AF668880

Marine Archaeological Perspective of the Indian Ocean

Marine Archaeological Perspective of the Indian Ocean

Editor
Alok Tripathi

Organising Committee
International Seminar on Marine Archaeology
&
Sharada Publishing House
Delhi

Proceedings of the International Seminar on Marine Archaeology 2003

The responsibility for the facts stated, opinions expressed or conclusions reached is entirely that of the contributors of the articles. The ASI and IN accepts no responsibility for them.

First Published 2004
Revised and enlarged edition 2008

ISBN: 978-81-88934-53-9

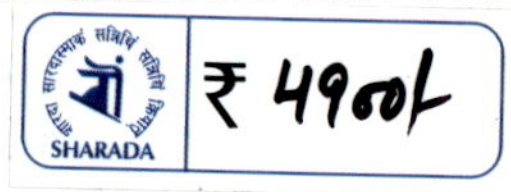

Published by:
Organising Committee
International Seminar on Marine Archaeology
New Delhi
&
SHARADA PUBLISHING HOUSE
111, Vidya Vihar Apartment
Sector-9, Rohini, Delhi-110 085 (INDIA)
Tel.: +91-11-3292 1560, 2755 2102
E-mail: bansal_sph@rediffmail.com

Printed at:
Salasar Imaging Systems
Delhi

Contents

Preface vii

Introduction ix

Contributors xiii

List of Figures xvii

Section - I

Welcome Address 3
Rear-Admiral R.F. Contractor

Keynote Address 7
Smt Gauri Chatterji

Inaugural Address 11
Admiral Madhvendra Singh

Closing Address 15
Vice-Admiral M.P. Awati

Vote of Thanks 23
Rear-Admiral R.F. Contractor

Section - II

Ships and Shipping in Ancient India 27
R.S. Bisht

The Later Indian Trading and Power Systems in the Indian Ocean (4th-12th Century AD) 47
Ranabir Chakravarti

The Arab Seagoing Traditions in the Indian Ocean (8th-15th Century AD) 93
John Carswell

The Relative Maritime Technologies at the time of European Encounter (15th-16th Century AD) 111
K. Raja Menon

Archaeological Investigation of *Princes Royal* off Bangaram Island - Introduction and Underwater Operations 143
A.K. Patnaik

Archaeological Investigation of *Princes Royal* - Excavation and Identification 161
Alok Tripathi

Traditional Boats and Ships of India Through Ethnographical Studies of Admiral Paris (1806-1893) 187
Eric Rieth

British Strategic Vision of the Indian Ocean (1857-1947) 219
Andrew Lambert

The Stellar Compass and the *Kamal* - An Interpretation of its Practical Use 245
Jose Manuel Malhao Pereira

The Indian and Pacific Oceans in World Maritime History 289
Joseph J. Krol

Maritime Heritage, Mutual Heritage - Research Beats Collecting 305
Thijs J. Maarleveld

Delegates 331

Index 341

Preface

The Indian Ocean, covering about 20% of the earth's water surface, is the third largest of the world's oceanic divisions. It has been an arena of migration, trade and cultural interchange since antiquity and its littoral has played an important role in world history. The historical and cultural dynamism of the Indian Ocean is yet to be studied properly.

India is centrally located in the Indian Ocean. With more than 7,500 km long coastline, her history and culture has been influenced with activities in the Indian Ocean. The researches aimed to understand the role of the sea, so far, have been limited and her over 5000 years old maritime history is yet to be written, systematically.

Realising the need for serious studies for writing definitive maritime history and sincere efforts to search, study and preserve underwater cultural heritage the Archaeological Survey of India and the Indian Navy have come together to organise joint fieldworks and International Seminar on Marine Archaeology (ISMA).

Limited copies of the first "Proceedings of the International Seminar on Marine Archaeology" were published in 2004 by the Organising Committee and they were distributed long before. It is nearly three years since demand for the book has been on increase. Meanwhile the Proceedings of the ISMA-2 were brought out where several things were standardized. During the organization of ISMA-3 the Organising Committee decided to bring out the revised version of the proceedings of the ISMA-1. Another decision was made to get it published in collaboration with a publisher to ensure wider distribution and regular supply.

Readers would find this revised edition in entirely new look. With the experience of earlier two publications several improvements have been made. Not only that, a presentation, which was made by Thijs J. Maarleveld during the seminar is an addition. A word-index has also since been prepared. All along it has been my earnest endeavour to make proceedings of ISMA collectable by all researchers working on Indian Ocean.

I am happy to place "Maritime Archaeological Perspective of the Indian Ocean" with the trust that it would fulfill the growing demand from scholars and researchers alike with its distribution. Collaboration between the Archaeological Survey of India and the Indian Navy is serving its purpose well and the outcome is fresh archaeological evidence and publication of reference material, the necessary steps towards writing a definitive maritime history of Indian Ocean in general and India in particular.

New Delhi **Alok Tripathi**

30 May 2008

Introduction

Marine archaeology lies at the heart of maritime history. India's maritime history can be compared to an iceberg. Just one-tenth is visible and has been recorded. The remaining exists as everyone knows, but it is yet to be written. To anyone involved with the sea, in India, the mismatch between our centralised continental preoccupation and the relative neglect of our maritime interests is only too apparent. Naval officers appearing for the Staff College entrance examinations still do not have an authoritative maritime history text to study, as a background material. Yet, maritime history is the foundation for maritime strategy, a subject that will engage the minds of these officers for the next two decades. The Indian history that students grow up with has few references to the sea and often none at all. Hence, the longest period of colonisation that the country underwent is often thought of as having begun at the battle of Plassey, with no explanation of how the British came to be in the district of Murshidabad, or Trichinopoly.

The handicap that Indian maritime history begins with is formidable, and has run its course for two hundred years. To rewrite Indian maritime history comparable to the history on land is a project that would take a national effort spread over a couple of decades. The Indian Navy and the Archaeological Survey of India has made a beginning with a series of seminars to attract the best brains in the field of maritime history and marine archaeology to revisit the entire narrative of India's maritime history. At the heart of these seminar proceedings lies the report on the dive over the *Princes Royal*, a shipwreck lying off Bangaram atoll. The seminar was only the visible portion of the first step of this great journey. Many more such ventures should follow, but more importantly, more collections of well-researched papers will add to the basic material from which a scholar in the future, could write a definitive maritime history of India.

Among the formidable papers presented at this seminar, the first is on the 'Ships and Shipping in Ancient India' as gleaned by the archaeological and literary evidence. Dr R.S. Bisht, a noted archaeologist and one of the foremost authorities on Harappan culture has thrown light on the maritime trade and contacts of Harappans with the Mesopotamians. He also discusses about their boats and navigation. The second paper is by Professor Ranabir Chakravarti on 'Later Indian Trading and Power Systems in the Indian Ocean'. Professor Chakravarti has listed both Western and less known Asian references in a comprehensive piece, that charts a record of Indian ports in international trade during 4th-12th century. At the end of reading his paper it becomes clear, that whatever evidence he has placed on record is so firmly drawn up that it clears up, much of the doubts on the volume and nature of the Indian Ocean trade during this relatively less researched period. Of particular interest is the material he has presented on maritime activities on the east coast from what is today Andhra, Orissa and Bengal and their links through Myanmar to Southeast Asia.

In the next paper, Professor John Carswell, who comes with a formidable reputation as an archaeologist, presents historical evidence through the device of excavated pottery and ceramics for the movement of goods across the Indian Ocean. Professor Carswell's emphasis is on what in India is a little known Sri Lankan port of Mantai and its tremendous importance as the sea exit for the prosperous kingdom of Anuradhapura. Favoured as a natural harbour, Mantai would perhaps qualify as the first major landfall in the Gulf - India to Southeast Asia and China trade. Another far-reaching point made by Carswell is the enormous fillip that Arab navigators gave to seaborne trade by taking the risk of starting their eastern voyage during the turbulent tail end of the monsoon, and thereby arriving in India with the entire winter sailing season still ahead of them to make their way farther east.

The paper presented by Rear-Admiral Raja Menon (retd.) looks at the relative technologies of the Western and Indian seafarers at the time of the European encounter. It is clear that the level of Indian technology was no match against the advances that Europe had instituted, and the effect of the imbalance is apparent in the ease with which India was conquered from the sea. Making a detailed comparison in the fields of shipbuilding, sailing outfits, cannon making, navigation and producing the national infrastructure to support a maritime industry, Menon concludes that some deep sociological factors prevented the Indian mariner from translating his individual skill to achieve effective ocean going practices. In the West, this was made possible by mathematicians, engineers, metallurgists, astronomers and scientists supporting the mariner, whose social standing was at par with the highest in the land. In India, the mariner was a social outcast left to fend for himself, except perhaps under the Marathas and the Cholas.

The report on the archaeological investigation of the shipwreck *Princes Royal* jointly carried out by the Archaeological Survey of India and the Indian Navy must be a first in India for the systematic manner in which the excavation was conducted. The excavation was conducted to depths up to 61 metres. The report on the actual dive was presented by Commodore Patnaik, the Principal Director of Special Operations and Diving, Indian Navy and the report on excavation and identification of the findings by Dr Alok Tripathi of the Archaeological Survey India.

The paper by Dr Eric Rieth of the Musee de la Marine in Paris, describes the activities of the French Admiral Paris and his pioneering work as the founder of nautical ethnography. Admiral Paris participated in three circumnavigation expeditions, and his monumental work on Extra-European boats is the best of its kind anywhere in the world. In three expeditions between 1826-1840, Paris established the plans of almost 250 extra European boats, including boats in use on coasts of India. Paris went on to become the curator of the Maritime Museum in Paris where he converted many of his drawings into models.

The presentation by Professor Andrew Lambert describes the British strategic vision of the Indian Ocean between 1857 and 1947, the years when the Crown took over the direct administration of India from the East India Company. Dr Jose Manuel Malhao Pereira who was a Captain in the Portuguese Navy gives an interpretation of the practical uses of the stellar compass and the *Kamal,* both of them navigational instruments that facilitated intertropical ocean navigation. Dr Pereira's paper eventually leads us to an understanding of the route navigation technique followed by Indian mariners, who followed a star to a particular destination, and used the *Kamal* to know how far south or north they had to go to arrive at the latitude of their destination. Rear-Admiral Krol of the US Navy who is posted on the staff of Commander-in-Chief Pacific went over the USA's current interests in the Indian Ocean.

The introduction would be incomplete without mentioning the brief talks and erudite summing up by the discussants, among whom the remarks of Shri M.C. Joshi, Dr Om Prakash, Professor Roddham Narasimha, M.K. Rasgotra, Professor B. Arunachalam, Captain Peter Hore and Mrs Lotika Varadarajan stand out in their moderation and wisdom. An excellent account of the attempts of Indian Kings to build modern warships was given by Mr Jean Marie Lafont of the French Embassy.

The collaboration between the Archaeological Survey of India and Indian Navy will go a long way to protect and preserve the underwater cultural heritage. This is not the last seminar of its kind in India, and each of the proceedings will make weightier reading along with the proceedings of other years, so that the country moves closer to the goal of collecting enough material to enable a maritime history of India to be written.

Editor

Contributors

R.S. Bisht
Joint Director-General (retd.)
Archaeological Survey of India
9/19, Rajendra Nagar-3
Sahibabad, Gaziabad - 201005
INDIA
rsbishtarch@gmail.com

Ranabir Chakravarti
Centre for Historical Studies
Jawaharlal Nehru University
New Delhi - 110067
INDIA
ranabir 1953 @yahoo.com

John Carswell
Cortijo Moya, 29710 Periana
Provincia de Malaga
SPAIN
carswell @activanetes

Joseph J. Krol
Associate Administrator for the Office
of Emergency Operations
National Nuclear Security Administration
U.S. Department of Energy
1000 Independence Ave., S.W.
Washington, DC 20585

Andrew Lambert
Laughton Professor
King's College, London
UNITED KINGDOM
andrew.lambert@kcl.ac.uk

K. Raja Menon
8 Shroff Apartments
2 Nizamuddin East
New Delhi - 110013
INDIA
rajamenon@vsnl.com

A.K. Patnaik
c/o Principal Director of Spl Ops & Diving
Integrated Headquarters of MOD (Navy)
New Delhi - 110011
INDIA
dirdiv_02rediffinail.com

Jose Manuel Malhao Pereira
R. de S. Gabriel n. 6
Alto do Lagoal
2760-170 Caxias
PORTUGAL
madrugada@sapo.pt

Eric Rieth
Musee National de la Marine
Palais de Chaillot, Dept. Archeologie Navale
17, Place de Trocadero
75016 Paris
FRANCE
e.rieth.cnrs@libertysurf.fr

Alok Tripathi
Superintending Archaeologist
Underwater Archaeology Wing
Archaeological Survey of India
Red Fort
Delhi - 110006
INDIA
alok.asi@gmail.com

Thijs J. Maarleveld
Professor of Maritime Archaeology
Institute of History and Civilization
University of Southern Denmark
Esbjerg
DENMARK
t.maarleveld@hist.sdu.dk

List of Figures

Colour

1. Chairman, Organising Committee, delivering Welcome Address
2. Director-General, Archaeological Survey of India, delivering Keynote Address
3. Chief of the Naval Staff, Indian Navy, delivering Inaugural Address
4. Director-General, Archaeological Survey of India receiving memento from Chief of the Naval Staff, Indian Navy
5. Audience at the seminar
6. Presentation of papers
7. Academic interaction during the seminar
8. Presentation on the first underwater excavation
9. Audience at the seminar
10. Chief Patron, Maritime History Society, delivering Closing Address
11. Chief Patron, Maritime History Society, receiving memento from the Chairman, Organising Committee of ISMA
12. Chairman, Organising Committee, delivering Vote of Thanks
13. Organising Committee
14. Delegates

15. *Princes Royal*: Buoy showing location of shipwreck off Bangaram Island
16. *INS Jamuna*: Survey ship
17. *INS Nireekshak*: Dive support ship
18. *Princes Royal*: Divers getting ready for dive
19. *Princes Royal*: Divers laying lines

20. *Princes Royal*: Diver transporting debries from the trench
21. *Princes Royal*: Pottery and copper objects exposed in excavation
22. *Princes Royal*: Divers retrieving a brown glazed jar
23. *Princes Royal*: Brown glazed jar of small size
24. *Princes Royal*: Iron anchor
25. *Princes Royal*: Cannons
26. *Princes Royal*: Diver excavating a cannon
27. *Princes Royal*: Cannon retrieved with lifting balloons
28. *Princes Royal*: Ballast stones after partial excavation
29. *Princes Royal*: Wooden planks of the ship
30. *Princes Royal*: Bronze bell inscribed with PRINCES ROYAL 1792

Black and White

1.1. Marine shell resources and distribution network
1.2. Mehrgarh: Shell objects, Period I
1.3. Mehrgarh: Shell objects
1.4. Mohenjo-daro: Representation of boat on a terracotta amulet
1.5. Mohenjo-daro: Representation of boat on a steatite seal
1.6. Ur: Harappan influenced objects
1.7. Tell Asmar: Elephant and Rhino seal impression
1.8. Kish: Indus type seal
1.9. Ur: Carnelian beads found in Royal cemetery
1.10. Susa: Etched-carnelian beads
1.11. Tell Agrab: Harappan influenced object
1.12. Umma: Harappan seal
1.13. Mohenjo-daro: Indus black-slipped ware

2.1. Old Goa: Hero stone in the Archaeological Museum
2.2. Ajanta: Ship painted in the cave no. 2

3.1. Mantai: Aerial view
3.2. Mantai: Fragment of burnished red pottery with Brahmi inscription

4.1. The Battle of Nagapattanam

4.2. A Ventian Galleas at the Battle of Lepanto

4.3. A scale model of the French warship *l' Ocean* on the basis of which the warship was actually built in 1775

4.4. Mandvi: Traditional shipbuilding in Gujarat

4.5. Mandvi: Naturally shaped wood of poor quality used as ribs

4.6. Mandvi: 14 inch nails and the heaviness of the planking relative to the ribs

4.7. Model of a two-mast lateen-rigged *Dhow*, with a small mizzen

4.8. Beypore: Modern Indian shipyard

4.9. Beypore: Handmade blocks for mainstay

4.10. Beypore: Mould frames produced by lofting drawings

4.11. Beypore: Mould frames produced by lofting drawings

4.12. A scale model of a 74 gun ship of the late 18th century

4.13. A model used for construction in a French shipyard with the side opened up to show the arrangement in the hold

4.14. The Miller Atlas by Lopo Homem, c. 1515 CE

4.15. The Mediterranean chart of Sebastao Lopes, c. 1555 CE

4.16. The Chart of the East Indies by Peter Goos, c. 1660 CE

4.17. *Malik-e-maidan* - made by a Turkish designer in Ahmednagar in c. 1548 CE

4.18. *Zafarbaksh* probably the best Mughal piece cast in Aurangzeb's time, c. 1674 CE

4.19. A cannon cast during Adil Shah's reign, Bijapur, late 16th century

5.1. Bangaram Island, Lakshadweep

5.2. *Princes Royal*: Rigid Hull Inflatable boat used in diving operations

5.3. *Princes Royal*: Preparation for the excavation during the passage to the wreck site

5.4. *Princes Royal*: Wooden boat to support diving and excavation activities

5.5. *Princes Royal*: Diving on site

5.6. *Princes Royal*: Conducting acoustic survey
5.7. *Princes Royal*: Laying of grids for excavation
5.8. *Princes Royal*: Briefing in progress
5.9. *Princes Royal*: *In situ* documentation
5.10. *Princes Royal*: Marking of objects

6.1. *Princes Royal*: Remote sensing survey
6.2. *Princes Royal*: Marking of exposed objects
6.3. *Princes Royal*: Artefacts lying on middle terrace
6.4. *Princes Royal*: Grids laid for excavations
6.5. *Princes Royal*: Underwater video camera used for documentation
6.6. *Princes Royal*: Constant monitoring of underwater operations on the surface
6.7. *Princes Royal*: Brown glazed jar
6.8. *Princes Royal*: Pottery lying on lower most terrace
6.9. *Princes Royal*: Sherd of blue and white Chinaware
6.10. *Princes Royal*: Copper vessel
6.11. *Princes Royal*: A pan of copper
6.12. *Princes Royal*: Thick and round through bolt
6.13. *Princes Royal*: Copper fastening
6.14. *Princes Royal*: Rotary quern
6.15. *Princes Royal*: Rotary quern
6.16. *Princes Royal*: Brick
6.17. *Princes Royal*: Keel of the ship
6.18. *Princes Royal*: Wooden planks
6.19. *Princes Royal*: Piece of a copper sheet bearing two stamps

7.1. Admiral Francois Edmond Paris
7.2. Plan of a *catimaron* (Catamaran)
7.3. Details of a *catimaron* showing the aft part of the logs with a lee-board used as a rudder
7.4. Plan of a *doni* outfitted with an outrigger

7.5. Plan of a coastal ship

7.6. Plan of sailing-ship with lugsail rounded by Paris at False Divi Point, coast of Golconde

7.7. Plan of a sailing tanker for fresh water carried in 14 jars, coast of Golconde, mouth of the Godavari river

7.8. The sailing tanker with a fore lugsail and an aft sprit sail

7.9. A *doni*

7.10. *Doni* of Karakal pictured at Pondicherry. The rigging of the aft shows European influence gaff sail, top mast table, shrouds with deadeyes

7.11. A sailing-ship rigged with lateen sails, Pondicherry

7.12. Model of a *chelinque* or *masula*: General view

7.13. Model of a *chelinque*: Details of stem

7.14. Model of a *chelinque*: Details of sewn joints

7.15. Model of a *catimaron*: General view

7.16. Model of a *catimaron*: Details of logs and mast step

7.17. Plan of *chelinque* or *masula*

7.18. Detail view of the fore part of the hull of a *chelinque*

7.19. Master frame of a *chelinque*

7.20. Water-colour drawing of *chelinque* on a beach

9.1. Intertropical areas at sea and European and Asian routes of navigation

9.2. The need for direction

9.3. Parallel sphere - Observer at the Pole, stars never rise or set, only half of the celestial sphere visible throughout the year

9.4. Right sphere - Observer at the Equator, stars trace circles perpendicular to the horizons, equatorial stars defining, during rising and setting, long periods of accurate bearing

9.5. Oblique sphere - Observer at 45^0 of latitude, stars trace circles oblique to the horizons, defining, during rising and setting, very short periods of accurate bearings

9.6. Stellar compass of the Indian Ocean

9.7. A graphical representation of the bearings of the stars

9.8. The north Indian Ocean area of navigation

9.9. The Little Bear, the pole-star and the main operation needed for navigating along a parallel

9.10. Two examples of *Kamal*

9.11. Geometrical principal of the *Kamal*

9.12. Some of the stars of the southern sky during the observations

9.13. The *Kamal* and its use at sea

11.1. Underwater archaeology in North Sea basin

11.2. UNESCO publication

11.3. Histogram showing maritime archaeological discoveries in the Mediterranean

11.4. Prehistoric boat found in silted estuarine environment, 11 m. below sea-level

11.5. Roman shipwreck find in Netherland, 6m. below sea-level

A Marine Archaeological Perspective of the Indian Ocean: A Gateway to the Continents

International Seminar on Marine Archaeology

Organised by
Indian Navy and
Archaeological Survey of India

14-15 March 2003
New Delhi

1. Rear-Admiral R.F. Contractor, Chairman, Organising Committee, delivering Welcome Address

2. Smt Gauri Chatterji, Director-General, ASI, delivering Keynote Address

3. Admiral Madhvendra Singh, Chief of the Naval Staff, delivering Inaugural Address

4. Director-General, Archaeological Survey of India receiving memento from Chief of the Naval Staff, Indian Navy

5. Audience at the seminar

6. Presentation of papers

7. Academic interaction during the seminar

8. Presentation on the first underwater excavation

9. Audience at the seminar

10. Vice-Admiral M.P. Awati, Chief Patron, Maritime History Society, delivering Closing Address

11. Chief Patron, Maritime History Society, receiving memento from the Chairman, Organising Committee of ISMA

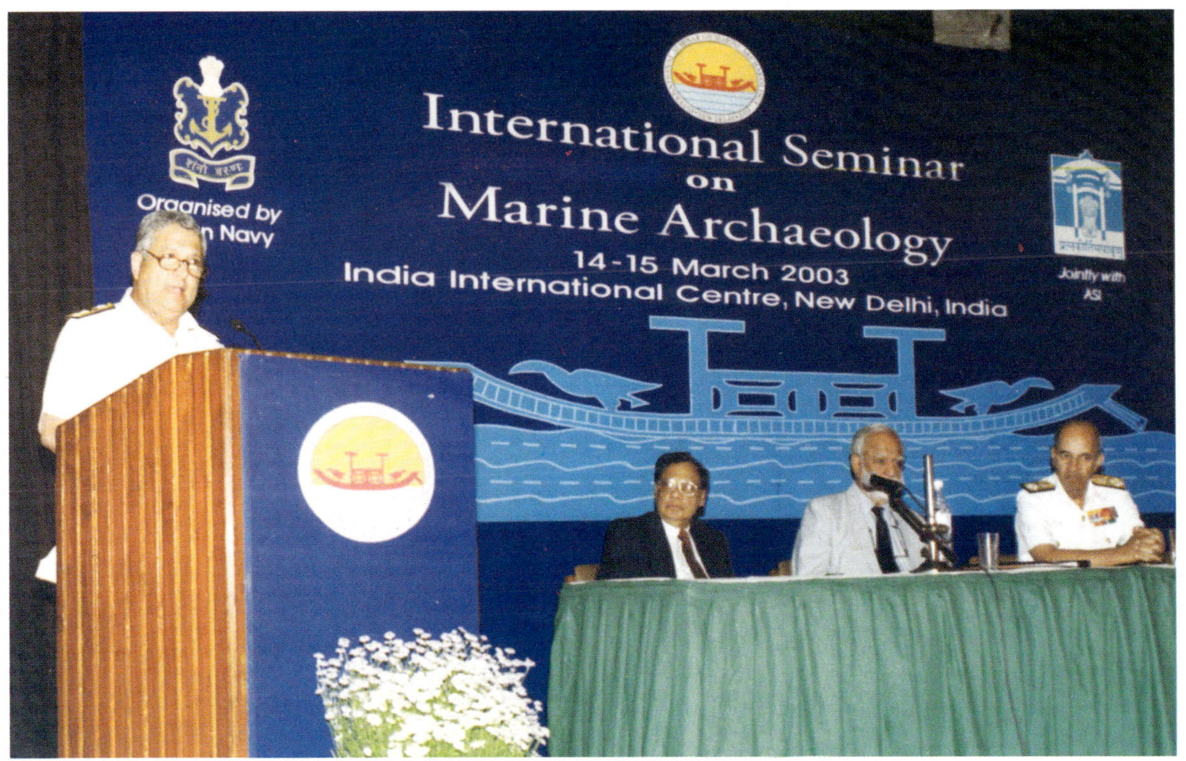

12. Chairman, Organising Committee, delivering Vote of Thanks

13. Organising Committee (L-R) Cdr A.K. Aukta, Cmde P. Chauhan, Cmde M.K. Thomas, Cmde A.K. Patnaik, Rear-Admiral K.R. Menon (retd.), Rear-Admiral R.F. Contrator, Dr Alok Tripathi, Cmde S.R. Nathan, Cdr V.S. Rana

14. Delegates

Sitting (L-R): Prof B. Arunachalam, Prof John Carswell, Admiral V.S. Shekhawat, Admiral J.G. Nadkarni, Prof Ranabir Chakravarti, Admiral R.H. Tahiliani, Admiral Madhvendra Singh, Rear-Admiral Joseph J. Krol, Captain Jose Manuel Malhao Pereira, Rear-Admiral R.F. Contractor

Standing (L-R): Commodore A.K. Patnaik, Shri M.C. Joshi, Rear-Admiral K. Raja Menon, Dr Jean Marie Lafont, Vice-Admiral M.P. Awati, Dr Alok Tripathi, Shri M. Rasgotra, Mrs Lotika Varadarajan, Dr R.S. Bisht, Vice-Admiral M.K. Roy, Prof Roddam Narasimha, Prof A. Lambert, Prof Thijs J. Maarleveld, Captain Peter Hore, Dr Eric Rieth

Section - I

Welcome Address

Rear-Admiral R.F. Contractor, NM
Chairman, Organising Committee
International Seminar on Marine Archaeology

Good morning, Ladies and Gentlemen, Admiral Madhvendra Singh, the Chief of the Naval Staff, Mrs Gauri Chatterji, the Director-General, Archaeological Survey of India and distinguished guests.

At the outset, I would like to inform this august gathering that Hon'ble Raksha Mantri Shri George Fernandes has conveyed that due to unforeseen developments, he would not be able to present in our midst today to inaugurate the seminar. However, he has wished all success for the seminar. It is indeed a great honour for me as Chairman of the Organising Committee to welcome Admiral Madhvendra Singh, the Chief of the Naval Staff, who has kindly consented to inaugurate the International Seminar on Marine Archaeology (ISMA).

As you all know, the secrets of the sea have always fascinated mankind and he has always been particularly attracted to the mysteries of human history and treasures that lie beneath its vast waters. Since times immemorial, man has used his ingenuity to invent means of underwater exploration. Aristotle, the famous Greek historian, dreamt of diving-bells, and sources reveal that Alexander the Great used a glass like cage to inspect the fortifications of the Port city of Tyre in Lebanon way back in 332 BCE. Leonardo da Vinci drew machines to facilitate breathing underwater, and the first real diving-bell appeared in the 16th century. However, it was not until the 19th century and the invention of the autonomous diving-suit that man could finally explore and operate in shallow waters. This landmark achievement marked the onset of crucial discoveries that laid the foundation for development of marine archaeology.

By the end of 1950s, marine archaeology became a scientific discipline transporting historians, archaeologists, engineers and divers in its wake. For a long time, the Mediterranean basin was the preferred excavation site, and it is only during the last 20 years that marine archaeological activities have gathered momentum and extended right from the Atlantic Ocean to the South China Sea.

Maritime history and marine archaeology have thus far been little researched areas in India and more so in the Indian Navy. As such, a wealth of scientific, cultural, historic and traditional data pertaining to our area of interest, namely the Indian Ocean, have remained unexplored. Archaeological evidence unearthed during excavations, the records chronicle in the ancient text as well as certain enduring maritime traditions which are practiced even today in the little villages that dot our Indian coastline, are testimony to this great Indian heritage. As you are all aware that a sizeable slice of this great maritime history lies buried under the waters of Indian Ocean. It is therefore natural that we attempt to unravel the mysteries that lie there. Since the century has been declared the Century of Oceans, the Indian Navy has taken up initiative in partnership with Archaeological Survey of India to highlight this very important subject and enhance its awareness amongst the people here and the international community. It was therefore imperative that we endeavor to examine the history of events that pervaded our neighbouring seas through the ages. This indeed has prompted the Indian Navy and the Archaeological Survey of India to take the lead in organising an International Seminar on this subject.

Existing data on the survey of wrecks off the Indian coast indicate that a plethora of information, including that dated to the 4th-5th century CE could become the missing interface between the past and the present. The investigation of these wrecks and subsequent analysis could reveal vital information on the shipbuilding, navigational and technical skills, handicrafts, trade and other aspects of maritime heritage and traditions that existed through the ages. A seminar to promote marine archaeology was thus mooted and, as pilot project, excavation of a known shipwreck off Bangaram Island in the Lakshadweep group of islands, was undertaken by the Archaeological Survey of India and Indian Navy, the findings of which would be presented during the seminar.

The theme of the seminar has very aptly been titled *A Marine Archaeological Perspective of the Indian Ocean — A Gateway to the Continents*, as the Indian Ocean with its geographic location connects most of the continents and yet its history has by and large remained unexplored. I am certain that the seminar would renew awareness of our ancient seafaring traditions and their influence on economic activities and development in the region.

This forum could explore and critically examine the existing material on the trade routes in the Indian Ocean through which silk, spices, pottery and other material were trans-shipped across the ocean. There has also been a confluence of technology, design and construction of sailing craft between the Arabs, the Chinese and the Indians, which needs to be explored more precisely. The presentation of papers by scholars and specialists working in the international arena on these diverse, yet related disciplines would hopefully throw light on our vital, albeit neglected, aspect of maritime history. If, in the light of this information, it is possible to have a clear picture of Asian maritime networks, the contour of the two way transmission process that took place in this part of the world would become apparently clear.

Finally, it would be only apt to say that we, the Indian Navy and the Archaeological Survey of India, have a long term interest in safeguarding our maritime heritage, and would contribute fully towards the development of marine archaeology.

I would be candid in confessing that though we are a late starter in the archaeological investigative field, I am sure that, in years to come this forum will hear more from us in terms of involvement, enhanced activity and contribution towards marine archaeology.

Keynote Address

Smt Gauri Chatterji
Director-General
Archaeological Survey of India

Ladies and Gentlemen, I feel honoured to address this 'International Seminar on Marine Archaeology' and would like to thank all the delegates and experts who have accepted our invitation and participated and enriched this seminar.

As marine archaeology is an interdisciplinary study, which acts as a bridge between the humanities and the sciences, the major objective are of course to search, study and preserve sunken cultural heritage. Submerged cities and shipwrecks of historical importance are, therefore, a major source of information in reconstruction of maritime history.

The antiquity of India's maritime activities can be traced back to as far as the 3rd millennium BCE. Ancient Indians were shipbuilders and navigators who set out on many voyages, mostly for trade. There are ample evidence to show that India was actively trading with the countries of Southeast Asia as well as Africa and West Asia. Studies of the Harappan culture reveal some pioneering feats in marine engineering and maritime activities. There are archaeological evidence to show that their ships carrying precious cargo traveled as far as the African and the Sumerian ports. This evidence is corroborated with the findings of the Harappan artefacts.

India has an enviable treasure of underwater cultural heritage that is perhaps, one of the richest and also the least disturbed. While some other countries, mostly in the western part of the globe have made a lot of progress in underwater archaeological study particularly excavation and search of shipwrecks in the last 30 to 40 years. We have started our underwater investigations comparatively recently.

Underwater archaeological investigation carried out off Dwarka drew a lot of interest from the historians, archaeologists and oceanographers and also from the public. In fact the mystery of the sea attracts a lot of public imagination and public interest. The vast sea around the country, our mighty rivers, ancient lakes and other water bodies we have a subcontinent, so we have a vast source of water bodies. These are storehouses of rich cultural heritage. Sunken archaeological sites on our coast form an important source of history for us. Their discovery and systematic study in the coming years may serve to bridge many gaps in Indian history to shed light on newer perspective.

Since the late 80's archaeologists have been involved in exploration of ancient sites on Indian coast. These finds have indicated fluctuations in the sea-level which have been responsible for the destruction or submergence of some of the ancient settlements that existed on shore. The ASI has taken up the efforts to develop this branch of archaeology in this country to preserve underwater cultural heritage of the nation. Underwater archaeology is highly scientific branch and these researches need to be carried out under the direction of trained persons.

Since the late 80's, archaeologists have been involved in exploration of ancient sites near Tranquebar on the Tamil Nadu coast. The finds have indicated a definite rise in the sea-level, which has been responsible for the destruction and submergence of settlements that existed in the region. This is further corroborated by the finds of terracotta ring-wells and storage jars. The archaeological remains extend up to Poompuhar which is 15 km. from Tranquebar. The marine archaeologists have also studied the ancient ports of Maharashtra namely the Sopara, Gharapuri, Chuel, Vijaydurg and some ports of the Karnataka coast namely Honnavar, Bhatkal and Malpe.

In 1498, the discovery of sea route to India by the Portuguese remains a major milestone in world maritime history. This led to an appreciable increase in the maritime activities in the Indian Ocean. The Portuguese being the pioneers, dominated the Indian Ocean throughout the 16th century and fought many battles against Afro-Asian powers as well as the British, French and the Dutch for more than hundred years to control the trade with India and the spread of Christianity. The frequent battles at sea fought during this period is one of the reasons for finding hundreds of shipwrecks known and unknown, strewn on the ocean floor. Also, a large number of ships sailing across the Arabian Sea, sank due to the hidden reefs and other natural factors while traversing between Africa, Arabia and the Malabar coast. While many of these wrecks are recorded in the archives, there are others lying buried unrecorded. The wrecks covered with sediments and preserved on the sea floors serve as excellent chronological landmarks and can prove to be of immense exploratory valve.

With this as a backdrop, the Archaeological Survey of India and the Indian Navy undertook a joint excavation of one of the wrecks around the Lakshadweep Islands, the result of which would be presented as one of the papers during this seminar.

I would like to add here that it is gratifying to find that the Indian Navy has been taking keen interest in excavating sunken ships and without their invaluable assistance it would have been impossible to achieve the results that we gathered from the recent excavatory dives in the Lakshadweep Islands. They also have a marine archaeological club in Mumbai and some officers have undergone training in marine archaeology. In years to come we hope to see more contributions from the Indian Navy in this field.

Finally, I would like to thank Admiral Madhvendra Singh, the Chief of the Naval Staff, distinguished guests and all delegates for sparing their valuable time to be present for the function.

Inaugural Address

Admiral Madhvendra Singh, PVSM, AVSM, ADC
Chief of the Naval Staff
Indian Navy

Smt Gauri Chatterji, Director-General, Archaeological Survey of India, distinguished delegates of the seminar, ladies and gentlemen. Jai Hind and a very good morning to all of you. At the very outset I must once again apologise on behalf of our Defence Minister who had to be called away for urgent work. I am grateful to the organisers for inviting me to inaugurate the International Seminar on Marine Archaeology (ISMA), which has been jointly organised by the Archaeological Survey of India and the Indian Navy.

As the country's premier maritime service the Indian Navy has been trying for some time now to focus attention and highlight our maritime history and maritime heritage. And towards this end people like Admiral Awati started the Maritime History Society about two decades ago that continues to do well. Admiral Ganesh when he was in south started the Maritime History Society, south India chapter, recently in Mumbai we founded the Maritime Heritage Foundation which is also coming of age and I am sure we will get better and better and it is in this same spirit that we thought that we should organise this seminar to focus on marine archaeology in our area.

Marine archaeology is a bit intriguing and I am sure that this gathering will be highly enlightened during the proceedings of the seminar. It is a modern science and in fact it is so modern that I tried to access it on the net and did not find many sites on it. I must admit of course, that I am not very good at finding things on the net, I am a bit of a novice, but I did find a clip obviously written by somebody who did not like archaeology or archaeologists. He says that he did not take to archaeology because he did not want his life in ruins. Of course I must say, as Mrs Chatterji has said, the mystique of the sea and things like that marine archaeology is certainly much more interesting as is everything to do with the sea including sailors. It deals with the remains of ancient civilizations, shipwrecks buried or lying on the seabed. Considerable amount of work has already been done world-wide on the archaeological excavations on land and underwater which throws light on the extent of development that took place during the periods to which the sites pertain. By analysing these artefacts recovered during the excavations, we are able to reconstruct the history to a large extent in terms of culture, value systems, traditions and the technological expertise that prevailed during those times.

India is perhaps one of the richest countries in the world, from the archaeological point of view, which has a history of civilization that can be traced back to over five thousand years. The excavations carried out at the sites of Harappa and Mohenjo-daro, most of which you all will be familiar with, showcase the progress that occurred in the field of art and architecture, housing and development of state of township and cities that existed several millennia before Christ.

'Archaeology' has often been referred to as the post-mortem operation of the civilizations and cities buried under the earth and the sea. The study of the past in its historic as well as natural context is essential for understanding the changes that have occurred in the earth's structure, evolution of life, man's impact on nature and nature's influence on the growth of civilization. It would be incorrect to merely confine archaeological activity to excavations as is so often thought, as it must bring out information dealing with the past, which could be corroborated with other parameters available to the archaeologists. Today, therefore, archaeology has a broader dimension and it has the potential and promise to unravel many mysteries of life and recollect the turning points of important civilizations of the past.

With the rapid changes in our ecological system, it would not be prudent to restrict the work of archaeologists only to excavation of the archaeological sites. Archaeologists should also be concerned with the preservation of the environment. With utter disregard to the environment, urbanisation and industrialisation has polluted and destroyed the coastal strips, wastelands, riverbeds, and coral reefs which is a matter of concern to us. But it is not a modern day phenomenon. It has happened earlier in the past. It is estimated today that nearly 25,000 plant species and more than a thousand vertebrate species are threatened with extinction, and these are alarming statistics. Therefore, archaeological research should not only help in discovering what was existing and destroyed, but should also, with the help of other scientific disciplines, be involved in rebuilding and preservation of the ecosystem.

Fortunately, in this seminar we have not restricted speakers to purely marine archaeology as such. We therefore have papers on a wide variety of subjects which deal with related issues and highlights some aspects of maritime heritage and history in this area.

Perhaps, the most fascinating and interesting aspect of underwater exploration is the excavation of ancient shipwrecks, which lie strewn along the coasts of many countries. These wrecks indicate the craftsmanship and also throws light on the shipbuilding technology as also the type of trade, etc., that was carried on at that time by various societies. After being wrecked, a ship in a sense, continues to travel perhaps not in space but certainly in time. Hidden from the sight of man, it continues its solitary and secret itinerary down through the annals of time. The finder of the wreck is responsible for bringing new life of this forgotten heritage and expose it to the light of the day and as also explain his findings and deductions.

Our ancestors have been engaged in seafaring activities and water borne trade in and outside the country from the very early days. Over the centuries the sea-level along the Indian coast have changed numerous times which have often submerged entire port cities and at times even vast stretches of land. The evidence of sea-level changes are preserved in the ancient literature, viz. the tamil literature of the Sangam age which refers to the bridge built by Lord Ram to go over to Lanka. The sea has also been a silent witness to the movement of ancient traders, travellers and adventurers many of whom unfortunately met with a watery grave.

India, by virtue of its geographic location occupies a key position from the marine archaeological angle, as it overlooks vast Indian Ocean that has a network of marine routes since times immemorial, with countries on the West as well as on the East. India boasts of a glorious past with well developed education system, splendid architecture, rich art and culture and above all a highly developed shipping industry which contributed to the spread of its art and culture to several countries of the world.

In ancient times, India did excel in the art of constructing vessels and many earlier western traders and travellers have expressed the same views. Infact Admiral Awati has done a lot of work on it and we are happy to have him here today. Thomas Bowrey, an English traveller who visited India during 1669-79, observes that many English merchants had their ships and vessels built at Madapollum which was a flourishing shipping centre on the east coast.

A Venetian traveller of 16th century, Cesare de Fedrici, while commenting on the east coast of India had observed that there was an abundance of material for shipbuilding in this area and many Sultans of Constantinople found it cheaper to have their vessels built in India than at Alexandria. Marco Polo, the famous traveller who visited Thane port in India had also written about the well developed shipbuilding industry that existed in India and highlighted the abundance of natural resources and technological skills available in the country.

We are and have been a seafaring nation and should take pride in our rich maritime culture and traditions. It is precisely in this context that today's seminar holds great significance for our maritime community at large and I must thank the Archaeological Survey of India for joining hands with the Indian Navy to conduct this seminar which will remind and educate us of our almost forgotten great maritime heritage.

I am confident that this forum would generate mutual exchanges and ideas that are intellectually stimulating. There are many eminent scholars who have come from distant lands and I would like to thank you and I have no doubt that you would enrich us with your vast knowledge and experience in marine archaeology and maritime history.

I sincerely hope that this seminar would throw light on the fascinating aspects of our history and the maritime heritage, and we gain considerably from the proceedings of the seminar. I look forward to interacting with all of you and I wish this seminar all success.

Closing Address

Vice-Admiral M.P. Awati, PVSM, VrC (retd.)
Chief Patron
Maritime History Society, Mumbai

Ladies and Gentlemen. This closing address was to have been delivered by the Minister of Culture. The Chief of the Naval Staff asked me if I would oblige. I have a strange feeling that at the end of this many among you are going to say 'I wish you had not'! I am a reluctant speaker this afternoon. I feel like the civilian gentleman who, driving around Willingdon Island entered the Island Gate of the Naval Base by mistake. Stopped by the naval sentry, the gentlemen rolled down the car window and said 'I am sorry, but I do not want to be here'. 'Sir', replied the sentry 'neither do I'.

Having thus clearly established my credentials as a reluctant performer let me get on with the inevitable — the Closing Address. I crave your indulgence. We have had two very interesting days discussing a range of subjects, which bear upon India's perspective on the Indian Ocean. Marine archaeology will play an increasing role in unearthing incontrovertible evidence which must establish many facts of our maritime history and in demolishing many legends and myths surrounding India's long association with the seas which wash her shores. The Director-General, Archaeological Survey of India told us in her keynote address that her department will join the Indian Navy in exploring and excavating submerged ports and offshore wrecks to help reconstruct the maritime history of India. The effort will also help in generating useful data for a study of sea-level fluctuations and the effects of the marine environment on underwater objects. That was a good beginning!

Underwater exploration is designed to explore and search for antiquarian remains, which might, but for such exploration, have remained submerged. All over the world, there are cities and ports, which have been, over a period of time, washed away by tides, floods and erosion of land. They constitute important archaeological sites, which reveal many facets of history and culture of once thriving communities living by the sea and its fruits. There is another aspect of underwater exploration; and this has to do with ships and boats. Our people have been engaged in seafaring and in water-borne trade in and outside the country from very early times. Battles were fought and fleets boats traversed the waters of the oceans and the rivers, laden with a variety of goods for a great many years. It is only to be expected that many of these vessels would have been hit by calamities, natural as well as man-made. Having battled them bravely, they would have gone to their final resting place on the bottom of the sea or on the beds of rivers. To search, to know and to tell others about the adventures of these mariners is surely a thrilling experience and this is the true domain of the marine archaeologist. So what is the Navy doing dabbling in archaeology when it should be studying, practicing and preparing for war. The short answer to that is there is always room and time for a pastime.

Explorations and excavations on land have revealed a plethora of relics of our past, which have enabled us to get to know and understand it better. India is one of the richest countries in the world from the archaeological point of view, with a history of civilisation dating back over five thousand years. If one were to scratch a little somewhere, I dare say, one would find something that would qualify as an antiquity. Of course, it is important not to get carried away by such possibility because we cannot be digging at just 'any old place'. That would be an extremely inefficient way of expending effort and resources.

But explorations and excavations are a slow, tedious and expensive process and need to be conducted in a systematic manner on the basis of clues that provide an indication of the possible existence of material of interest, lying beneath a spot or a mound. These clues could be gleaned from ancient coins and seals, or from pilot excavations, paintings and epigraphs, even from ancient literature. It is true that some really startling discoveries are made quite by accident — for instance, the unearthing of the most ancient and important of our cultural vestiges, namely, those at Harappa and Mohenjo-daro, was accidental. However, it is prudent to bear in mind that such fortuitous accidents do not occur often. It is usually hard and consistent effort; dig, dig, dig! — till you hit pottery sherds, bricks or whatever.

In the West, a great deal of work has been done in archaeological excavations, both on land and underwater. This has added a great deal to our knowledge and understanding of these civilizations and their interfaces with other, more distant ones. It is the careful study of the artefacts recovered from such excavations that has enabled us to reconstruct the history of their past and our own from interpolating our relationships with them, with a fair degree of accuracy.

Ours is an ancient civilisation, a past with a well developed social system, architecture, culture and art. India occupies a key position in the world, sitting astride one of the most important sea-lines of communication, which connect the West to the regions of South and Southeast Asia. Maritime activities in this part of the world were fairly well developed in ancient times and a large number of ships of various shapes, sizes and nationalities, have traversed these waters. Quite a few of these have gone down to "Davy Jones Locker", and, the data available with us, meager as it is, confirms that there are a number of wrecks in our waters, many of them of considerable historical interest. Excavations of these wrecks will very likely provide valuable information which would enrich our knowledge about ourselves. So much then for marine archaeology!

As Professor Carswell tells us, the Arabs were the most dominating influence at sea in the pre-European Indian Ocean. They dominated it through trade. The scientific knowledge of and seamanship skills of the Arab mariner once linked the civilised world, from the Atlantic to the Pacific. We must acknowledge our debt to the Maghrabin traveler, Ibn-e-Batuta for our knowledge of medieval geography and history. Later, it was a distinguished *Muallim* from the Nejd, Ahmed bin Majid, who guided Vasco da Gama across the Arabian Sea in May 1498. That voyage turned out to be epoch making. It gave a new, quite unexpected, turn to our history.

The Bedouin of the treeless wastes of the Hadramawt, Dhofar, Mahar, Sur and Oman had little going for him except sand and the occasional date palm. He was forced to turn to the sea for food, early in his history. Fishing over open beaches needs a very different mental approach and a physical endurance not demanded of the estuarine fisherman. The Arabs became the incomparable seamen and navigators of the Indian Ocean. It was in the fitness of the things that they should graduate to becoming the carriers of trade of the ocean littoral and beyond, using the God given seasonal wind system of the north Indian Ocean for outward and homeward voyages of his *Dhow*. The word *Dhow* is the general term for the Arab wooden sailing-vessels, which have sailed the eastern seas for at least two millennia, scarcely changing their shape or settee rig. The Arabs have never used the word *Dhow*. Instead, they have or I should say had, many names to describe their ships of trade and of commerce. Today, these *Dhows* are almost extinct, or on the point of extinction, victims of the changing times and of the great God — oil — black gold.

We Indians too had displayed a remarkable ability at sea, for ocean navigation and for shipbuilding. The Arab, however, showed vigor and an ability to improvise which was unmatched. The great forests, which backed the coasts, especially in the West supplying quality shipbuilding timbers like teak, punnai and karimaradu, gave the Indian shipwright an advantage, which he used to the full, in building ships and craft of quality, for Indian and Arab shipowners.

My view on Arab sailing traditions is grounded, or anchored, to use a maritime term, on two factors. The first is geographical, the other cultural-historical. The Arabs of the littoral were literally pushed out to sea by the geography of his homeland. Later, the new desert religion motivated them to travel, to voyage, to proselytise. Religion usually followed trade, that was the pattern. This daring and positive approach to seafaring by the Arab was hobbled among Indians, by a curious and inexplicable fear of crossing the seas by the time we come to the beginning of the 8th century. No historian, I know of, has commented on how and why this superstition established itself in our caste conscious society. It did and it had an adverse effect on our competence at sea. Thus, for example, in pre-European times, numerous littoral rulers appointed Arab port officers or administrators, the *Shah Bandars*, obviously because of an absence of competent locals. I do not wish to labour the point. Not long ago, I came across a book by Thor Hayerdahl of Kon Tiki fame on archaeological findings in the Islands of Nilandu and Fua Mulaku in the Middle Maldives. These excavations point to a flourishing pre-Arab civilization based on oceanic trade. Professor Carswell has told us of similar findings in his diggings at Mantai in northwest Sri Lanka. There is much evidence to show that the Arabs profited from generations of Persians. Similarly, through their contact with Chinese mariners trading with Kerala, the Arabs — and the Persians — became familiar with the magnetic needle. By the end of the eleventh century they had adopted it to their own use as a floating compass, the *da'yra*. It is interesting to me that whatever came to the West as maritime lore and wisdom, of mathematics and astronomy is usually attributed to the Arabs. There is little or no speculation as to where the Arabs might have picked up their knowledge of ocean navigation, by the use of heavenly bodies. My own experience is that the Arabs are not particularly fond of the sea. They put to sea from sheer necessity. They are certainly great trader and the Bedouin in them make them a compulsive wanderer, inured to hardships qualities, essential in all great seamen! All in all, the Arab sailing traditions in the Indian Ocean is a great legacy in the history of this ocean. It impinged on the character and the culture and the beliefs of the people of the entire littoral.

Ancient and medieval India was a competent builder of ships. Sewn plank construction was preferred. Dr Eric Rieth spoke on Admiral Paris's contribution to the ethnology of the shipbuilding communities of India through a study of the architecture, the method of construction, the rig of the boats and vessels they built.

Admiral Raja Menon bemoaned a general reluctance among the shipbuilders to learn from others, in absorbing better building technologies to improve their products. They preferred verbalisation. The consequences of this rigid attitude towards change were predictable. The ship forms change little over the years, indeed ages. There could be no standardisation. Method of construction remained archaic. The Europeans forced some changes. Nails came in for bonding in place of the sewn planks. The transom stem became a fashion. The fore and aft rig however remained. Whatever new technology was invented, it was not backed by the State.

I have long been fascinated by the *Dhow*, the great trader and voyager of our ocean for over two thousand years, perhaps even longer, scarcely changing its shape or its rig over those millennia. Surely, an amazing record for any investigator or researcher to look into, to unravel the level of culture and technology of those who conceived them and built them. In living memory these fascinating crafts were contemptuously dismissed as 'country crafts' by our erstwhile rulers. The terminology remains in the vocabulary of the successor, Brown Sahib! Much credit then to investigators like Admiral Paris who, more than a century ago began a search for their ancestry. He carefully scrutinised their architecture, the methods of construction, their sailing and sea keeping qualities. He wrote about them in his essay 'On Naval Constructions of non European Peoples' illustrated with detailed and accurate line-drawings to scale and beautiful sketches in water-colour. I have a photocopy of that essay and of the drawing. It is a prized possession. After him there were others, Hornell and Villiers, Ricciardi and Hawkins who, fascinated by these crafts, both Arab and Indian, wrote about them and sailed in them.

It is a matter of regret that no Indian, except perhaps, K.B. Vaidya, has devoted any time to researching or writing about these and other indigenous crafts of our coasts and seas.

Around the end of the Gupta period, a treatise on shipbuilding appeared. It was titled *Yuktikalpataru*. It was written in classical Sanskrit by one Bhoja Deva and was, evidently, meant for the learned, certainly for the literate. Perhaps the shipwrights of the period were literate, if superstitious. For there is great deal of the superstitions in that book. For a start the treatise classifies timber according to caste on the analogy of the established *varnashrama dharma*. Mixing of types of timber in the construction of vessels is expressly forbidden by the book, for a hybrid vessel would soon rot and sink before its time! Similarly, iron fastenings were not recommended as magnetic rocks would attract them and break up the ship, so, it goes on. It talks about unlucky ships. Whatever influence it may or may not have had on the art of shipbuilding in India of that period, it is a book of considerable interest and value to the nautical ethnologist. Did the ships of that period reflect the cultural values of the shipbuilder? Surely they must have.

The great days of the *Dhow* are now history, except for the motorised variety. The *Dhow* is now almost extinct. I discovered that during my search along the coasts of India in 1997-98. Despite the fact that the *Dhow* and kindred indigenous craft offered the cheapest mode of bulk transport of cargo along the coasts, they fell victim to the demands of speed and schedule. To societies in a hurry, time is all important. To the poor *Dhow*, time was never a concern. Its movements depended on tides and winds.

There is now a revival in the West of interest in and therefore a study of, local boat forms, as they constitute the best means of knowledge of 'the man of the water', of how those forms are influenced by geographical, historical, techno-economic factors and thereby be a document of value to the ethnologist. The massive ship of Tuticorin, the *Thoni*, is, as I was to discover, witness to the cultural and religious mores of the Roman Catholic Rayer community of Portuguese descent. They own these magnificent, if lumbering, sailing-vessels. Their relations with the neighbouring Hindu Nader community as well as the equally Hindu but Malyalee builders of these vessels, the *mistries* of Beypore and Mangalore are worthy of study by the nautical ethnologist. At the other end of the scale I found the stitched plank, *Masula* and the impoverished, open-to-the-sea, *catamaran* of the coasts of Tamil Nadu and Andhra Pradesh perfectly suited to the purpose for which they are built. I have seen the later being driven by a discarded *mundu* or a dhoti doing the duty of a sail! The now extinct *Baghla* and *Batel*, the still-in-service, double ended *Kotia* of Gujarat in its motorised form, the once aristocrats of the *Dhow* world, the decrepit *Machwa* and the elegant *Patamar* of the Bombay coast were and are boat forms which tell us a good deal of those who built them and of those others who sailed them.

The concept of sea power remained alien to the Indian mind until well after the arrival of the Europeans on our coasts. I would go so far as to say that sea power or maritime power, the constituents of which make for sea power are, even today, not fully understood by Indian political leadership, nor indeed by a very large section of the educated classes. Sea power is mistaken for high handedness, with aggression, with gunboat diplomacy. To them the Indian Ocean need never be *mare nostrum*. Indeed the reason for such an attitude to one's own front yard comes naturally to a people who have been brought up in the tradition of *vasudhaiva kutumbakam*, the 'World is a Family'. Everyone was welcome to trade with us, so long as they paid their dues. And they all did, for millennia. That changed with the arrival of the Portuguese, with their high freeboard, square-rigged ships mounting cannon along the broadsides. The Samudri of Calicut had seen nothing like them before even in the days of the great Admiral Cheng Ho. But he remained unreceptive. The European agenda, however remained uncompromising. They demanded and obtained a monopoly in goods they wished to trade in.

The Indian Ocean thereafter became a bone of contention among the successive European maritime powers with Indian littoral powers reduced to the status of onlookers. The final was played between Britain and France. The rest is modern history of India on the back of Pax Britannica and now the fledgling Pax Americana.

The Indian Navy and the Archaeological Survey of India have for the first time jointly undertaken a full fledged underwater excavation. This effort really needs to be viewed as a pilot project for the subsequent excavations which we intend to undertake within the country in the near future. It should be our endeavour to uncover the hidden treasures that lie beneath the ocean surrounding us so as to reconstruct the history surrounding them. This effort in the public domain must be applauded.

Finally, I would like to laud the efforts of the organisers for having been able to get together such a large body of scholars who are working in this branch of knowledge and for meticulously planning and conducting this seminar just concluded.

I am sure that the deliberations during the seminar have enlightened many of us and would also help the future generations to establish their roots and to chart the future with the confidence born of knowledge and awareness. So until we meet again, *au revoir* and *bon voyage*.

Vote of Thanks

Rear-Admiral R.F. Contractor, NM
Chairman, Organising Committee of the International Seminar on Marine Archaeology

Vice-Admiral Awati, Admiral Madhvendra Singh, the Chief of the Naval Staff, distinguished delegates, ladies and gentlemen.

As this International Seminar on Marine Archaeology draws to an end, it is my distinct honour and privilege to propose a vote of thanks.

At the outset, I would like to thank Admiral M.P. Awati who has consented to preside over the closing session of the seminar.

I am also extremely grateful and express my appreciation to all our distinguished guests and delegates who have made it convenient to congregate here from different parts of the country, and from abroad, to participate in this seminar and present their papers. The papers have indeed been very enlightening and educative, and I would like to place on record our appreciation to all the delegates for their valuable contribution. We are also grateful to the audience for their active participation and the keen interest they have evinced in the proceedings.

I would also like to mention the useful roles played by *Indian Naval Ship Jamuna*, a survey ship, in conduct of the preliminary survey, and that of *Indian Naval Ship Nireekshak*, a dive support vessel in actual conduct of the joint archaeological investigation. I also take this opportunity to acknowledge the contributions of Reef Watch Marine Conservation, Mumbai, a Non Government Organisation who were associated in providing us with precious inputs on marine flora and fauna in the area.

I would be failing in my duty if I did not acknowledge the useful assistance provided by the Lakshadweep administration to our teams during the period of excavation as also in providing local knowledge of the islands.

I must admit here that we received unstinted support from the Staff of Embassies of France, Portugal and the United States of America and the British High Commission in coordinating with the delegates from their countries and we really appreciate their contribution.

At the end of the two days of gruelling sessions, I am sure all of us have gained immense scientific, cultural, historic and traditional data pertaining to the Indian Ocean. The presentation of the papers has thrown a new light on the seafaring activities prior to and after the arrival of the Europeans in the Indian Ocean.

I would like to make a special mention of Rear-Admiral K.R. Menon who played the key role of Honorary Consultant and rendered professional guidance for the seminar.

Finally, I would like to acknowledge the tireless efforts of my colleagues in the Organising Committee who have ensured the smooth and successful conduct of the event.

Section - II

1
Ships and Shipping in Ancient India

R.S. Bisht

Man's association with the Ocean has been ever since the evolution of the man as such. He migrated from one place to another and went across the continents. There were periods of marine transgression and recession. During the period of recession some ridges appeared, that's how men went across the continents. Bering Straits, Gulf of Gibraltar, Palk Strait, area between Southeast Asia and Australia are some such major land crossings.

Remains of man have practically been found on all continents except for Australia but aborigines were there too. They were harvesting the marine resources, a good evidence of which is found at several places. In Oman piles of fish bones and shells are found which he must be collecting from the sea for food. However there is not much evidence to prove how he was harvesting these marine resources.

Civilization process in India started around 7000 BCE. Excavations conducted by French archaeologist Jarrige and his team at Mehrgarh unearthed archaeological evidence from 7000 BCE to early phases of civilization period. More than 7 meters of occupational deposit of the Neolithic period is found at Mehrgarh. The period when metal or pottery was not known is called aceramic Neolithic culture or pre-pottery Neolithic culture. In Neolithic period man had created an economy in which he had tried to win over nature by domestication of plants and animals. Mehrgarh has given a very good account of beginning of settlements, construction of houses and domestication of plants and animals. During early periods there were more wild animals than domesticated. In due course of time the number of domesticated animals increased and the same is the case with the plants. During this period they were cultivating three or four types of wheat also. This site is very important because a large number of marine shells also have been found. Mehrgarh is located in the Kachi plain on which the Bolan river debouches from the Baluchi hills in Pakistan.

As it is far off from the sea it means that man was going to the sea and collecting the shells and later using them for making ornaments.

The exploitation of shells from various areas like rocky and sandy beaches, lagoons and estuaries, shallow areas of reef indicate large-scale procurement of marine products (fig. 1.1). The site also indicates that it had linkages with other contemporary archaeological sites. Mehrgarh is located close to the Mohenjo-daro site however at that time the city of Mohenjo-daro did not exist. Pieces of ornaments, beads, pendants, necklaces, etc. were made of marine shells (fig. 1.2) and were found there. Although very systematic' expression had not arrived but man's progress across time, from pre-pottery Neolithic culture through pottery making and slowly refinement in culture, economy, arts and crafts are seen clearly.

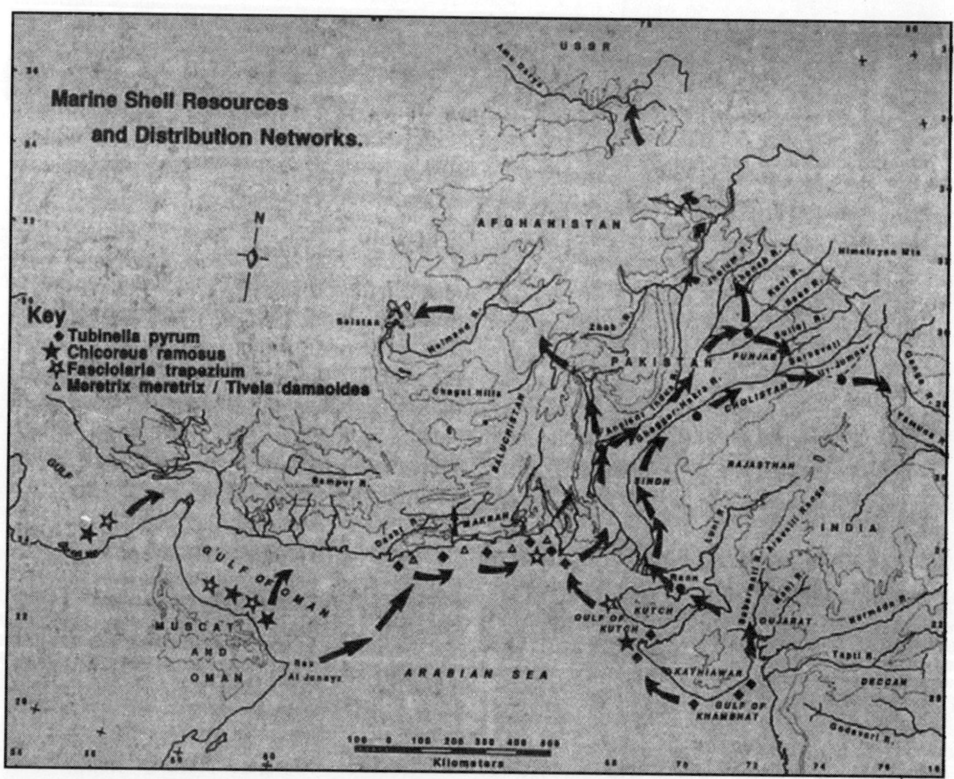

1.1. Marine shell resources and distribution network

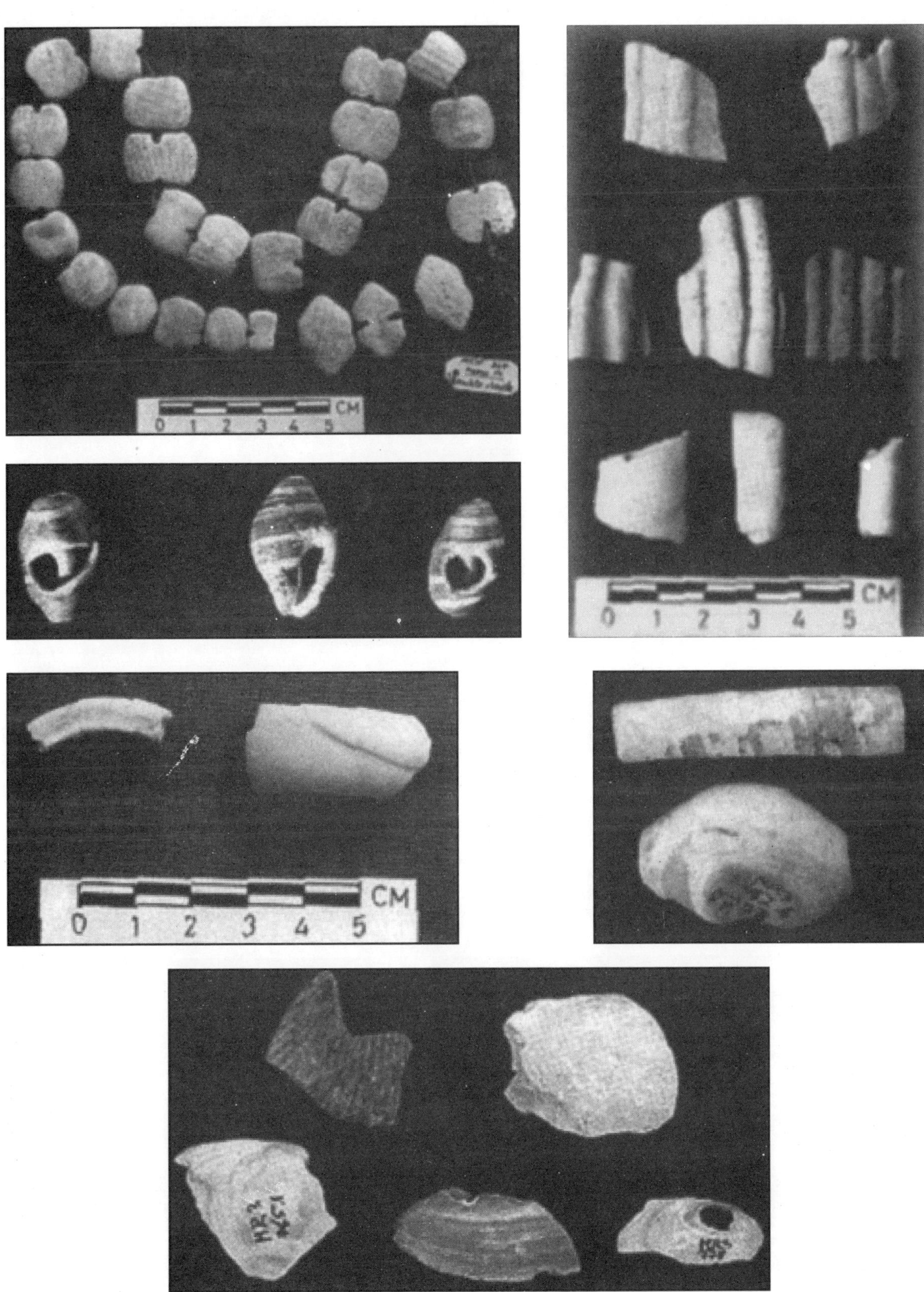

1.2. Mehrgarh: Shell objects, Period I

Settlements at Mehrgarh are of different periods. Jarrige had divided these cultural deposit into seven periods. Period 7 is the period when urbanisation started growing but beyond that period we find the use of marine shells (fig. 1.2-1.3), which shows intimate association with the sea. Then comes the Copper Age, around 5000 - 4500 BCE. The period of incipient urbanism that started around 3000 BCE is very significant. It is not confined only to India but that is a period when similar development is seen in Mesopotamia, Iran, Egypt and many areas of West Asia.

Here a passing reference may be made about the importance of

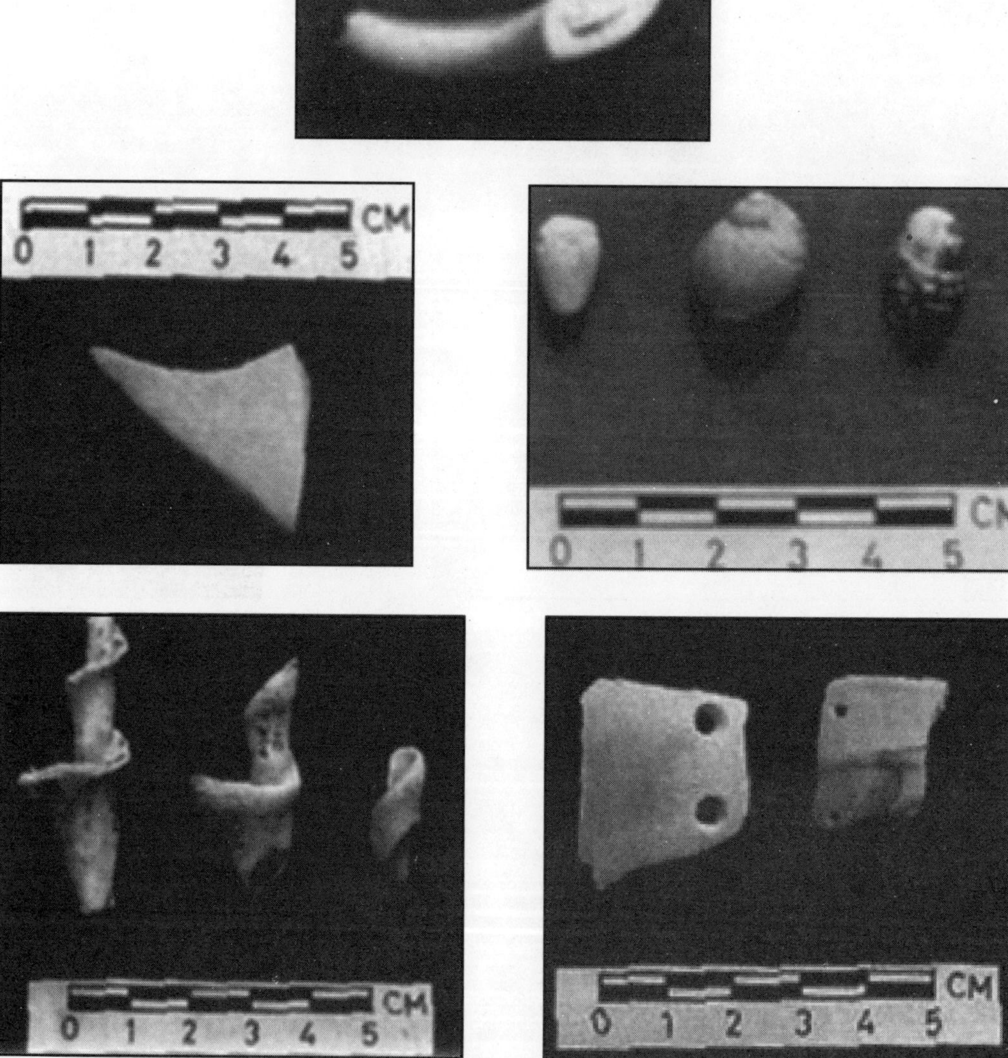

1.3. Mehrgarh: Shell objects

3100 BCE in our history. That is the period when Bronze Age started and man knew the technique of alloying copper with tin and other metals. It is well known that the copper is a very soft metal but if mixed with tin, it becomes harder, bronze. The period when bronze was used is known as Bronze Age. That is the period when urbanisation started growing all over the world. This significant period in India is remembered in different ways. It is popularly believed that it was the period when Mahabharata ended or Parikshit was born. Kashmiri era also starts around that period. Some how it is also believed to be the beginning of Kali era. The truth is somewhere in the legends but one thing is certain, that man in India remembered that 3100 BCE was very important date and tried to associate it with different events.

The Harappan Age witnessed a highly advanced urban civilization. Harappans were great pioneers, great architects, great builders and they achieved a phenomenal growth in art, crafts, trade and commerce. They also embarked on international trade. Harappan material has been found in Oman peninsula, Bahrain, Failaka, Mesopotamia, even in Central Asia and on certain sites in Iran. Harappans may have covered some of these areas over the land but they had to cross the sea to reach Oman, Bahrain, Failaka. The boat on the logo of this conference with a cabin in the centre and *dishakak,* the land finding crows or birds (fig. 1.4), is an evidence of their seafaring. The terracotta amulet has three faces. On one of the face there is a crocodile and on the third there is Harappan legend. A boat engraved on another seal show a cabin in the center (fig. 1.5). It is interesting to note that more Harappan material has been found in Oman, Mesopotamia and Iran. A variety of objects including seals, weights, beads, gold beads, etc. have been found in Mesopotamian sites, Oman, Bahrain and other places but they are negligibly small in quantity and quality. Since there was frequent trade among Mesopotamia, Harappa and Persian Gulf these three different regions had seals which they used for sealing of documents, sealing of cargo, bales and consignments.

1.4. Mohenjo-daro: Representation of boat on a terracotta amulet

1.5. Mohenjo-daro: Representation of boat on a steatite seal

The seals of these three different regions are different from each other. Where Harappan seals are generally square in shape, Mesopotamian seals are cylindrical in shape and Persian Gulf seals are circular in shape. In Harappan context we have found all of them, but cylindrical seals seems to be locally manufactured not imported from India. This means that Harappans were more active on the sea borne trade. That is why Sargon of Akkad in the middle of the 24th century BCE boast that the ships from, Meluhha, Magan and Dilmun are anchoring at the quay of the city. So often the scholars identify Meluhha with the Harappan Empire. The Harappan Empire does not mean the political empire, it may be economical empire or cultural empire. It is more an economical empire having cultural entity in the form of pottery, architecture, ornaments, seals, etc.

The Harappans were spread by and large in the riverine plains that was very poor in natural resources. They had to collect their material like gold, silver, copper from different sources. This may possibly be one of the reason for the Harappans for going to Oman or having trade contacts with Oman to have copper from there. Ancient copper workings have been found there and Oman could have been a source for copper for a long period of time. The other source was Baluchistan. It is doubtful how much copper they exploited from Indian sources. Although there are certain areas in north and south Rajasthan which are rich in copper. Direct evidence of collecting, mining and smelting copper from Indian sources is yet to come. Since copper was a very useful metal because that was a period when Iron was not known and copper was the only metal, which was needed for making tools and also ornaments. Good quantity of copper from Oman was coming to India by ships. A boatman community lives very close to Mohenjo-daro and they are called Mohana. They say that Mohenjo-daro is named possibly after them.

A rectangular large brick made basin at Lothal, situated very close to Gulf of Khambat, is identified by its excavator as the dockyard. Here an inlet channel was made through which the water during high tides was coming into the dock and bringing the smaller boats here. Larger boats must be anchoring somewhere far away in the Gulf. There was also a system of sluice gates to let the surplus water out. It may be mentioned here that many archaeologists feel that it was not a dockyard but a fresh water reservoir because that is an area where ground water is saline.

Dilmun, Magan and Meluhha are the three important countries east of Mesopotamia, which are referred to in the Mesopotamian accounts, including that of Sargon of Akkad who unified Mesopotamia. The Mesopotamian accounts talk about Meluhha, because it was a foreign tongue for them. They needed interpreters and they contributed to the temple activities and maintenance of temples and all that they were associated with the social life of that area as usual. Objects made of chlorite and alabaster is found from various sites from Mesopotamia. Material like alabaster has been used for making a variety of things particularly vessels, the cosmetic things and the great manufacturing centre for them was Saurashtra.

The Persian Gulf type of seal bearing Indus signs (fig. 1.6) indicates interaction between these areas. Elephant and Rhino shown on a Mesopotamian seal (fig. 1.7) must have been carved by a person who had deep knowledge of these animals that are obviously Indian during that period of time. A typical Harappan seal has been found from Kish (fig. 1.8), a Mesopotamian city.

As carnelian beads, particularly long cylindrical beads, manufactured by Harappan have been found in the Royal Cemetery of Ur (fig. 1.9). They must be highly valued there. Chanhu-daro near Mohenjo-daro was one of the manufacturing centers. Lothal and many other cities also were engaged in bead making. It must be a very thriving industry for domestic consumption as well as for international trade. The etched carnelian beads are found from Susa (fig. 1.10), southeast Iran and from, other archaeological sites in West Asia.

1.6. Ur: Harappan influenced objects

1.7. Tell Asmar: Elephant and Rhino seal impression

1.8. Kish: Indus type seal

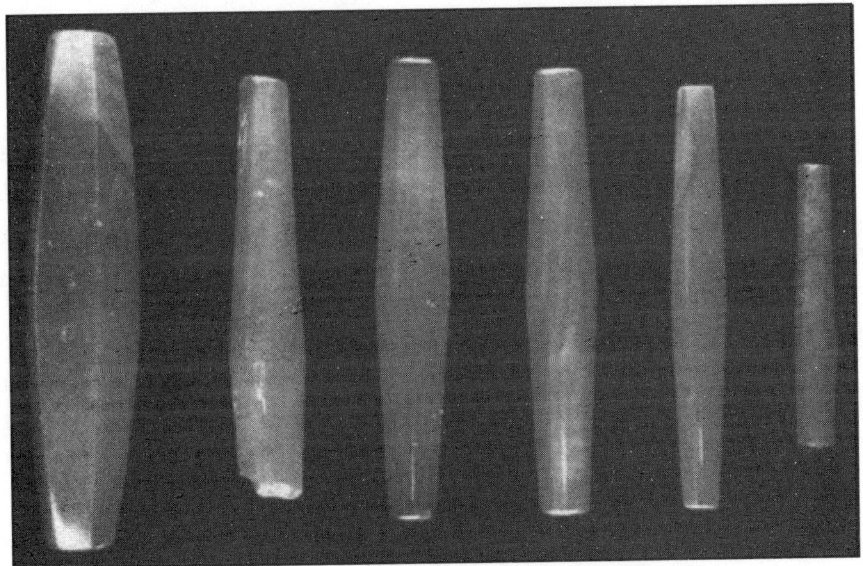

1.9. Ur: Carnelian beads found in Royal cemetery

1.10. Susa: Etched-carnelian beads

1.11. Tell Agrab: Harappan influenced object

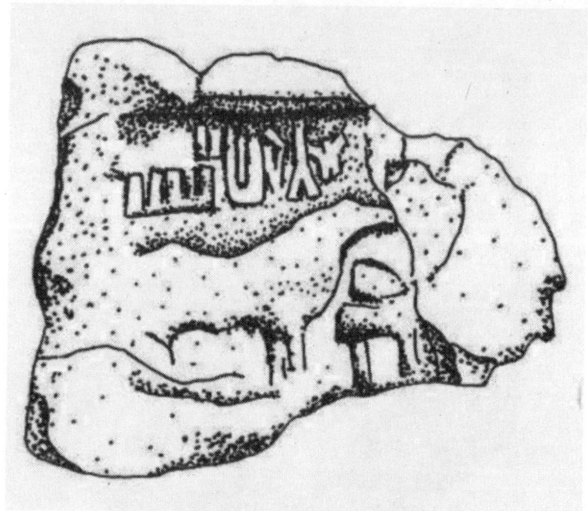

1.12. Umma: Harappan seal

The tangible items which can last longer (fig. 1.11-1.13), but we do not know how much of other things were being transshipped or exchanged or traded or sent there for trade. Items such as medicinal plants, incense, fragrance material, cloth, wood, etc. which were regularly sent to Mesopotamia from the eastern countries of Meluhha and Magan as their accounts say. A variety of incense (*google*) was found in Kachchh and Saurashtra area. This must be a regular item of export from India and also from Oman peninsula. Timber must be in great demand as it is not found there. Mesopotamians were getting timber from forests in Lebanon and Harappans would have exported teak, as they also required ships and river crafts.

The *Periplus of the Erythrean Sea,* a manual written by a Greek for the mariners who were trading between Red Sea and going up to Bay of Bengal in the 1st century CE, enumerates a large number of things of trade. Some of these are incense, fragrance, cloth and medicinal herbs. These medicinal herbs from Kashmir and Himachal Pradesh have been highly praised in Vedic literature also. They were used for medicine, for making scents and also have aphrodisiac properties.

The process of urbanisation had started growing from 3000 BCE onwards and about 2500 BCE it reached the zenith. The period between 2600 BCE to 1900 BCE, was certainly the time when the Harappans remained prosperous, having large cities like Harappa, Mohen-jo-daro, Rakhigarhi, Dholavira and very impressive array of smaller cities and towns.

Around 2100 BCE Mesopotamia was in political turmoil. It is the

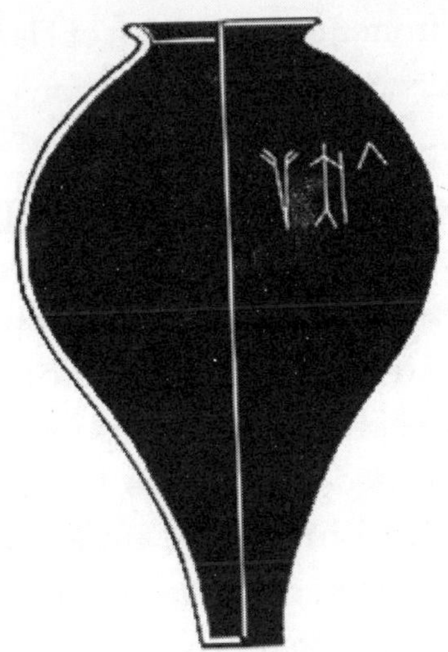

1.13. Mohenjo-daro: Indus black-slipped ware

period when Harappan cities started showing signs of decline. International trade, which was bringing in a lot of wealth to the Harappan Empire dwindled and was under a lot of stress. Archaeological evidence have been found to confirm that cities were under decline, industries were under stress and there seems to be a period of recession. That is why they could not maintain big cities and started moving out to the countryside and went back to their most dependable profession that was agriculture and animal herding. Harappans made a valiant effort to revive their trade between 2000 BCE to 1800 BCE and during that period Bahrain played a very important role of an intermediate port. Probably the Harappans may not be going up to the Mesopotamia. They were going up to Bahrain where they were exchanging their wares. That period of revival is best documented in Kachchh and coastal areas of Saurashtra. After that a period of about 1000 years, is called a period of darkness as far as economic prosperity and trade is concerned.

Rigveda which is a collection of hymns, is the earliest literary composition, possibly, of humanity. It was generally believed that Aryans were barbarians who were snatching land and destroying the cities of adversaries. On one side you find Harappan cities and on the other side the Aryans say that with the help of Indra, their warlord, they had destroyed 100 *puras.* Such situation is very ideal to identify the Aryans with the destroyers of the Harappan cities, so Mortimer Wheeler said that the Indra stands accused. But later on when this evidence was challenged by American archaeologist Reeve, Wheeler said I exonerate Indra from the destruction of Harappan cities because we do not find any evidence that any outside force destroyed Harappan cities.

On the other side we do not know who authored the Harappan civilization, whether they were Sanskrit literate or Dravidians or some other. So far as the *Rigvedic* period is concerned we do not know what was their material culture. No archaeological culture has been related with the early Aryans, i.e. Rigvedic Aryans. Effort are made to study the Vedas, and other related literature to try and see some linkages between archaeology and literature. Unless we dovetail both, a complete picture cannot emerge, so Harappans remain nameless. The name Harappan culture is given because it was first discovered in Harappa in 1921. It was also called the Indus Valley Civilisation because the early cities were located in the Indus plains.

"May our Ship embark to all quarters of the earth" is mentioned as the motto of the Rigvedic seafarers. Another is "May my going abroad be pleasant and may my return be pleasant" (*RV* X.24.6). Contrary to the earlier general belief archaeological finds made it clear that the Rigvedic people did not destroy the cities but they founded many such cities and in fact lived there. A large number of adjectives given to *pura* these fortified settlements and many of them qualify as ports. They were trading with manufacturing people, trading people, city dwelling people and navigators in rivers as well as in sea and going across the sea for wealth. There are numerous references, which show that the Aryans had a craze for foreign goods. They talk about the wealth corning from across the sea and going across the sea for wealth, glory and undertaking naval operations.

The standard term for the ship or the boat was *nau* from which navigation is derived. Other words for boats found in the *Rigveda* are - *tari* (X.69.7), *tarani* (III.29.13; VII.26.4; VII.67.8), *pratarani* (V.46.1), *plava* (I.182.5), *drun* (VIII.85.11). Word *nau* comes very frequently and we find a graphic description of *nau*, a seagoing vessel, having oars sometimes as many as 100. Hundred in *Rigveda* means many. There are several references, which talk about ship or boats being fitted with oars. A hundred oared ship is a favourite theme which is repeatedly narrated in the *Rigveda*. It seems that ships had three masts and each mast was fitted with two sails. Sails are compared to horses, birds, eagles, swans and it seems that each boat had three wheels, which might be for loosening sails or dropping of the anchor.

Varuna was the lord of the water, but the Ashwin, the heavenly stars, were the real guarding force. They were the presiding gods over trade or sea voyages. When Bhujyu, the son of Tugra, met with an accident, a shipwreck, Ashwin Kumar saved him in their boat.

Another important goddess, who is very intimately related to sea voyages, is Ushas, the dawn. There are also several words which are used for handling of ships at sea such as *niyana* (passing over with carriage or anchoring), *nyayana* (docking), *parayana* (sailing abroad), *sanyayana* (commissioning), *avartana* (homeward journey), *nivartana* (returning or reaching), etc.

In an episode Vasishta is shown swinging in a ship in the midst of an ocean. Thus, the acquaintance with the seas and seafaring of the Aryans is well attested.

During the early historic period the stage changes from the Indus Valley and peripheral areas to the Gangetic plain. The Gangetic plain emerges as the hub within its sphere slowly encompassing the whole of India.

Some of the things which were used in the *Rigveda* continued to be used later but some underwent changes. The early Buddhist literature are full of accounts of sea voyages, shipwrecks, carrying of goods and even missionaries. *Jatakas* gives a lot of information about ships, shipbuilding, water traffic schemes, river traffic schemes, going across the seas, etc. They even talk about Manimekhla which later on inspired the famous article in the *Sangam* literature. Manimekhla is said to be the daughter of heavenly gods, who protected mariners, sailors and merchants at sea.

In *Jatakas* there are references to *Nala-mala Samudda* (Arabian Sea), *Nila-mala Samudda* (Bay of Bengal), *Agina-mala Samudda* (Sea around Andaman and Nicobar Islands), *Kura-mala Samudda* (Persian Gulf or Red Sea?) and many other such *Samudras.* The Arabian Sea was the *Neel Malai Samudra* because it was along the *Neel Malai* (blue mountains), the western ghats. Bay of Bengal was the Nal Malai because it was along the eastern ghats which are called Nal Malai or Nilgiri even today. *Agni-Mala Samudra* should be the south China Sea, 10 degree channel all that because that was the area of interaction during the historical period.

There comes a reference of Nal and Neel who were great engineers who constructed the bridge from Rameshwaram to Sri Lanka, so that the army of Ram could enter Sri Lanka. This is the area possibly where hill ranges of Neel Malai and Nal Malai meets. The Adam's bridge is dedicated to them and the likewise the *Ramayana* and *Mahabharata* give a number of direct and indirect metaphorical references of naval expeditions and sea trade. The *Puspak Viman* of Ravana appears nothing but a ship, if one look into the details in literature. It was not an aircraft but a ship because in the sea where the horizon meets the sea all looks alike. Likewise if one look into the accounts of the attack by Salva on Dwarka it becomes very clear that the Salva brought a fleet of warship to invade Dwarka.

In the southern recension of *Mahabharata* we find averment to *Roma* (Rome), *Antakhi* (Antioch in Syria) and the city of *Yavanas*. It seems that they are talking about the cities and ports of Mediterranean Sea. It is significant that south India witnessed vigorous Indo-Roman trade during early centuries of common era that is why these cities are mentioned there.

From these archaeological and literary evidence discussed here it is quite clear that the ancient Indians were great mariners and courageous seafarers. I conclude by saying that Navy was one of the four essential limbs of the military, which was always cherished by any powerful king. Kautilya in his *Arthasastra* talks about *nau- adhyaksha* the Chief of the Naval Staff.

Comment

M.C. Joshi

Dr Bisht has given a very good idea of the growth of shipping and use of ships in antiquity. He has largely brought before you, the archaeological material and references from the literature, particularly from the Vedic literature. Because it is felt by most of the historians, that the Aryans never knew the sea and they were confined to the upper part of India, whether they did reach *sindhu* or not, is a problem. Sometimes, some rivers are also mentioned as a large water body. It is also stated that the *samudra* which is sometimes mentioned in *Rigveda* stands for a large water body and *sindhu* may not refer to the ocean itself. He has given good ideas and I don't have much to say.

So far as the Harappan culture is concerned, that is not the area of my specialisation. I know a little about it and may offer a few comments on that particularly since he has already touched upon the arrival of gold. Indeed gold and silver have also been found over there and he was talking about the copper ingots which may have been brought from outside. But what was the idea of urbanisation. The idea of urbanisation is the movement and urbanisation is connected with the concept of money. When it was introduced replacing the barter system then the key was of using any scarce material to determine value. That was gold because of its rarity. What could be the source of gold? S.R. Rao stated that it could be from Karnataka but it appears to be a little doubtful. Hence, the only source of gold may be West Asia or more likely Egypt but I doubt whether at that early stage we had direct connections with Egypt. It appears even from the later references that the Indian mariners or traders were familiar with the Persian Gulf or the Gulf of Oman. And from there they were going to Mesopotamia and Babylon. And the same route was followed by Alexander's army.

When Alexander's army returned, it was divided into two parts. One section returned from the land route including Alexander himself who died in Babylon and another section returned from the sea under Niarchus one of his Commanders. Most of the movement of ships till the discovery of trade winds was only along the coast and this is supported by the naval logbooks or the shipping logbooks belonging to the ships of 15th-16th century written in Gujarati. About six such logbooks have been acquired from Gujarat by the National Museum at the instance of the Archaeological Survey of India. They also have charts which were primarily utilised for hajj purposes.

The trade between Gujarat and Kerala was also along the western coast. So far as the depiction of ships is concerned, rosewood may have been utilised by the Harappans. But I think bulk of the ships might have been made by using reeds. This was also the system in Egypt. Now to what extent the mast and sails were used is a matter of how they were fitted. It would need further studies. In literature apart from the Vedas we have the great story of Manu. The biblical flood which is also recorded in the Puranic literature, that how humanity was saved with all its associates including animals by Manu. So that was the utility of boats when there was flood on the land or the sea. In India we have identified a number of places which in the Biblical story they point to the western part of the old world.

The later literature is more useful. There was de-urbanisation between 2000 BCE to up to 1800 or 1600 BCE. We do not exactly know what happened in the early historical period before 600 BCE as rightly pointed out by Bisht. There is no documentation of anybody going westwards by sea except by the land route till the beginning of the Common Era. Then of course our literature mentions *Baveru Jataka* that is again connected with Babylon but we have more interesting references in later literature. *Supparak Jataka,* which is a Sanskrit *Jataka,* gives very interesting details about the dangers of the sea. The ship is led by Bodhisatva and what they refer to as the manager of the ship is called *Potavanij*. We have got a number of names for boats. As Dr Bisht has pointed out *Tari, Tarani,* were sea faring vessels which could also be identified as *Pota,* which could not be used in the river, whereas *nau, nava* were river crafts. *Potavanik* can be called the owner or manager and perhaps the captain was known as *Karnadhar* because he was operating the sails. Sails were known as *Poksha.* The dangers of the sea were categorised in the ascending order as follows *Khurmali, Dadimali, Agnimali, Kushmali, Nalmali* and *Bharwamukh.*

It is interesting to know how these *Potavaniks,* the owners of the merchant ships were operating. At a port they used to announce a few days earlier that such and such ship would be sailing and the announcement was being made using gongs and the bells and then they would ask where would you go and they would collect the money and then pass either eastwards or westwards. In one of the stories, in one of the *Avdanas* they reach an Island where they see a mountain, and a desert and the name is referred to as Badradweep. Badra is also a Sanskrit word which stands for berries. But here because the journey is westward from the western coast the Badar may be an Arabic name. Normally we have two words for moon in Arabic literature, *Kamar* and *Badar. Kamar* is the crescent moon and *Badar* is full moon and so it appears that this was a kind of semi circular or circular Island about which exact direction and location cannot be fixed. But these are the ideas and apart from this we have a lot of later literature.

Though Dr Bisht says that there was a kind of navy as one of the organs of Indian system of the Army, but we don't have specific references. We have a very elementary depiction of a large boat in one of the paintings of Ajanta datable to 500 CE, where a huge boat with elephants, horses, warriors is depicted in a cave. The painter of Ajanta was not familiar with the kind of regular ship or boats fitted with weapons. There are certain passing references in the Indian epigraphy, which show the capture of the coastal forts. To what extent we could develop our shipping industry on the west coast on our own or whether we learnt anything from the Chinese or Arabs, this is a matter for investigation.

2

The Later Indian Trading and Power Systems in the Indian Ocean (4th-12th Century AD)

Ranabir Chakravarti

India's past in the pre-modern period, spanning several millennia, is increasingly illuminated with the availability of diverse data — both archaeological and literary — and also by fresh inputs to the already known data. From the middle of the twentieth century, the historiography of early India (from the earliest times to 1300 CE), is marked by a significant shift in historical enquiry. The previous penchant for traditional dynastic history began to be replaced by incisive analyses of socio-economic institutions and cultural practices. The result was the portrayal of the possibilities of changes in traditional society and economy outside and irrespective of dynastic shifts; certain forces of continuity were also discerned. Historians of early India have offered significant explanations of the changes and continuities in social, economic and cultural life. The other important aspect of the changing perspective is the attempt by historians to establish linkages among society, economy, polity and cultural activities, which are not treated in a compartmentalised manner. Another methodological shift is that nowadays the historian does not entirely depend upon the literature for an understanding of early Indian society, economy and polity. It is now well recognised that the religious literature, however important they may be, provide norms relevant to Brahmanical ideals. These precepts were often at variance from many practices which are evident from other sources, e.g., archaeological material, epigraphic data, impressions left behind by foreigners, accounts in creative literature in Sanskrit, Buddhist and Jaina texts largely in Pali and Prakrit.

These preliminaries will help situate better a particular aspect of early Indian socio-economic life, namely India's role in the maritime network of the Indian Ocean in the early period. It is impossible to deny that the material milieu of Indian society has been agricultural over millennia and the bulk of the Indian population were engaged in agriculture. This has naturally paved the way for a rich literature on the agrarian history of India. Without belittling the significance of agriculture in India, one also has to acknowledge that trade *(vanijya)* was regularly recognised as one of the major ingredients of economic life from remote times. One must make it clear that commercial activities of the pre-modern times are to be situated in the overall agrarian material milieu of the subcontinent. While nobody can lose sight of the vast continental landmass of India, it is also unmistakable that the subcontinent is washed on three sides by the Indian Ocean. The other inescapable geographical fact is the almost central position of India (along with Sri Lanka) in the Indian Ocean. The Indian Ocean, the only enclosed ocean in the world, occupies almost 20% of the maritime space of the earth. The map of the National Atlas and Thematic Mapping Organization shows that the Indian Ocean includes in it two important sea-lanes in the west — the Red Sea and the Persian Gulf — and washes the east coast of Africa; its eastern sector is marked by the Bay of Bengal (but not the Java and the China Seas); it stretches up to the Antarctica in the south (Gotthold 1987). Though the Atlantic and the Pacific Oceans are larger than the Indian Ocean, the Indian Ocean has been acting as a bridge among numerous communities of Asia and Africa over a very long period of time with rich historical consequences. Franz Broeze rightly argues that a long-range view, say spanning over the last two millennia, would project Asia and not Europe as the leading maritime zone of the world (Broeze 1989: 8). The Indian Ocean undoubtedly dominates the sea-face of Asia. It was the seminal researches of Fern and Braudel that established that a civilization can be related to a maritime space. He repeatedly emphasised that maritime studies do not merely throw light on the sea, but drive home the inseparable unity between a given maritime space and its related landmass (Braudel 1972-73). Ashin Das Gupta further elucidates that neither the sea nor the continent has a history; it is the history of the activities of people who depend on the sea and whose functions also reach out into the inland areas, often away from the direct shores and waves of the sea. Maritime history, therefore, is not an arena for 'at sea' historians since it does not remain confined only to naval battles, naval policy and diplomacy (Das Gupta 2001; Das Gupta and Pearson 1987). Sea-borne trade is certainly one major facet of maritime history but it is not the sole area of thrust. Maritime history addresses the question of the attitudes and orientation of a given society and polity towards its relevant maritime space and takes a close look at those whose life depend on the sea, e.g., the sailor, the ship-captain, the shipbuilder, the fisherman and even the pirate.

The existing historiography of India's position in the Indian Ocean affairs suggests three broad strands of academic interests:

(a) The maritime linkages of the Harappan civilisation (2500-1750 BCE) with Sumer and Akkad through the Persian Gulf;

(b) India's flourishing sea-borne commerce with the Roman empire through the Red Sea (late first century BCE to mid-third century CE) and

(c) The sea change in India's maritime situation with the arrival and consolidation of Europeans in the trade and politics of the Indian Ocean (1500-1800 CE).

The last aspect has received greater scholarly attention as the historians of these centuries had relatively easy access to the massive quantitative data of the Portuguese empire and the English, French and the Dutch East Indian Companies. The study of these three crucial centuries highlight how India and the Indian Ocean became gradually incorporated into the capitalist world economy and ultimately succumbed to the colonial expansion of north Atlantic powers. In the long chronological range of India's role in the Indian Ocean network one can easily see a historiographical gap ranging from 4th to 13th century CE. This paper looks into this chronological phase. This phase in Indian historiography is generally denoted as early medieval, the label implying that these nine centuries or a millennium witnessed the passage from the early historical to medieval. In at least one genre of historical research, the early medieval phase of Indian history is portrayed as one experiencing crises in social, economic and political spheres, ushering in feudalism in India from 300 to 1200 CE. One of the salient markers of the economic formation under Indian feudalism is the emergence and consolidation of self-sufficient, enclosed and stagnant village economies which were hardly conducive to trade, especially long-distance overseas trade. The assumed paucity of trade over long distance is said to have brought about widespread urban decay in India and the attendant 'monetary anaemia'. The historiography of Indian feudalism highlights the relegation of trade, especially long-distance maritime trade, to marginalia especially during the period from 500 to 1000 CE. The historian of Indian feudalism suggests some revival of trade, including maritime trade, in India from 1000 CE onwards (Jha 2000; Kosambi 1956; Sharma 1980; 1987; Yadav 1973).[1] The above portrayal of a sharp decline of overseas commerce of India during the early medieval times has resulted in the much cherished notion of a relatively languishing role of India in the affairs of the Indian Ocean. The above formulation, however, has been effectively contested both on conceptual and empirical grounds (Chakravarti 2001a; 2002a: 58-171; Chamapakalakshmi 1995: 266-307; Chattopadhaya 1994; 1995: 308-346; Sircar 1982). K.N. Chaudhuri (1985; 1990) in his two monographs presents the patterns of commerce in the Indian Ocean from the rise of Islam to 1750 CE after which the situation vastly changed because of the domination of the European powers. Chaudhuri's contribution along with some other leading scholars leave little room for doubt that the rise and spread of Islam was a fillip to Indian Ocean commerce (Hourani 1951).

The very coinage of the expression *Mari Indicum* (the Indian Sea) by Pliny as early as the second half of the first century CE clearly suggests the familiarity of merchants, sailors and shippers at least with the western sector of the Indian Ocean (Rackham 1942: 381). While classical authors were less familiar with the eastern Indian Ocean (the Bay of Bengal) than its western counterpart, the rise and spread of Islam — with a pronounced orientation to commerce and urban life— resulted in a more intimate association with the Indian Ocean from late eighth and early ninth century onwards. A perusal of the geographical texts and travel accounts of Arabic and Persian authors reveals the coinage of the expression *Al-bahr-al-Hind*, the near equivalent of the later English expression the Indian Ocean. In the *Al-bahr-al-Hind* was included *Bahr-al-Larvi* (the sea of Lata or Gujarat), corresponding to the Arabian Sea. The Arab authors show a clear awareness of the *aghbab* or the channel separating southern part of the peninsula from Sri Lanka; this seems to be the same as the Palk Strait. Further to the east the maritime space was designated as *Bahr-al-Harkal* (or *Harkand)*. Harkal is clearly the Arabicisation of the name Harikela, identified with the present Chittagong, Noakhali, Comilla regions of Bangladesh (south-eastern most Bangladesh to the east of the river

Meghna) (Mukherjee 1975: 115-119). The description of the *Bahr-al-Harkal* in Arab accounts (especially in the Persian treatise *Hudud-al-Alam* written by an anonymous author in 982 CE) closely matches the coinage of a Sanskrit term, *Vangasagara* (the sea of Vanga) in a Sanskrit inscription of 971 CE found close to Dhaka. The coinage of terms denoting the Indian Ocean or different segments of it, especially by Arabic and Persian authors, cannot but demonstrate their familiarity and acquaintance with the maritime space in question, obviously in the wake of the burgeoning commerce in the Indian Ocean. It is true that contemporary Indian were content with merely using the terms *purva* (east) and *paschima/apara* (west) to suggest their awareness respectively of the Bay of Bengal and the Arabian Sea (Basak 1952: 51-58; Chakravarti 2002b; Sircar 1958: 337-339). Chinese annals also speak of maritime linkages with India and various areas of the Indian Ocean. The Chinese authors often used a generic term to point to the Indian Ocean; it was known as *Hsi-hai,* the western sea. At least on one occasion the Chinese account speaks of the *Silan-hai* or the sea of Ceylon (Sri Lanka). This term, distinct from *Hsi-hai,* may correspond to the Bay of Bengal (Ray 1987: 109-124). The above exercise in the study of nomenclatures of the Indian Ocean is a pointer to the frequent sailings across the vast maritime space.

No less significant is observation regarding another geographical feature of the Indian Ocean, the immense influence of the monsoon wind system on the sailing patterns in this maritime space. Prior to the arrival of steamships, shipping in the Indian Ocean was largely guided and shaped by the more or less predictable alterations of the south-western (June to September) and the north-eastern monsoon (October to January) wind system. The term monsoon is derived from Arabic *mawsim* (plural *mawasim)* which means season. However in the celebrated navigational manual of Ibn-e-Majid (fifteenth century), the greatest navigator in the Indian Ocean prior to the arrival of the Portuguese, the term *mawsim* stands also for a sailing season to reach a particular destination across the sea (Tibbetts 1971: 34-36). When the Indian Ocean network was at its zenith, its western terminus was situated in Siraf (in the Persian Gulf) under the Abbasid Caliphate and Al-Fustat (Old Cairo) under the Fatimid Caliphate in Egypt. The eastern reach of this network stretched to the ports of maritime Southeast Asia and China. As it would be impossible for a traditional ship to complete a round trip between the western and the eastern termini in a single year because of the alterations of the monsoon winds, the two seaboards of India — dotted with numerous ports — were ideally suited as stopovers and points of transshipment. This must have proved immensely advantageous to the overseas commerce of India which was also involved in brisk export and import transactions (Chaudhuri 1985).[3] Contrary to the image of a languishing overseas trade in the early medieval times, evidence available from a wide range of sources strongly suggest India's active and important role in the Indian Ocean commerce. The documentation in use here covers inscriptions, indigenous literary texts, a large number of Arabic and Persian accounts, Chinese texts, the accounts of Marco Polo in the late thirteenth century, and last but not the least, the letters of Jewish merchants (Genizah documents) trading with India from the eleventh to the thirteenth century (Ahmed 1960; 1989; Beal 1983; Broadhurst 1952; Elliot and Dowson 1939; Ghosh 1990a; 1990b; Gibb 1929; Goitein 1954: 247-257; 1973; 1980: 43-66; 1987: 448-464; 1967-96; Grennville 1980; Hirth and Rockhill 1911; Legge 1974; Mills 1970; Minorsky 1937; 1942; Reif 2000; Sachau 1910; Shaked 1964; Weinstein 2001; Yule and Cordier 1903).[4] The information offered by these sources is hardly adequate, often scattered and stereotyped and offers little statistical data. These belong, instead to the category of 'qualitative sources'.[5] The letters of the Jewish merchants, highlight the activities of *Musafirun-ul-Hind* (travelers to India), whom S.D. Goitein immortalised under the label, 'India Traders'. These unique letters enable us to hear the voices of the merchants who actually participated in the Indian Ocean trade; they also occasionally provide some statistical information.

The Indian Ocean, especially its western sector, came into great prominence during the first three centuries CE, on account of India's brisk trade with the Roman Empire. The maritime network of the western sea-board of India through the Red Sea plays an important role in bringing India closer to the Mediterranean commerce. This far flung commerce, however, experienced a slump after c. 250 CE as there was a lesser demand of the Indian products in the Roman world in the wake of the gradual decline of the Roman Empire. The rise of the Sasanid Empire in Iran and the eastern Roman Empire with its capital in Byzantium again brought the western Indian Ocean to prominence after 400 CE. The maritime commerce of the eastern Roman Empire with India is evident in the coin hoards, largely from south India, yielding many coins of the Byzantine rulers. It was maintained largely through the Persian Gulf, which now overshadows the Red Sea network. The rise of the Sasanids of Iran coincided with their distinct interests in the commerce of the Persian Gulf. In fact one may discern some rivalries between the Byzantine and the Sasanid powers over the Persian Gulf trade with India. The Syrian Christian monk Cosmas Indicopleustes who visited India in the late sixth century spoke of the importance of a few ports on the Konkan coast like Kalliene (Kalyan) and Sibor (Chaul). However, his accounts particularly underline the great prominence of Sri Lanka in the trade of the western Indian Ocean. The linkages of the western sea-board with the northern part of the Persian Gulf are also brought out by the distribution of sites yielding red ware. These sites are mostly coastal, located in Gujarat, the Indus delta, the Makran coast and the littorals of southern Iran; the archaeological evidence support the impression left behind by Cosmas (Whitehouse and Williamson 1973: 29-49). Hsuan Tsang (travels in India from 629 to 645 CE) informs us about the importance of Bhrgukachchh on the mouth of the Narmada, though the port probably was past its great glory during the early centuries of CE. The Indus delta was witnessing the rise of a new port, Debal, mentioned as Daibul in early Arab accounts. Early Arab accounts recounting raids on this port in c. 627 CE attest to its importance. The descriptions of the first Arab invasion of Sind in early eighth century, narrated in a later source, the *Chachnama*, point to the direct voyages between Daibul and Sri Lanka. That the Sasanid Empire was interested to maintain linkages with South Asia is seen in the account of the embassy sent by Chalukya Pulakesin II (c. 610-642 CE) to the court of Khusru, his Sasanid counterpart (Majumdar 1962). To Pulakesin II also goes the credit of conquering Puri, generally identified with the Elephanta Island near Mumbai, by his own fleet.[6] One is not sure if the Chalukyas of Badami (near Bijapur) really maintained a regular navy, as this was essentially a land-based polity. But his conquest of the Elephanta Island may point to his awareness of the commerce in the Arabian Sea and the Persian Gulf. This gains ground in the light of his diplomatic contacts with the Sasanid ruler.

Ports on the east coast of India appear to have been less affected by the waning of the Roman contacts with India. The premier port in the Kaveri delta was certainly Kaveripattinam, identified with the archaeological site of Puhar or Poompuhar. Kaveripattinam had already figured in the Sangam poems as an important port. But it is in the later two famous Tamil epics, the *Silappadikaram* and the *Manimekhalai* (ascribed to c. 5th century CE) that Kaveripattinam attains its celebrated status. Excavation has yielded remains of a wharf at this port, while the *Silappadikaram* describes the facility of showing signals to ships even at night — a forerunner of a lighthouse. The graphic description of a specific quarter in the port town earmarked for the *Yavana* merchants indicates the regular arrival of non-indigenous traders from the West to this celebrated port (Champakalakshmi 1996; Chattopadhyay 1997: 181-208). To the north of the Coromandel coast stood the littorals of Vengi, situated in the deltas of the Krishna and the Godavari. The *Periplus exo Thalasses* (distinct from the well known *Periplus tes Erythrus Thalasses),* datable to the fourth-fifth century CE, reveals to us the existence of an *aphaterion* or a point of departure of ships in the deltas of the Krishna. The *aphaterion* was suitable for ships bound for Chryse Chora and Chryse Chersonesys (comparable to *Suvarnabhumi* and *Suvarnadvipa* of early Indian tradition), sought to be located in mainland Southeast Asia (Chakravarti 1986; Miller 1962).

In the Bengal coast the outstanding port was Tamralipta, identified with the archaeological site of Tamluk on the river Rupnarayan in West Bengal. Its importance as a port of international commerce is well driven home in the Chinese accounts. When Fa Hsien, the well known Chinese Buddhist pilgrim (travels in India from c. 399 to 414 CE), left India for Sri Lanka he boarded a merchant vessel from Tamralipta. The account of the voyage clearly shows that it was undertaken in winter months, taking advantage of the north-east monsoon to reach Sri Lanka directly. This was not a coastal voyage, but was a high sea journey. The Chinese pilgrim then boarded another vessel from Sri Lanka bound for Java from where he finally sailed for the coast of China. The description not only highlights the maritime linkages between the Bengal littorals and Sri Lanka but also points to the interconnectedness of the Bay of Bengal with Southeast Asia. This point is further illustrated by the accounts of Itsing (travels in India from c. 675 to 695 CE) who reached Tamralipta by sailing from the Malay peninsula. The other Chinese pilgrim, Hsuan Tsang, also did not fail to observe the significance of Tamralipta (Chakravarti 2002b; Ray 1990: 1-14). Hsuan Tsang is also the first to inform that *San-mo-ta-ta* (Samatata area, identified with the Noakhali-Comilla region of Bangladesh) maintained contacts with Shi-li-chi-ta-lo (Srikshetra in Myanmar with its capital Prome on the Irawaddy), *Kia-mo-lang-kia* (Kamalanka, identified with

Pegu), To-lo-po-ti (Dvaravati the famous Myanamrese kingdom), I-shung-na-pu-lo (Isnapura to the east of Dvaravati), Mo-ho-chen-po (Mahacampa in Vietnam) and Yen-niona-chen (Yamanadvipa, identification uncertain) (Watters 1960: 200). It is not unreasonable to infer that these areas in mainland and maritime Southeast Asia could be reached from Samatata in the south-eastern most part of the Bengal delta, though it must have been overshadowed by the premier port of Tamralipta. The importance of the Bengal coast as a major point of sea-borne contacts with Southeast Asia also figures in a sixth century Sanskrit inscription from the Malay peninsula. It speaks of *mahanavika* (the master mariner) Buddhagupta who was a native of *Raktamrittika* (usually located in the Murshidabad district, West Bengal) and who evidently went to the Malay peninsula, obviously in connection with his profession as a master mariner.[7]

The above survey, synoptic and hurried, does not suggest a sharp decline in the maritime trade of India in the centuries succeeding its commerce with the Roman Empire. It is however true that it does not match in scale with the commerce between the eastern Mediterranean and India. One cannot also deny that around the eighth century premier ports of the previous epoch, like Bharuch, Kaveripattinam and Tamralipta, gradually faded out. This may have prompted the historians to infer a period of slump of India's maritime commerce. On the contrary, in the historiography of India the voluminous evidence available in the Arabic and Persian texts (several of them being composed prior to 1000 CE) has been consistently ignored and sidelined. Impressions left behind by Sulaiman (c. 851 CE), Ibn-e-Khuradbeh (c. 882 CE), Al-Masudi (c. 915 CE), *Kitab-ul-Ajaib-ul-Hind* (c. 955 CE), the anonymous author of the *Hudud-al-Alam* (c. 982 CE), Al-Beruni (1030 CE), Al-Marvazi (1120 CE), Al-Idrisi (c. 1162 CE), Ibn-e-Batuta (early fourteenth century) speak in volume of the over-seas contacts with India, the importance of Indian ports and the transaction of commodities. To this can easily be added the evidence of Chau ju Kua (1225 CE) who as an officer supervising foreign trade under the Sung dynasty of China wrote a valuable account of commodities and ports of India. Towards the end of the thirteenth century Marco Polo also informs us about India's role in the commerce of the Indian Ocean. We shall also utilise here the evidence of the Jewish business letters. The Jewish merchants' commercial ventures spanned from Tunisia and Levant to Egypt,

Aden from where they regularly sailed to ports on the west coast of India.

The establishment of the Abbasid Caliphate in Baghdad paved the way for the growth of trade in the Persian Gulf where stood the premier port of Siraf. After the decline of Siraf since late eleventh century, Kish or Qays came to great prominence. It is from the thirteenth century one notes the rise of Hormuz as a port of immense significance in the western Indian Ocean commerce. The Red Sea network seems to have taken a backseat during the period from circa seventh to the tenth century CE. The rise of Fatimid Caliphate in Egypt in 965 CE once again swung the pendulum in favour of the Red Sea. The Red Sea network further offered linkages with the vibrant Mediterranean maritime trade through Egypt, which acted like a hinge between the Indian Ocean and the Mediterranean with far reaching consequences. The rise of Islam indeed provided a fillip to the maritime commerce of India.

This brings us to another survey of the ports in both the seaboards of India, active during the period from circa ninth to thirteenth century. In Sanskrit sources ports are called *velakulas,* in Tamil as *velapuras* and *pattinam/pattanams.* In the Indus delta the port of Daibul certainly prospered. It seems to have been well connected with Multan and Mansura on one hand and with the Makran coast on the other. Its remains are found in the excavations at Banbhore (Dani 1988; Kervan 1999: 70-153).

On the Gujarat coast Bharuch is replaced with Stambatirtha/ Stambhaka which is the same as the famous port of Khambat. It figures prominently in the Arab accounts under the name Kanbaya and Kambaya. The rise of the Chalukyas as a regional power of Gujarat paved the way for the integration of the coast with the extensive interior by a number of overland routes (Jain 1990). Early medieval Gujarat experienced remarkable agricultural growth also. In western India, especially in Gujarat, Rajasthan, Malwa and western part of Uttar Pradesh, emerged a new type of exchange center in the early medieval period. These are known as *mandapikas* (modern *mandis*) which provided vital linkages between the rural level market centres *(hatta/ hattika)* and larger markets at urban centres and ports *(pattana/velakula)* (Chakravarti 2002b; Chattopadhyaya 1994). These were instrumental in the efflorescence of the ports of Gujarat, particularly Khambat. The port maintained coastal linkages with ports in the Konkan and Malabar littorals, as will be evident from the accounts of the Arab writers. Two early medieval Indian texts, *Vastupalamahatmayam* and *Jagaducaritam,* bear clear impressions of the contacts between Khambat and Ardrapura, which stands for Hormuz in the Persian Gulf (Buhler 1892a; Sandesara 1953). The availability of an extensive hinterland and the agricultural prosperity of Gujarat in the early middle ages certainly contributed to the sustenance of Khambat; it was also well served by a few smaller ports in Gujarat and northern Konkan. These relatively smaller ports acted as feeder ports for the premier Gujarat port. One may enlist in this context Mangalapura (Mangrol in the Kathiawad coast), Diu, Somnath and Ghogha. One of the earliest mentions of Diu is available in a twelfth century Jewish business letter which speaks of a voyage from Al-Manjrur (Mangalore) to Al-Divi (Diu) in Gujarat from where the ship sailed to Aden (Goitein 1980; Chakravarti 2001a). Diu was thus connected with both Kanara coast and Aden. Though Somnath is principally celebrated in Indian history as a great Saiva sacred center, it was also a port of some prominence, located at the confluence of the Sarasvati with the sea *(Sarasvtisagarasamprayogavibhusita)* (Buhler 1892b: 271-287). Al-Beruni was one of the earliest authors to have appreciated the importance of Somnath as a port; he states that it was a vantage point of departure for Zanz or Zanzibar in East Africa (Sachau 1910). We have however no further corroborative evidence in support of Al-Beruni's mention of voyage from Somanth to the East African coast. That Somnath did receive ships from Hormuz is proved beyond doubt by a remarkable bilingual (in Arabic and Sanskrit) inscription of 1264 CE from Somnath (Chakravarti 2000a; Desai 1961-62: 17-24; Hultzsch 1882: 241-245; Ray et al. 2000; Sircar 1983: 402-403). *Hurmuji-vahana* or ships from Hormuz also reached the port of Ghogha *(Ghogha velakula)* in Gujarat.

To the south of the Gujarat coast was situated the narrow strip of Konkan, repeatedly figuring in the Arab accounts as Kunkan, Kamkam and Makamkam (Nairnar 1942). At the northern most fringe of Konkan, almost touching southern Gujarat, was the port of Samyana, Arabicised as Sanjan/Sindan. A port to the south of Sanjan was Sristhanaka or Thana. Thana seems to have come to prominence later than Sanjan. A Jewish business letter of 1145 CE narrates a coastal voyage from Al-Manjrur to Thana wherefrom the ship further proceeded to Barus (Bharuch). It appears in greater prominence under the name Tana in the accounts of Marco Polo in late thirteenth century. Thana during that period seems to have served as the major port of north Konkan under the Yadavas of Devagiri. Cemulya or Chaul, situated to the south of present Mumbai, was a significant port of Konkan. Already known as Sibor in the sixth century accounts of Cosmas Indicopleustes, Chaul appears as Saimur in the Arab accounts. Three ports in the southern part of Konkan rose to some prominence in the early centuries of the second millennium CE. These are Balipattana (modern Kharepatan), Candrapura (Sindabur in Arabic texts) and Gopakapattana (Old Goa). While Balipattana was located in the kingdom of the Silaharas of southern Konkan, Candrapura (Chandore) and Gopakapattana belonged to the realm of the Kadambas of Goa. A close perusal of Sanskrit inscriptions, Arab texts and Jewish business letters brings out a few features of the Konkan ports. Though geographically the Konkan ports were located in the most conducive areas for harbour building, they were less prosperous and prominent than the ports in Malabar and Gujarat. These ports in Konkan neither had rich agricultural hinterland nor commanded regular linkages with the interior because of the geographical barrier of the Sahyadri range between the coast and the mainland. The ports of Konkan witnessed regular settlements of Arab merchants who figure in Sanskrit inscriptions under the category of *Tajjikas.* Al-Masudi noted the presence of large number of Omani, Sirafi and Baghdadi merchants at the port of Saimur or Chaul. The long presence of these Arab merchants explains how and why their names sometimes appear in Sankritised forms in inscriptions: e.g. Mahumata (Muhammad), Alliya (Ali), Shariyarahara (Shahriyar), etc. A close study of two inscriptions of the Silaharas of south Konkan, one of 1008 CE and the other of 1094 CE, strongly suggests that ports of Konkan were engaged in looping coastal commerce. This coastal linkage further extended to Gujarat and Malabar. Impressions of long coastal voyages from around Goa to Somnath in the Kathiawad coast are available from the inscriptions of the Kadamba rulers of Goa.[8]

The Arab accounts leave a impression that the ports of north Konkan, especially Sanjan and Saimur, derived considerable advantages from their linkages with the Persian Gulf commerce. A few Jewish merchants undertook a voyage from Aden to Sindabur, close to Goa, as known from a correspondence of a Jewish merchant. Sindabur thus seems to have experienced a gradual expansion of its network up to

Aden in the twelfth century. To the south of Goa the most important port in the Kanara coast was Al-Manjrur, sometimes mentioned in the Arab texts, but repeatedly figuring in the business letters addressed to Abraham bin Yiju Yishu, a Jewish merchant of Al-Mahdiyya or Tunisia who lived in this port town for more than seventeen years (1132-1149 CE).[9]

The southern most segment of the western sea-board is famous as Malabar, known as Malibar, Manibar, Al-Mulaybar in the Arabic texts and the Jewish business letters. Three ports deserve special mention: these are Kulam Mali (Quilon), Panatalayani Kollam (Fandarina) and Calicut. The Malabar ports could be reached from both the Persian Gulf and the Red Sea ports in around 30 days by utilising the south-western monsoon. These ports earned their celebrity for handling the invaluable cargo of spices, especially the pepper of Kerala aptly described as 'black gold'. Calicut does not seem to have attained its great glory prior to the early fourteenth century. Ibn-e-Batuta is one of the first authors to have described its immense importance. He noted that the huge Chinese junks did not sail further west of Calicut and Calicut provided excellent transshipment facilities for Chinese and Arab vessels. Calicut reached its zenith in the fourteenth and fifteenth century. The famous Chinese voyages under Admiral Cheng Ho/Zeng He (from 1404 to 1433 CE) touched Calicut as many as eight times (Appadurai 1936; Sastri 1972).

In the east coast the Tamil merchants regularly frequented the Bay of Bengal to reach Sri Lanka, Sumatra, Thailand, lower portions of Myanmar and even China. A number of Tamil inscriptions found in these countries bear eloquent testimonies to the sea-borne ventures of the Tamil merchants. This is also reflected in the distinct interests of the Chola rulers in maintaining cordial relation with the Sung court of China and the kingdoms of Srivijaya (Palembang) and Kadaram (Kedah, Malay Peninsula). Three Chola rulers, Rajaraja (Lo-ts-lo-tsa), Rajendra and Kulottunga I (Ti-hua-kie-lo), sent embassies to China respectively in 1014, 1033 and 1070 CE. The premier port in the Tamil coast in the eighth and ninth century was Mahabalipuram, south of Chennai, under the Pallavas. The establishment of the Chola realm as the most formidable power of early-medieval south India (850-1200 CE) witnessed a shift in the site of the premier port, from Mahabalipuram in Tondaimandalam to Nagapattinam in the Kaveri delta, the very core area of the Chola power (Hall 1980; Sastri 1955).

To the north of the Coromandel coast, in the Andhra-Kalinga coastal segment of the eastern sea-board, emerged a new port in the second half of the eleventh century. This port is Visakhapatanam, obviously the same as the homonymous modern port in the northern part of the Andhra coast. The very name Visakhapatanam figures as early as 1068 CE in an inscription. It seems to have been under the eastern Chalukya kingdom of Vengi or the eastern Ganga rulers of Kalinga. The most interesting point about Visakhapatanam is that sometime between 1068 and 1090 CE it was formally named Kulottungacholapattanam, though the earlier name Visakhapatanam was also retained in inscriptions. The renaming was done after the reigning Chola king Kulottunga I (1070-1120 CE). Kulottunga who originally belonged to the eastern Chalukya house of Vengi, ascended the Chola throne in 1070 CE by virtue of a marriage alliance. This led to the merger of the Vengi kingdom into the Chola territories. This is a unique instance in the entire history of the Chola dynasty that the name of an existing port was renamed after a reigning king. By doing so, Kulottunga seems to have officially enhanced the status of a port which belonged to his original seat of power. In and around 1090 CE Kulottunga also remitted tolls and customs (hence his title, *Sunganadavirttachola),* obviously to promote long-distance maritime trade in the Bay of Bengal. Inscriptional evidence may suggest his interests in diplomatic contacts with the ruler of Pagan which lay on the other side of the Bay of Bengal, across the Andhra coast (Chakravarti 1995: 57-78). Despite the efforts of Kulottunga, Visakhapatanam/ Kulottungacholapattanam did not enjoy long lasting prominence. It faded out after twelfth century. In the Andhra coastal area another port came into considerable prominence in the thirteenth century. This is Motupalli, figuring in an inscription of 1245 CE, issued by Ganapatideva (1199-1261 CE), the greatest ruler of the Kakatiya house of Warangal. Ganapati's inscription is categorically labeled as an *abhayasasana,* a charter of security to merchants *(samyatrikas)* from distant areas across the sea *(dvipantara)* who had earlier been harassed by rampant piracy. He put an end to this menace and declared fixed levies, in cash and kind, on different commodities carried in ships to the port of Motupalli. These measures appear to have proved beneficial, as Motupalli, under the name Mutfili, is much praised by Marco Polo as a fine port in late thirteenth century. It appears that the two ports of the early medieval Andhra-Kalinga coast did not attain their full promise and potential in spite of the considerable political and administrative support provided to them from the apex political authority in the interior. The littoral in question did not probably develop adequate linkages with the hinterland in the interior. Kalinga and Andhra have a long history of agricultural settlements, but these were largely in the coastal belt. Neither western and central Orissa nor the Telangana area in Andhra offered much scope of agrarian expansion and craft production in the early middle ages. It is only from around the time of the Kakatiyas of Warangal that agrarian expansion became visible in the relatively drier areas of central and western Andhra. The same period also witnessed the emergence of *pentha/pintha/pemta* (modern *peth)* as market place providing crucial linkages between the rural level exchange centers *(addas)* and larger markets at urban areas and ports (Chakravarti 1995: 57-78; 2001c: 99-120; 2002b: ch.X; Hultzsch 1913-14: 188-197).

As one moves along the eastern sea-board to the north of Orissa, one enters the Bengal coast spanning from the coast of West Bengal in India to the easternmost section of it around Chittagong in Bangladesh, almost touching the Arakan littorals. The premier port in Bengal was Tamralipta till the eighth century after which it gradually faded out mainly due to the siltation of the navigable riverine channel on which it stood. From the middle of the ninth century onwards a large number of Arab authors began praising a port in the Bengal delta, named Samandar. The name itself seems to have been derived from *samudra (samudra, samundar, samandar)* or sea. Most historians prefer to locate it in and around present day Chittagong in south-easternmost Bangladesh. Al-Idrisi in the second half of the twelfth century informs that it was situated on an inlet like formation *(khawar)* which would have facilitated the ingress and egress of sailing crafts. Close to Samandar was an island where flocked merchants in large numbers from different lands. This island is generally identified with Sandwip near Chittagong. The port was known to Ibn-e-Batuta under a different name in the fourteenth century, Sudkawan, apparently based on the indigenous name Cattagrama. Early medieval Arab authors speak of the voyages between Samandar/Sudkawan on the one hand and Uranshin (Orissa coast), Kanja (Ganjam in the southernmost part of coastal Orissa or Kanchipuram in Tamil Nadu) and Silandib/Serndib (Sri Lanka). This new port thus seems to have been well integrated in the Bay of Bengal sea-borne network. The Bengal coast has a general orientation to Southeast Asia. In the ninth century the Sailendra ruler of Java, Balaputradeva, requested his contemporary Pala king Devapala to grant land in favour of the famous Nalanda monastery. This speaks of the linkages of the Bengal coast with maritime Southeast Asia. Interestingly enough, Ibn-e-Batuta enlightens us on the sailing between the Maldives and Sudkawan, implying a further expansion of network of this Bengal port with the south-western sector of the Indian Ocean. It is difficult to perceive a slump in the maritime commerce in the Bengal coast following the decline of Tamralipta. It however cannot escape our attention that the location of the major sea-port of early-medieval Bengal shifted from the western part of the delta to its south-easternmost corner. A close perusal of the Arab accounts amply bears out that products of the extensive interior were brought to Samandar. This impresses upon the hinterland commanded by the port in question. The Bengal delta was the only effective outlet to the sea for greater parts of the land-locked Ganga valley. The fluvial routes, so typical of an active delta, must have facilitated the communication of Samandar with the mainland. Thus Ibn-e-Batuta undertook a riverine journey from there along the Blue River (the Meghna) as far north as Habang (Habiganj in Bangladesh). The flowering of this port to prosperity is also evident from the accounts of the celebrated Ming voyages under Admiral Cheng Ho (1404-1433 CE). The Chinese fleet, as Ma Huan reports, reached Chittagong as many as on four occasions by direct voyages from Southeast Asia through the Malacca Strait. All these underline the integration of the Bengal coast with the Indian Ocean network of the early middle ages. If the early Portuguese accounts hailed Chittagong as the 'porto grande' in early sixteenth century Bengal, its foundation was laid in the first-half of the second millennium (Chakravarti 2002b; Mukherjee 1981: 65-83; Ray 1993).

However great was the port of Samandar, it could not have stood in isolation, without being served by a few feeder ports. Such smaller ports, less prominent than Samandar, can be located in the interior of the delta. The fluvial routes in the delta provided excellent opportunities towards the making of a few inland riverine ports. One such port, Devparvata, stood in Mainamati-Comilla. Three inscriptions, ranging in date from late seventh to early tenth century CE, graphically describe Devaparvata as being encircled by the river Ksiroda (modern Khirnai) on which regularly plied many boats manned by expert sailors *(navika)*. These leave behind fine portrayal of an inland riverine port.

An inscription of 971 CE reveals the existence of another riverine port to the north of present Dhaka. This port, located on the river Bangshi, offered warehousing facilities and therefore designated as a *sambhandariyaka.* Though situated in the interior and away from the coast, this *sambhandariyaka* seems to have been connected with the sea through the existing inland riverine communication system and hence was given the epithet *Vangasagara-sambhandariyaka* (a riverine port with warehousing facilities and connected with the Bay of Bengal or *Vangasagara).* Towards the very end of the twelfth century (1196 CE) one comes across a place called Dvarahataka in the vicinity of Gangasagara. The place was so named as it was a rural market center (*hatta*) located as it were at the gate *(dvara)* to the Bay of Bengal. More or less similar in character was Betaddacaturaka, figuring in a thirteenth century inscription, its eastern limit being marked by the Ganga *(purve Jahnavisima).* This is the same as Betore close to Howrah and mentioned as a convenient anchorage in the sixteenth century Portuguese accounts. Early-medieval Bengal not only experienced the rise of Samandar as the premier port, but also was a witness to the proliferation of several smaller inland riverine ports in different areas of the Bengal delta (Chakravarti 2002b).

These ports on the two sea-boards of India were involved in the exchange of commodities. The sources mentioned above also throw light on the items of exports and imports. Among the perennial export items of India were various types of textiles for which India attained legendary fame. Textiles of Gujarat, Malwa and Bengal deserve particular mention in this context. Textile products ranged from the famous and expensive muslins of Bengal to ordinary cotton cloth meant for daily necessity (Gopal 1965). There was also the expected craze for precious gems and stones in abroad. The pearl fisheries of the Tuticorin coast in the Pandyan realm and the diamond mines in Andhra under the Kakatiyas figure prominently in the accounts of Marco Polo (Appadurai 1936; Sastri 1972). Among spices, the pepper of Malabar was highly prized. The Genizah papers of Jewish merchants are replete with references to regular shipping of pepper in large quantities from Malabar to Aden, and from there through the Red Sea to Egypt.[10] From Egypt pepper finally reached Italian ports by voyages across the Mediterranean. At the port of Aydhab on the western shore of the Red Sea Ibn-e-Jubayr, an Andalusian traveller, saw huge heaps of Indian pepper in 1184 CE.[11] Pepper was a forest product procured in Kerala but assumed the character of a luxury as it reached the markets of the Red Sea and the Mediterranean, fetching very high prices. Two other forest products, aloes wood and teak had wide demand in West Asia, according to the Arabic and Persian texts. The best two varieties of aloes were Multani (from Multan) and Kamaruni (from Kamarupa or upper Assam). Indian teak was regularly exported to Persian Gulf and Red Sea ports as the basic raw materials for building sea-going vessels for Arab merchants who did not have a local source of durable wood for boat-building. Two highly prized exotic items were rhinoceros horns and yak tails, both available in the north-eastern part of India.[12] These were obviously luxury items catering to the taste of the rich, but a relatively small number of clientele.

Along with high value, portable, small quantity luxuries, were shipped daily necessities in bulk quantities. The Genizah letters repeatedly tell us about the regular shipment of various types of iron from Al-Manjrur and other ports of Malabar to Aden during the period from eleventh to the thirteenth century. There appears to have been a steady demand for Indian iron in the West, possibly for making weapons like swords.[13] A combined testimony of Ibn-e-Batuta and Ma Huan would reveal Bengal shipped rice to Maldives which in return sent out to Bengal cowry shells. Cowry shells functioned as a major currency not only in eastern India, but also in the East African port in Kilwa and the West African kingdom of Dahomey.[14] Huge quantities of cowry shells certainly functioned as small exchanges and also as ballast in the maritime network of the Indian Ocean. This clearly negates the widely held notion that the use of cowry currency in early-medieval Bengal was symptomatic of languishing trade and signified only local level transactions. Cowry currency was well integrated with the Indian Ocean commercial economy of the pre-modern times. There is a distinct possibility of a regular supply of coconut coir (mentioned as *qinbar* in Arab accounts) from the west coast of India to the ports of the Red Sea and the Persian Gulf. Coconut coir was an indispensable ingredient in the wooden ships of the Indian Ocean, which did not use iron nails to fasten the planks. The west coast of India, noted for the regular plantation of coconut since at least second century and also exported this item in bulk to the shipbuilding centres in the Islamic world (Dionisius 2002: 174-208; Gopal 1970: 108-122). Besides the demand for Indian cotton textiles in the Persian Gulf and the Red Sea ports was closely linked the regular export of Indian indigo to the West. The

Genizah documents inform about a eleventh century Jewish merchant, Ibn-e-Awkal who maintained meticulous details of correspondence with his agents. While his seat of business was at Al-Fustat (Old Cairo), his agents spread as far as Al-Mahdiyya (Tunisia) and the Levant. His correspondence speaks of his dealings in Sanjani indigo; this is certainly the indigo produced in western India and shipped from the port of Sanjan in north Konkan. That the Sanjani indigo competed well with indigo from Amta (in Palestine) and Kirman (south-western Iran) is clearly impressed in the correspondence of Ibn-e-Awkal. One is therefore not surprised to come across another merchant, involved in the 'India trade', specifically dealing in indigo; hence he is named Sheikh Alib Muhammad Nili *(nil* = indigo) (Goitein 1973: 190). This is another instance of the importance of bulk items in the maritime trade of India. It effectively rejects the stereotyped portrayal of India's export of merely high value, small quantity, exotic, luxury or prestige goods.

India received exotic spices and gems from Sri Lanka and Southeast Asia as imports many of which were further shipped to the Western destination. This suggests the involvement of Indian ports in transit trade. A case in point is the importation of cinnamon, cardamom, camphor and sandal; the same can be said of the import of ruby from Sri Lanka. Indian imports included precious metals, especially gold and silver, though the actual quantum cannot be ascertained. In at least one Genizah letter merchants preparing for a sailing to India from the Red Sea port of Aydhab are said to have preferred silver to gold as the form of payment to purchase Indian commodities. In another letter of 1139 CE the import of silk instead of gold from Aden to Al-Manjrur is clearly mentioned (Chakravarti 1996: 18-30; Stillman 1971: 15-88).[15] That copper was brought to India from West Asia is amply borne out in a few Jewish letters. Among India's imports, a particular commodity demands our attention. This is the horse from Arabia and Fars. There was a constant demand for imported war horses. This demand seems to have greatly increased in the early middle ages when there were numerous regional and local powers, engaged in endemic warfare. One of the regular sources of good quality war horses was from the Central Asian steppe regions, called Tatari horses in medieval sources. These were brought to India by overland routes through the north-western borderlands of the subcontinent. With the onset of the second millennium, excellent horses from Arabia and Fars began to be sent to India, by overseas transportation. These horses were therefore labeled as *bahri* (sea-borne) in Arabic and Persian sources and these fetched the maximum price. Marco Polo speaks of the shipping of ten thousand horses per year to the Pandyan realm alone from the ports like Shihr, Kish, Hormuz and Aden. Marco Polo's accounts, when read along with those of Ibn-e-Batuta, Barani and Abdulla Wassaf, impress upon the enormous price of these horses, each fetching a price of 220 gold coins. The most significant point of disembarkation of these *bahri* horses was the Malabar coast. This coastal zone, significantly enough, was the arena of great activities of Tamil horse-dealers, called *kudiraicettis* of Malainadu *(kudir*=horse, *cetti*=merchant). They were possibly engaged in the distribution of the imported horses to political powers inland, for instance, the Pandyas of Madurai. No less interesting is the observation of Marco Polo that the king around the port of Tana (Thana) colluded with the pirates on the condition that he would receive all the horses from the pirated ship, while the rest of cargo would remain with the pirates. Polo impresses upon the desperate act of a coastal authority in order to capture by force and fraud the shipload of imported horses which apparently were too costly for him to afford regularly (Chakravarti 1991: 159-182; 1999; Digby 1971).

Now we briefly deals with an overview of the ships and vessels, the carriers of men and merchandise. The main impediment to our understanding is the lack of specific data in available literature. Sanskrit texts often use the term *nau* to denote any watercraft, from a sea-going vessel to a river-craft. The *Yuktikalpataru* of Bhoja speaks of two broad classification of vessels: *samanya* (river-craft) and *vishesa* (sea-going ship). These are further subdivided into *deergha* (long) and *unnata* (tall) (Mookerji 1912; 1957). Though the data contained in this text are often cited by historians of ancient Indian shipping, it is not beyond doubt whether these describe actual ships and boats. Inscriptions are not familiar with these terms and on the other hand mention *pravahana* and *vahitra.* A close examination of inscriptions from the Konkan coast suggests that *pravahana* were coasters plying along the coastal routes. The *vahitra* generally figures in association with the descriptions of overseas voyages *(dvipantara).* There are literary images in classical Sanskrit texts of *yanapatras* or sea-going ships carrying many passengers (*samyatrikas*) (Chakravarti 2002c: 28-31).

Some iconographic information is available on Indian ships in a few twelfth and thirteenth century sculptures. One such panel at Borivali near Mumbai depicts three scenes of a sea-battle. One particular

2.1. Old Goa: Hero stone in the Archaeological Museum

sculpture in the Archaeological Museum at Old Goa (fig. 2.1), shows a large ship being gradually lowered into the water from a shipbuilding area (Deloch 1987: 165-184; 1996: 199-224). The Chinese vessels plying in the Indian Ocean and visiting Indian ports came under the general category of Junks. The most and frequently occurring tenn for a ship in Arabic texts is *markab.* At least in one twelfth century Jewish letter occurs the Arabic term *jahaz,* which however relates to the preparations of fitting out a ship for a long voyage. Three types of ships, *burrama, shaffara* and *jashujiyat* were employed by the rapacious king of Kish during a naval blockade around Aden in the early twelfth century, according to a Jewish Genizah letter (Goitein 1954: 247-257). The *jilab* was in frequent use in the Red Sea area; Ibn-e-Jubayr saw how the *jilab* was always overloaded with passengers in the port of Aydhab.

The most important figure on board the Arab ship was the *muallim,* navigator or captain, who figures very prominently in the shipping manual of Ibn-e-Majid, himself a great navigator. Right from the beginning of the preparation of a voyage till its reaching the ultimate destination, the *muallim,* according to Ibn-e-Majid, was the final authority in the running and maintenance of the ship.[16] Two subordinates under him were the *kardar*[17] and the *tandel.* The *muallim* of Ibn-e-Majid closely corresponds to the *sasaka/mahanavika* of Sanskrit sources. Interestingly enough, the Buddha appears as a master mariner with great expertise in bringing in and taking out the ship from harbour (*aharana* and *apaharana*). Fascinating stories of sailing, surviving shipwrecks and great dangers of the sea voyage are narrated in the *Kitab-ul-Ajaib-ul-Hind* of Buzurg Ibn-e-Shahriyar (955 CE) who himself was a famous ship-captain. The helmsman figures under the category *karnadhara*. The men who actually manned merchantmen came under the category of *navikakarmakara* (literally engaged in sailing). The term *navikakarmakara* has a correspondence to *al-askar* in Arabic, meaning a sailor in general.

If the dependence on the alteration of the monsoon wind was a perennial feature of shipping across the Indian Ocean, the other fact of unity lies in the construction of traditional ships of the Indian Ocean with the help of wooden planks, held together by coconut coir. These crafts did not have any iron nails, hence the expression *nilloham* in Sanskrit texts as an epithet of the traditional ship. Though the term, *sitapatra* (*siyapata* in Prakrit) stands for sail of the ship, visual representations of early Indian ships rarely display sails (Schlingloff 1987). One notable exception to this pattern is the famous painting of the ship in the Ajanta caves (fig. 2.2). There is little account available of the facilities for passengers. Nahray bin Alan, a prominent Jewish

2.2. Ajanta: Ship painted in the cave no. 2

merchant, wrote about his satisfaction having been provided with a *bilij* or cabin in a ship bound for India (Goitein 1973: 197). Ibn-e-Batuta was mostly very critical of the arrangements of living in the ships of his time; he however spoke in approving terms of Chinese Junks where he got a separate cabin for an aristocrat like himself.

Attention now may be paid to merchants engaged in the maritime trade network of the Indian Ocean. The most important figure in this context are the ship-owning merchants. It is from around 1000 CE that shipowners stole the limelight in the ports on the west coast of India. Sanskrit inscriptions coined the term *nauvittaka* to denote a person whose wealth (*vitta*) was rooted to his ships (*nau*). This exactly corresponds to the Persian word, *nakhuda* (*nau*= ship and *khuda*=lord), denoting ship owning merchant. In some Sanskrit inscriptions the two terms were also abbreviated respectively into *nau* or *nakhu;* sometimes the two terms were also used as synonymous. All these suggest considerable familiarity of the coastal society with the shipowners. Largest number of such shipowners come to our notice in the Jewish Geniza letters. A leading Jewish merchant, Madmun bin Hassan, (the representative of merchants / *wakil-ul-tujjar* at Aden and a close friend of Abraham bin Yiju, the Tunisian Jewish merchant residing in Mangalore) wrote in 1130 CE of a ship ordered by him for carrying commodities and passengers from Aden to Sri Lanka. This is probably the same craft mentioned as 'the ship of Elder Madmun' in which Nahray bin Allan, another prominent Jewish merchant, sent his goods in 1141 CE. In 1149 CE Bin Yiju informed his brother in Al-Fustat that he had sent some presents for him in the ship of Madmun. These letters, taken together, indicate that Madmun's ship was operational between Aden and the south western sea-board of India (also with Sri Lanka) for nearly two decades. Though Madmun was not primarily a shipowner he seems to have been successful in his shipping business. That is why in a letter to Bin Yiju (around 1140 CE) he expressed his desire to launch another ship in Aden in collaboration with four Indian merchants, *viz.* Sus Siti, Kinbati, Isha and Isaq. Rich and exhaustive data on shipping and shipowning merchants prompt us to take a close look at a letter of 1139 CE. Khalaf bin Isaac bin Bundar, a cousin and also a close business associate of Madmun acknowledged the receipt of two *bahars* and one third of 'refurbished iron' sent by Abraham Yiju to Aden in the ship of *nakhuda* Ibn-e-Abi'l-Kataib. Madmun also informed him of the arrival of *nakhuda* Joseph at Aden from Dahabattan or Vallarapatnam (in Malabar) in the ship of Ibn-al-Muqaddam. Besides Jewish and Muslim *nakhudas,* two ships of an Indian shipowner, Fatanswami (Sanskrit *pattanaswami*) also figure prominently in Khalaf's letter. Khalaf informs the safe arrival of the smaller ship of the Fatanswami to Aden but announces the terrible news of disaster striking the bigger ship. Pepper in that ship was completely lost, but portions of the iron was salvaged by employing professional divers of the port of Aden. Khalaf continues in his letter to inform that he sent some presents for Bin Yiju in the ship of Fadiyar who, according to Goitein was another Indian shipowner. In another letter from Madmun, Bin Yiju was informed of the shipment of a bag of copper of twenty-three pieces sent in the ship of Ramisht of Siraf, the merchant millionaire of the port of Siraf, also separate consignments of copper in the ships of Al-Muqaddam and Nambiyar, the last one being certainly an Indian merchant from Malabar (Chakravarti 2000a: 34-64). These letters are thus replete with reliable images of close interactions and linkages among Jewish, Muslim and Indian merchants. Abraham Yishu seems to have used the ships of Ramisht on several occasions, though ill luck occasionally struck both Yiju and Ramisht in the form of shipwrecks. In a letter dated in late 1130s Joseph Bin Abraham broke the news of shipwrecks of two ships of *nakhuda* Ramisht; in those ill-fated ships Bin Yiju sent some consignments to Aden. Joseph consoled Bin Yiju : "Do not ask me, my master, how much I was affected by the loss of the cargo belonging to you" (Goitein 1967-96; 1973: 193; Stern 1967: 10-14).

Jewish merchants also figured in the coastal network along the western sea-board as is visible in a letter of 1145 CE. Mahruz Bin Jacob, a *nakhuda* often plying between Aden and Malabar, wrote to his brother-in-law Judah Bin Joseph Ha Kohen (sister's husband), who was representative of merchants at the Egyptian capital. The letter was sent through Bin Yiju. Mahruz was greatly disturbed to learn that Judah was attacked by pirates during his coastal voyage from Al-Manjrur to Tana (Thana near Mumbai). The letter was originally sent to Tana but Judah had already left that port for Baruz or Bharuch at the mouth of the river Narmada. Mahruz wrote "The boats start presently from your place, from Kambayat and Tana; please set out immediately so that you reach Mangalore with vessels which god willing will soon reach Malibarat, Kayakannur and Mangalore" (Goitein 1973: 63). The Geniza letters thus showers a flood of light on the stellar role of Jewish India traders in the trade with India especially the sea-borne commerce.

From the thirteenth century onwards the Jewish merchants and *nakhudas* gradually faded out from the shipping scenario in the western Indian Ocean in the face of stiff competition from the Islamic *Karimi* merchants who seem to have had a stronger mercantile organisation. Goitein attributes the gradual lessening of the Jewish seafaring mainly because of the ever stricter interpretation and observation of the precepts of Jewish religion. The desecration of the Sabbath was unavoidable for sailors on voyage. Their employers were held responsible for all religious transgressions arising out of their employment. For the scrupulously religious Jews shipping became an increasingly less attractive avenue. The merchants in the Geniza letters were certainly well off, some of course very rich; but contemporary long-distance trade and shipping in the West was largely financed by the Islamic ruling groups having far greater resources. To this was added the organised business ventures of Al-Karim from the thirteenth century onwards (Goitein 1958: 175-184; 1967-96). The prominent role of the Jews in India trade had to give way for other more enterprising and resourceful competitors. It is not difficult to visualise the active presence of several *nakhudas* on the Gujarat coast, especially around Khambat, as revealed by epitaphs. Most of them were Iraqi or Iranian by birth, but settled in and breathed their last in port towns of Gujarat in 14th-15th century. Eminently successful and possessing enough wealth, they were leading (*sultan*) merchants and shipowners and great patrons of the *hajj* (Desai 1957-58: 12-13; 1963; Pearson 1994).

The point to take particular note of is the presence of Indian shipowners whose ships sailed across the Arabian Sea along with those of the Jewish and Muslim shipowning merchants. The general perception that Indian shipowners did not participate in the overseas voyages now needs a revision. Jagadu, a well known merchant of Gujarat, figures in the *Jagaducharita* as having sent his ships to Ardrapura or Hormuz from Khambat (Buhler 1892a). He is not described as a *nauvittaka* category of merchants and appears to have invested a part of his fabulous wealth in shipping business. There were also Indian shipowning merchants in the west coast of India whose ships mainly operated in coastal voyages. At the port of Balipattana (Kharepatan in the southern Konkan littorals) were active members of a *sresthi* family over three generations whose vessels operated in the Konkan sector, touching the ports of Cemulya, Nagapura, Candrapura and Sristhanaka. They are not categorically described as *nauvittakas* or *nakhudas* in the eleventh century inscription from Balipattana. It is quite evident that this *sresthi* family, residing in the port town of Balipattana, were interested in coastal shipping business and invested a part of their wealth in that (Chakravarti 1998: 97-124). In 1145 CE one comes across *nakhuda* Tinbu whose ships plied between Al-Manjrur and Thana. His constant interactions with Jewish India traders are writ large in a business letter of 1145 CE. This is true not merely in the western sector of the Indian Ocean. A Tamil inscription from Barus in Indonesia (1088 CE), records the presence of *marakkalam* (ships) and *marakkala-nayan* (shipowners) (Karashima 2002: 235-236; Subbarayalu and Shanmugam 2002: 23-24). The inscription proves beyond doubt the presence of Tamil shipowning merchants in Indonesia. There is a distinct possibility that the *marakkala-nayan* evolves into the famous *Maraikkayan(r)* or *Maraikkar* group of merchants, especially active on the Malabar coast in the sixteenth century.

One is not exactly sure how many of these shipowners did actually take to the sea; many of them could have been merchants ashore rather than afloat. But many other Indian merchants did undertake overseas voyages by the lure of gain. This is particularly evident in the presence of Tamil mercantile groups like *Manigramam, Nanadesi* and the 500 *Swamis* of Ayyavole in various parts of Southeast Asia and Sri Lanka. The most eloquent testimony to the activities of Tamil merchants in distant lands is offered by Tamil inscriptions. Out of many such instances, only a handful number of these are placed here as typical examples. As early as the ninth century the *Manigramam* had already reached Takua Pa in lower Myanmar. An inscription from Wat Boroma That of early eleventh century points to the presence of the same group in Thailand. That the famous mercantile organisation, 500 *Swamis* of Ayyavole, brought Sumatra within the orbit of their far-flung operation becomes evident from an inscription at Baros (1088 CE) recording the presence of ships and shipowners also. More or less during the same time another mercantile group, the *Nanadesi,* established linkages with Pagan in Myanmar. The Pagan inscription records the name of merchant Vinnagar, who belonged to the *Nanadesi* group and hailed from Mahodayapuram in Malaimandalam. Mahodayapuram in Malaimandalam is located close to present Cranganore in Kerala. The merchant in question is thus from the west coast of India who eventually arrived at the eastern shores of the Bay of Bengal. The regularity of the voyages of Indian merchants to maritime Southeast Asia is amply demonstrated by the engraving of a bilingual inscription (1256/1258 CE) in Javanese and Tamil, now in the collection of the Jakarta Museum. It is likely that at least some Indian merchants also reached China by sea from and via Southeast Asia. In 1281 CE Champanda Perumal alias Tavaccakkaravartikal donated an image of Siva at the famous Chinese port of Quanzhou. Another pointer to the interaction of south Indian ports with China is the archaeological discoveries of a large number of Chinese ceramics from sites in or near the coastal tracts. One of the most significant features of these Tamil epigraphic records is the regular eulogy of the widespread commercial activities of *Manigramam, Nanadesi* and especially the 500 *Swamis* of Ayyavole.[18] There is generally no eulogy of the political authority in these Tamil inscriptions, while Sanskrit inscriptions from different parts of Southeast Asia devote considerable space to court-poetical praise for kings. The merchant, especially the itinerant trader is the major recipient of laudatory accounts in Tamil (Ramesh 2002). The brief and hurried survey of voyages undertaken by Indian merchants both in the Arabian Sea and the Bay of Bengal highlights the hollowness of the notion that Indians rarely took to the sea because crossing the sea was forbidden in the *shastras* and brought under the category of *kalivarjya* (proscribed activities during the *Kali* Age). While the Sastric injunctions may at the most speak of Brahmanical ideals and norms, these were not insurmountable and were transgressed. The shipowning merchants were undoubtedly the key players in the maritime network in the Indian Ocean.[19] The traditional trade of India did not merely revolve around the ubiquitious pedlars, as Van Leur argued so strongly. In the *nakhudas, nauvittakas* and *marakkalanayans* one may find the forerunners of Mulla Abdul Ghafur of Surat, the greatest shipowning merchant of the eighteenth century (Das Gupta 1979).

We have already discussed two major points of unity in the world of the Indian Ocean, *viz.* the influence of the alterations of the monsoon winds on the shipping networks and hence on the transactional activities in the ports and their respective hinterlands; and the widespread use of ships made of wooden planks held together by coconut coir without the use of iron nails. A third point of commonalty is visible in the role of sea and the maritime merchants as bridges among diverse communities in widely dispersed territories. Merchants and their ships were not merely carriers of commodities, but purveyors of their cultural traditions too. This explains the spread of Buddhism and later the popularity of Vaisnavism and Saivism in many countries in Southeast Asia; the same is applicable in the use of Sanskrit and Tamil in Southeast Asia in inscriptions. The most eloquent examples are there in the Jewish business letters. These are replete with accounts of rivalries, court cases and complaints against merchants involved in the India trade. But nowhere in the entire range of the Genizah documents one comes across religious intolerance among Indian, Jewish and Islamic merchants active in the maritime trade of the western Indian Ocean. In a letter of 1145 CE the Jewish trader Judah Ha Kohen assured his brother-in-law Mahruz, who suffered an attack by pirates near Thana, to seek any financial assistance from the Hindu shipowner Tinbu, because between Kohen and Tinbu were 'inseparable bonds of friendship and brotherhood' (Goitein 1973: 64). There was little difficulty for Bin Yiju to marry a local woman from the Kanara coastal society during his stay at Mangalore for seventeen years (Ghosh 1990a). Al-Masudi noted in the early tenth century that Muslim merchants in the Konkan coast were given excellent support by the Rastrakuta rulers (Balhara kings in the Arab accounts, derived from the dynastic title of the Rastrakutas as Vallabharaja). The result was the construction of mosques in the port towns of Konkan. No less interesting is the information that Muslim Arab merchants were present as witnesses and important members of the coastal community during grants of land to a Durga temple in Sanjan by the Rastrakuta and Silahara rulers (Chakravarti 2001b). The most celebrated instance in this regards, however, comes from Somnath in Gujarat. A bilingual inscription (Sanskrit and Arabic, Sanskrit version more elaborate and Arabic version synoptic) of 1264 CE tells about the arrival of a pious Muslim *nakhuda* Nuruddin Firuz from Hormuz to Somnath. The texts of both the version leaves a clear impression that Firuz probably had made several previous voyages to Somnath and therefore, became well acquainted with the port town and its mercantile population, largely Hindus. With the active support of his Hindu merchant friends, he caused the construction of a *mijigiti (masjid* or mosque) in Somnath where were regularly observed a number of Islamic festivals (*vishesapujamahotsava*), typically associated with *nakhudas, nauvittakas* (shipowners) and *navikakarmakaras* (sailors). These were *Baratisabi* (*Sabe-barat*) and *Khatamrati* (the recitation of the Quran in a single night). The town council of Somanth, headed by a Saiva *Pashupatacharya*, approved of the procurement of land on which stood the mosque described in Sanskrit as a *dharmasthanam*. This was then endorsed by the provincial authority Maladeva and finally approved by Baghela Arjunadeva, the king of Gujarat. The Sanskrit inscription recording the construction of the mosque begins with a salutation to Allah (*Om Namastute*) who is given four epithets, *Viswarupa* (universal), *Viswanatha* (lord of the universe), *Sunyarupa* (formless/aniconic) and *Lakshyalaksha* (visible and being anicionic, also invisible). This is truly emblematic of the role the merchant communities in the Indian Ocean played as bridges among different ethnic groups and religious faiths (Chakravarti 2002b; Chattopadhyaya 1998).

The foregoing discussions indicate that the Indian Ocean had already become a much frequented maritime zone much before the advent of the Portuguese in the late fifteenth century. Vasco da Gama's voyage to Calicut cannot, therefore, be judged as a marker of the 'age of discoveries' in the Indian Ocean. Such a notion expresses an obvious Eurocentricism and justifies the European claim of inherent supremacy. One also cannot miss that prior to 1500 CE the Indian Ocean was a theatre of merchants, sailors and even of pirates. But only on very rare occasions did a political power consider the vast maritime space or a part thereof as an arena for establishing or even showing their political and military prowess. The famous Chinese voyages during the Ming period under the Admiral Cheng Ho (1404-1433 CE) spanned the greater parts of the Indian Ocean at least six decades before the arrival of the Portuguese in the Indian Ocean. But no territorial gains were made, neither were established any naval stations nor was made claim that the ocean belonged to the Chinese crown. Though the official Chinese annals described that Cheng Ho collected 'tributes' from the countries and ports he visited, a careful reading of the account cannot but suggest that the Chinese official missions were sent with a view to enhancing China's sea-borne commerce. A major exception to this pattern occurred in south India under Rajaraja I (985-1014 CE) and Rajendra (1012-1044 CE). The Cholas sent several maritime expeditions to conquer and annex Sri Lanka, by conquering which Rajaraja and Rajendra were eulogised as greater heroes than Ram. The Chola fleet also conquered '12000 islands in the open sea' which are usually identified with the Lakshadweep in the Arabian Sea. In 1025-1026 CE took place the most daring of the Chola maritime campaigns, the Kadaram campaign. As many as twelve important ports and centres of maritime commerce were conquered by the Cholas in the Bay of Bengal and maritime Southeast Asia. These include Talaittakkolam (Takua Pa on the

Myanmar coast), Manakkavaram (Nicobar Islands), Kadaram (Kedah in the Malay Peninsula) and Srivijaya (Palembang in Sumatra). No less significant is the fact that this naval campaign was immediately preceded by Rajendra's Gangetic campaign (1022-1024 CE) when the Chola army reached up to the Bengal coast from Coromandel through the coastal tracts of Andhra and Orissa (Sastri 1955). One may venture to suggest that there was a distinct design to establish the stamp of the Chola military superiority over the entire eastern sea-board before the naval campaign across the Bay of Bengal was launched. It is difficult to explain the motive of this rare maritime campaign of the Cholas in the Bay of Bengal. Except in the case of Sri Lanka, none of the areas in Southeast Asia was annexed to the Chola Empire. It is difficult to agree with George Spencer that the Chola maritime campaign was prompted by 'plunder dynamics', that too as a desperate measure because the Cholas had a weak material resource base (Champakalakshmi 1981: 411-426; 1996; Chattopadhyaya 1994; Sharma 1994; Spencer 1983; Stein 1980). The plunder dynamics can hardly be perceived as a potent explanatory model of a maritime policy which was in operation for nearly one hundred and thirty-five years (985-1120 CE). In view of the well established maritime linkages of Tamil merchants in Southeast Asia and the sustained Chola interests in maintaining contacts with the Sung rulers of China it is likely that the Cholas attempted to diminish the importance of many maritime centres of Southeast Asia which often played the role of a commercial intermediary between South Asia and East Asia. The Chola aggressive designs in the Bay of Bengal provide an exception to the general rule that political powers of pre-modern Asia and Africa did not consider the maritime space as an arena for political contestation. This is not symptomatic of any apathy to the sea, but certainly demonstrates an attitude to the sea different from that associated with the Europeans.

The Portuguese, the Dutch, the French and the English merchants did arrive in the Indian Ocean scenario in the post-1500 phase initially for the lure of profit from trade in spice and textiles which they ardently sought to import to Europe from the Indian Ocean countries; but on the other hand, the principal export of pre-industrial Europe to the world, including the Indian Ocean world, was violence (Prakash 1998: 165-182). A sea-change in the trading world and politics of the Indian Ocean was in the offing in the centuries succeeding 1500 CE.

NOTES

1. The relative absence of long-distance trade during the early middle ages is one of the major planks of the formulation of Indian feudalism which is said to have brought about self sufficient, enclosed and stagnant village society with little impetus to the non-agricultural sector of the economy.

2. The term *Vangasagara* occurs in the Madanpur Copper Plate of Srichandra of CE 971 edited first by Basak (1952: 51-58) and later by Sircar (1958: 337-339).

3. Chaudhuri (1985) has highlighted the commercial advantages of segmented voyages. He also suggests that from around c. 1000 CE emporia trade grew to the advantage of the Indian Ocean ports, including those in India.

4. Early medieval inscriptions are mainly concerned with political eulogies to rulers and grants of landed property to religious personalities and institutions. Information on trade, especially on seafaring merchants, appears in these inscriptions only in the shape of incidental notices. The Chinese texts used here are: Fa hsien - *Fu Kuo Chi*; Hsuan Tsang - *Ta tang His Yu Chi*; Itsing; Chau-Ju Kua - *Chu fan chi*; Ma Huan - *Ying-yai-Shen-lang*. The Arabic and Persian texts include: accounts of Sulaiman and Ibn Khurdadbeh; Buzurg Ibn Shahriyar - *Kitab-ul-Ajaib-ul-Hind*; Al-Beruni - *Kitab-ul-Hind*; Al-Idrisi; Ibn-e-Jubayr; Ibn-e-Batuta - *The Rihala*; *Hudud-al-Alam*. The letters of the medieval Jewish merchants were originally deposited in the genizah (meaning a store house) within a Synagogue in Al-Fustat (Old Cairo) in Egypt. Now the genizah is completely emptied, the documents having been preserved in various collections. The largest collection lies in the Cambridge University Library. Goitein (1954: 247-257; 1973; 1980: 43-66; 1987: 448-464; 1967-96) devoted considerable attention to the Jewish 'India traders'. The author is grateful to Professors Mark R. Cohen, A.L. Udovitch and Mordechai Friedman for allowing him to use and study the files of the unpublished India Book by S.D. Goitein now in the Dept. of Near Eastern Studies, Princeton University, Princeton.

5. That qualitative data are no less important than statistical information is argued for by Udovitch (1970: 37-62).

6. Puri is described in the Aihole *prasasti* of Pulakesin as the goddess of Fortune in the western sea (*aparajaladherlaksmi*) (Sircar 1983).

7. Sircar (1965: 497) opines that a *mahanavika* could also denote one in charge of a large ship (*mahanau*). Such a large ocean-going ship is

called *mahavakasa* (with large space in it) in the Jaina text, *Angavijja*, ascribed to the fourth century CE.

8. The inscriptions mentioning the port of Samyana and the presence of Arab merchants there were edited by Sircar (1962). For other inscriptions of the Silaharas of Konkan see, Mirashi (1978); Inscriptions of the Kadambas of Goa are available in Moraes (1931); analyses of these sources for the study of maritime trade have been made by Chakravarti (1998: 97-124; 2001b; 2002b: 153-161).
9. For a fascinating life-story of Abraham bin Yiju see Ghosh (2001). A Jewish business letter clearly describes Al-Manjrur situated on the 'great sea' (Weinstein 2001: 80).
10. Among many such letters, a letter of 1139 is particularly interesting for information on shipping of pepper from Malabar to Aden (Goitein 1973: 185-192). Export of pepper also figures prominently in another letter of 12th century (Goitein 2001).
11. Ibn-e-Jubayr was so impressed by the sight of profuse quantity of Indian pepper at Aydhab that 'so numerous as to seem to our fancy to equal in quantity only the dust' (Broadhurst 1952: 61).
12. See the accounts of Sulaiman, Khurdadbah, Masudi, Al-Idrisi and the anonymous author of the *Hudud-al-Alam* for the demand for Indian aloes wood, teak, rhinoceros horns and yak tails.
13. Letters mention information on shipment of various types of Indian iron. Some of the letters also show the urgency of Jewish merchants to bring Indian iron to Aden as quickly as possible, as there was a steady demand for Indian iron at Aden and it fetched good price (Goitein 1973: 185-192; 2001).
14. The widespread use of cowry currency in pre-modern West and East Africa is discussed by Newitt (Belshaw 1965: 70; Heiman 1980: 48-69; Newitt 1987: 201-202). The immense importance of cowry currency in the long-distance commerce of the Bengal coast is highlighted by Mukherjee (1981).
15. Sanjani indigo was exported from Sanjan to Al-Fustat and eventually to Al-Mahdiyya (Tunisia).
16. The Red Sea *jilab* was often captained by the *rubban* who however did not enjoy the status and prestige of a *muallim* in the manual of Ibn-e-Majid.
17. A kardal (kardar) figures in a Jewish trade letter of the 1130s (Goitein 1973: 193).
18. *Ayyavole* and *Manigramam* were two medieval merchant guilds of

south India (Abraham 1988). The 500 *Swamis* of *Ayyavole* was a larger body of merchants, while *Manigramam* was a smaller one. The term *Manigramam* was derived from Sanskrit *vaniggramam* (a body of merchants) the proper significance of which was first pointed out by Kosambi (1955: 50-71) on the basis of a first century BCE inscription from Karle.

19. For the sastric injunctions on *Kalivarjya*, see Kane (1962). It is impossible to accept Ramesh's (2002: 153) indefensible assertion - Indians had never displayed interest in seafaring activities and in building and plying their own ships, as can be inferred from the loud silence about the seas in Indian inscriptions, except in conventional contexts.

BIBLIOGRAPHY

ABRAHAM, Meera (1988) *Two Medieval Merchant Guilds of South India*, Manohar Publications, New Delhi.

AHMAD, S. Maqbul (1960) *India and the Neighbouring Territories*, Leiden.

AHMAD, S. Maqbul (1989) *Arab Classical Accounts of India and China (Accounts of Sulaiman and Ibn Khurdadbeh)*, Shimla.

APPADURAI, A. (1936) *Economic Conditions in South India AD 1000-1500*, 2 Vol., Madras.

BASAK, R.G. (1952) Madanpur Plate of Srichandra; Year 44, *Epigraphia Indica*, XXVIII, pt. II, Manager of Publications, Delhi.

BEAL, S. (1983) *Hsuan Tsang*, New Delhi.

BELSHAW, Cyril (1965) *Traditional Exchange and Modern Markets*, London.

BRAUDEL, Fernand (1972-73) *The Mediterranean and the Mediterranean World in the Age of Phillip II*, 2 Vol., (trn.) S. Reynolds, London.

BROADHURST, R.J.C. (1952) *The Travels of Ibn Jubayr*, London.

BROEZE, Franz (ed.) (1989) *Brides of the Sea*, Kensington.

BUHLER, G. (1892a) *Jagaducarita of Sarvananda*, Wein.

BUHLER, G. (1892b) The Cintra Prasasti of the Reign of Sarangadeva, *Epigraphia Indica*, I, Superintendent of Government Printing, India, Calcutta.

CHAKRAVARTI, Ranabir (1986) *Warfare for Wealth: Early Indian Perspective*, Firma Klm Pvt. Ltd., Calcutta.

CHAKRAVARTI, Ranabir (1991) Horse Trade and Piracy at Tana (=Thana, Maharashtra, India): Gleanings from Marco Polo, *Journal of the Economic and Social History of the Orient*, XXXIV.

CHAKRAVARTI, Ranabir (1995) Rulers and Ports: Visakhapattanam and Motuppalli in Early Medieval Andhra, In: K.S. Mathew (ed.) *Mariners, Merchants and Oceans*, Delhi.

CHAKRAVARTI, Ranabir (1996) The Export of Sindani Indigo from India to the West in the Eleventh Century, *Indian Historical Review*, XVIII.

CHAKRAVARTI, Ranabir (1998) Coastal Trade and Voyages in Konkan: The Early Medieval Scenario, *Indian Economic and Social History Review*, XXXV.

CHAKRAVARTI, Ranabir (1999) Early Medieval Bengal and the Trade in Horses: A Note, *Journal of the Economic and Social History of the Orient*, XLII.

CHAKRAVARTI, Ranabir (2000a) Nakhudas and Nauvittaks: Ship-owning Merchants in the West Coast of India (1000-1500), *Journal of the Economic and Social History of the Orient*, XLIII.

CHAKRAVARTI, Ranabir (2000b) Candrapura/Sindabur and Gopakapattana: Two Ports on the West Coast of India (c. 1000-1300), *Proceedings of Indian History Congress*, Diamond Jubilee Session, Calicut, Aligarh.

CHAKRAVARTI, Ranabir (ed.) (2001a) *Trade in Early India*, Oxford University Press, Delhi.

CHAKRAVARTI, Ranabir (2001b) Monarchs, Merchants and Matha in Northern Konkan (900-1053), In: Ranabir Chakravarti (ed.) *Trade in Early India*, Oxford University Press, Delhi.

CHAKRAVARTI, Ranabir (2001c) Between Villages and Cities: Linkages of Trade in India (c. AD 600-1300), In: Georg Bekemer, Tilman Frasch, Hermann Kulke and Jurgen Lutt (ed.) *Explorations in the History of South Asia*, Delhi.

CHAKRAVARTI, Ranabir (2002a) Politics and Society in India AD 300-1000, In: K. Satchidananda Murty (ed.) *Culture, Thought and Society in India AD 300-1000*, Delhi.

CHAKRAVARTI, Ranabir (2002b) *Trade and Traders in Early Indian Society*, Delhi.

CHAKRAVARTI, Ranabir (2002c) Seafarings, Ships and Ship Owners: India and the Indian Ocean (AD 700-1500), In: David Parkins and Ruth Barnes (ed.) *Ships and the Development of Maritime Technology in the Indian Ocean*, London.

CHAMPAKALAKSHMI, R. (1981) Peasant State and Society in Medieval South India: A Review Article, *The Indian Economic and Social History Review*, XVIII.

CHAMAPAKALAKSHMI, R. (1995) State and Economy: South India AD 400-1300, In: R. Thapar (ed.) *Recent Perspectives of Early Indian History*, Popular Prakashan Pvt. Ltd., Bombay.

CHAMPAKALASHMI, R. (1996) *Trade, Ideology and Urbanization: South India 300 BC-AD 1300*, Oxford University Press, Delhi.

CHATTOPADHYAYA, B.D. (1994) *The Making of Early Medieval India*, Oxford University Press, New Delhi.

CHATTOPADHYAYA, B.D. (1995) State and Economy in North India: Fourth Century to Twelfth Century, In: R. Thapar (ed.) *Recent Perspectives of Early Indian History*, Popular Prakashan Pvt. Ltd., Bombay.

CHATTOPADHYAYA, B.D. (1997) The City in Early India: Perspectives from the Texts, *Studies in History*, XIII.

CHATTOAPADHYAYA, B.D. (1998) *Representing the Other? Sanskrit Sources and the Muslims*, Manohar Publishers, New Delhi.

CHAUDHURI, K.N. (1985) *Trade and Civilization in the Indian Ocean from the Rise of Islam to 1750*, Munshiram Manoharlal, New Delhi.

CHAUDHURI, K.N. (1990) *Asia Before Europe — Economy and Civilization in the Indian Ocean from the Rise of Islam to 1750*, Oxford University Press, Cambridge.

DANI, A.H. (1988) *Recent Archaeological Discoveries in Pakistan*, Centre for East Asian Studies, UNESCO, Tokyo.

DAS GUPTA, Ashin (1979) *Indian Merchants and the Decline of Surat 1700-1750*, Franz steine verlage, Weisbaden.

DAS GUPTA, Ashin (2001) *The World of the Indian Ocean Merchants, 1500-1800*, Oxford University Press, New Delhi.

DAS GUPTA, Ashin and PEARSON, H.V. (ed.) (1987) *India and the Indian Ocean 1500-1800*, Oxford University Press, Calcutta.

DELOCH, J. (1987) Konkan Warships of the XIth-XVth Centuries as Represented on Memorial Stones, *Bulletin de L' École Francaine d' Éxtreme Orient*, LXXVI, Hanoi and Paris.

DELOCH, J. (1996) Iconographic Evidence on the Development of Boat and Ship Structures in India (2nd century BC - AD 15th century), In: H.P. Ray and J.F. Salles (ed.) *Tradition and Archaeology: Early Maritime Contacts in the Indian Ocean*, Manohar, New Delhi.

DESAI, Z.A. (1957-58) Inscriptions from the Prince of Wales Museum, Bombay, *Epigraphia Indica, Arabic and Persian Supplement*, Archaeological Survey of India, New Delhi.

Desai, Z.A. (1961-62) Arabic Inscriptions of the Rajput Period from Gujarat, *Epigraphia Indica, Arabic and Persian Supplement*, Archaeological Survey of India, New Delhi.

Desai, Z.A. (1963) Inscriptions of the Gujarat Sultanate, *Epigraphia Indica, Arabic and Persian Supplement*, Archaeological Survey of India, New Delhi.

Digby, Simon (1971) *Warhorses and Elephants in the Delhi Sultanate*, Oxford.

Dionisius, Albertus Agius (2002) Classifying Vessel-types in Ibn Battuta's Rihala, In: David Parkin and Ruth Barnes (ed.) *Ships and the Development of Maritime Technology in the Indian Ocean*, Routledge Curzon, London.

Elliot, H. and Dowson, J. (1939) *The History of India as Told by Its Own Historians*, Vol. I-III, Aligarh.

Ghosh, Amitav (1990a) *In an Antique Land*, Delhi.

Ghosh, Amitav (1990b) The Slave of Ms H.6, Occasional Paper, Centre for the Studies in Social Sciences, Calcutta.

Gibb, H.A.R. (1929) *The Travels of Ibn Battuta in Asia and Africa*, London.

Goitein, S.D. (1954) Two Eye Witness Reports on the Invasion of Aden by Kais in the Twelfth Century, *Bulletin of the School of Oriental and African Studies*, XVI, London.

Goitein, S.D. (1958) New Lights on the Beginning of Karim Merchants, *Journal of the Economic and Social History of the Orient*, I.

Goitein, S.D. (1967-96) *A Mediterranean Society*, 6 Vol., Los Angeles and Berkeley.

Goitein, S.D. (1973) *Letters of Medieval Jewish Traders*, Princeton.

Goitein, S.D. (1980) From Aden to India: Specimens of Correspondence of India Traders of the Twelfth Century, *Journal of the Economic and Social History of the Orient*, XXI.

Goitein, S.D. (1987) Portrait of a Medieval India Trader: Three Letters from the Cairo Geniza, *Bulletin of the School of Oriental and African Studies*, XLVIII.

Goitein, S.D. (2001) From Aden to India: Specimens of Correspondence, In: Ranabir Chakravarti (ed.) *Trade in Early India*, Oxford University Press, Delhi.

Goitein, S.D. (unpublished) *India Book*, in the Dept. of Near Eastern Studies, Princeton University, Princeton.

Gopal, Lallanji (1965) *Economic Life of Northern India AD 700-1200*, Motilal Banarsidas, Delhi.

GOPAL, Lallanji (1970) Indian Shipping in the Early Mediaeval Period, In: Lokesh Chandra, Devendra Swarup, Sitaram Goel and S.P. Gupta (ed.) *India's Contribution to World Thought and Culture*, Vivekananda Kendra Prakashan, Chennai.

GOTTHOLD, Julia (1987) *A Bibliography of the Indian Ocean*, Oxford.

GRENNVILLE, G.P.S. Freeman-(1980) Buzurg Ibn Shahriyar, *Kitab Ajaib ul Hind* (trn.), London.

HALL, K.R. (1980) *Trade and Statecraft in the Age of Cholas*, Abhinav Publications, New Delhi.

HEIMAN, J. (1980) Small Exchange and Ballast: Cowry Trade and Usage as An Example of Indian Ocean Economic History, *South Asia*, III.

HIRTH, F. and ROCKHILL, W. (1911) Chau-Ju Kua, *Chu fan chi* (trn.), St. Petersburg.

HOURANI, G.F. (1951) *Arab Seafaring in the Indian Ocean in Ancient and Early Medieval Times*, Beirut.

HULTZSCH, E. (1882) A Grant of Arjunadeva of Gujarat, Dated AD 1264, *Indian Antiquary*, XI, Bombay.

HULTZSCH, E. (1913-14) The Motupalli Pillar-inscription of Ganapatideva; A.D. 1244-45, *Epigraphia Indica*, XII, Superintendent of Government Printing, India, Calcutta.

JAIN, V.K. (1990) *Trade and Traders in Western India AD 1000-1300*, Munshiram Manoharlal, New Delhi.

JHA, D.N. (ed.) (2000) *The Feudal Order*, Delhi.

KANE, P.V. (1962) *History of Dharmasastra*, V, Poona.

KARASHIMA, Noboru (ed.) (2002) *Ancient and Medieval Commercial Activities in the Indian Ocean: Testimony of Inscriptions and Ceramic Shards*, Tokyo.

KERVAN, Monique (1999) Multiple Ports in the Mouths of the River Indus: Barbarike, Deb, Daybul, Lahori Bundar, Diul Sinde, In: H.P. Ray (ed.) *Archaeology of Seafaring*, Delhi.

KOSAMBI, D.D. (1955) Dhenukakata, *Journal of the Bombay Branch of the Royal Asiatic Society*, XXX.

KOSAMBI, D.D. (1956) *An Introduction to the Study of Indian History*, Popular Book Depot, Bombay.

LEGGE, J. (1974) *The Travels of Fa Hsien*, Delhi.

MAJUMDAR, R.C. (ed.) (1962) *The Classical Age*, Bhartiya Vidya Bhawan, Bombay.

MILLER, K. (1962) *Historia Graeco Minoricae*.

MILLS, J.Y.G. (1970) Ma huan, *Ying-yai-Shen-lang* (trn.), Oxford.

MINORSKY, Y. (1937) *Hudud al Alam* (trn.), London.

MINORSKY, Y. (1942) *The Accounts of Al Marvazi*, London.

MIRASHI, V.V. (ed.) (1978) *Corpus Inscriptionum Indicarum*, VI, Delhi.

MOOKERJI, R.K. (1912) *Indian Shipping*, Bombay.

MOOKERJI, R.K. (1957) *Indian Shipping: A History of the Seaborne Trade and Maritime Activity of the Indians From the Earliest Times*, Orient Longman, Calcutta.

MORAES, G. (1931) *The Kadambakula*, Bombay.

MUKHERJEE, B.N. (1975) The Original Territory of Harikela, *Bangladesh Lalitkala*, I.

MUKHERJEE, B.N. (1981) Commerce and Money in the Central and Western Sectors of Eastern India (750-1200), *Indian Museum Bulletin*, XVI, Calcutta.

NAINAR, S.M.H. (1942) *Arab Geographers' Knowledge of Southern India*, Madras.

NEWITT, M.D. (1987) East Africa and the Indian Ocean Trade 1500-1800, In: Ashin Das Gupta and H.V. Pearson (ed.) *India and the Indian Ocean 1500-1800*, Oxford University Press, Calcutta.

PEARSON, M.N. (1994) *Pious Passengers: The Hajj in the Earlier Times*, New York.

PRAKASH, Om (1998) European Corporate Enterprises and the Politics of Trade in India 1600-1800, In: Rudrangshu Mukherjee and Laksmi Subramaniam (ed.) *Politics and Trade in the Indian Ocean World*, Delhi.

RACKHAM, J. (trn.) (1942) *Pliny, Naturalis Historia*, London.

RAMESH, K.V. (2002) Texts and Translations of Indian Inscriptions in Southeast Asia, Appendix I, In: Noboru Karashima (ed.) *Ancient and Medieval Commercial Activities in the Indian Ocean: Testimony of Inscriptions and Ceramic Shards*, Tokyo.

RAY, Haraprasad (1987) China and the Western Ocean in the Fifteenth Century, In: Satish Chandra (ed.) *The Indian Ocean: Explorations in History, Commerce and Politics*, Delhi.

RAY, Haraprasad (1993) *Trade and Diplomacy in India-China Relations*, Delhi.

RAY, H.P. (1990) Seafaring in the Bay of Bengal in the Early Centuries AD, *Studies in History*, VII.

RAY, Niharranjan, CHATTOPADHYAYA, B.D., CHAKRAVARTI, Ranabir and MANI, V.R. (2000) *A Sourcebook of Indian Civilization*, Hyderabad.

REIF, S. (2000) *A Jewish Archive from Old Cairo*, London.

SACHAU, E. (1910) Al Biruni, *Kitab ul Hind* (trn.) London.

SANDESARA, B.J. (1953) *Literary Circle of Mahamatya Vastupala, and its Contribution to Sanskrit Literature*, Bhartiya Vidya Bhavan, Bombay.

SASTRI, K.A. Nilakantha (1955) *The Cholas*, Madras.

SASTRI, K.A. Nilakantha (ed.) (1972) *Foreign Notices of South India from Megasthenes to Ma Huan*, University of Madras, Madras.

SCHLINGLOFF, D. (1987) *Studies in the Ajanta Paintings – Identifications and Interpretations*, Ajanta Publications, New Delhi.

SHAKED, Shaul (1964) *A Tentative Bibliography of Geniza Documents*, Paris.

SHARMA, R.S. (1965) *Indian Feudalism (300-1200)*, University of Calcutta, Calcutta.

SHARMA, R.S. (1980) *Indian Feudalism*, Delhi.

SHARMA, R.S. (1987) *Urban Decay in India AD 300-1000*, Delhi.

SHARMA, R.S. (1994) The Segmentary State and Indian Experience, *The Indian Historical Review*, XVI, Delhi.

SIRCAR, D.C. (1958) Madanpur Plate of Srichandra, Year 46, *Epigraphia Indica*, XXVIII, pt. VIII, Manager of Publications, Delhi.

SIRCAR, D.C. (1962) *Epigraphia Indica*, XXXII, Manager of Publications, Delhi.

SIRCAR, D.C. (1965) *Select Inscriptions bearing on Indian History and Civilisation*, I, Calcutta.

SIRCAR, D.C. (1982) *The Emperor and His Subordinate Rulers*, Shantiniketan.

SIRCAR, D.C. (ed.) (1983) *Select Inscriptions*, II, Delhi.

SPENCER, George W. (1983) *Politics of Expansion: Chola Conquest of Sri Lanka and Sri Vijaya*, New Era Publications, Madras.

STEIN, Burton (1980) *Peasant State and Society in Medieval South India*, Oxford University Press, Oxford.

STERN, S.M. (1967) Ramisht of Siraf: A Merchant Millionaire of the Twelfth Century, *Journal of the Royal Asiatic Society*, London.

STILLMAN, Norman A. (1971) The Eleventh Century Merchant House of Ibn Awkal, *Journal of the Economic and Social History of the Orient*, XIII.

SUBBARAYALU, Y. and SHANMUGAM, P. (2002) Tamil Inscriptions in South East Asia and China, In: Noboru Karashima (ed.) *Ancient and Medieval Commercial Activities in the Indian Ocean: Testimony of Inscriptions and Ceramic Shards*, Tokyo.

TIBBETTS, G.R. (trn.) (1971) *Arab Navigation in the Indian Ocean Before the Coming of the Portuguese*, London.

Udovitch, A.L. (1970) Commercial Techniques in Medieval Islam, In: D.S. Richard (ed.) *Islam and Trade of Asia*, Pennsylvania.

Watters, T. (1905) *On Yuan Chwang's Travels in India*, Vol. II.

Watters, T. (1960) *On Yuan Chwang's Travels*, Delhi.

Weinstein, Brian (2001) Jewish Traders in the Indian Ocean - Tenth to Thirteen Centuries: A Review from Published Documents from the Cairo Geniza, *Journal of Indo-Judaic Studies*, IV.

Whitehouse, David and Williamson, Andrew (1973) Sassanian Maritime Trade, *Iran*, XI.

Yadava, B.N.S. (1973) *Society and Culture in Northern India in the Twelfth Century*, Central Book Depot, Allahabad.

Yule, H. and Cordier, H. (1903) *The Travels of Ser Mara Polo* (trn.), 2 Vol., London.

Comment

Om Prakash

I would like to begin by complementing Prof Chakravarti for writing an excellent paper and making a clear presentation covering a long period and the entire Indian coastline. The period of India's involvement in the Indian Ocean trade from the 4th to the 13th century CE that Chakravarti has talked about in his paper is today the least illuminated of the principal time periods that are usually covered in the literature that makes his efforts all the more worthwhile as well as challenging. In this brief comment all I can be able to do is to pick out some of the principal points that Chakravarti has made in the paper and subject them to a certain amount of scrutiny.

The central point in the paper relates to the debate on the overall question of the status and the scale of India's participation in the Indian Ocean trade during 400 to 1300 CE.

I greatly welcome the unequivocal position that Chakravarti takes that contrary to the widely held belief that these centuries particularly the period until about 1000 CE were categorised by overall stagnation had indeed declined in the volume and value of India's high seas trade in the Indian Ocean trade but these centuries might indeed have witnessed low decline and possibly even growth. As Chakravarti has mentioned those of us working in this period took advantage in so far as there is indeed an overwhelming a lot of documentation in the form of Portuguese, English, Dutch, French and Danish company records dealing in extensive details with the volume and value of European

Corporate Enterprise as well as private merchant trade in the Indian Ocean. Indeed a good part of this documentation deals not only with the relevant European corporate enterprise and private traders but also with the Indian and other Asian merchant traders in the ocean. This is more true of Dutch East India Company documentation than almost of any other European corporate enterprise but the documentation of each one of them has a good deal to say about the Indian and other Asian merchants. In the context of availability this vast body of information there need be no room for ambiguity regarding at least the broad trend in the volume and value of trade in respect of different regions and participating groups involvement in the Indian Ocean trade. Since that unfortunately is the case for the period prior to 1500 CE, it is all the more critical that as detailed and judicious use of the limited archaeological epigraph as well as written documentation be made as possible to delineate at least broad orders of magnitude. That however has not always been the case. A good deal of the literature does not go into this kind of evidence. A case in point is the feudalism debate in India for the period from 300-1200 CE which Chakravarti has thrown one's attention. The argument is made that feudalism involves emergence and consolidation of self-sufficient enclosed and stagnant village economies. It then follows, that this must by definition have involved a decline in trade particular long distance trade. This I submit is not only an academically unsound reasoning but indeed a potentially grossly misleading one, what needs to be done is marshalling of whatever evidence is available and then engage in careful evaluation of that evidence as indeed Chakravarti has done. The evidence is not exhaustive or always definitive but it certainly warrants broad conclusions that he draws.

I will indeed go a step further to suggest that detailed evidence available in respect of a later period may also be used on a selective and limited basis. Let me give an example. We now know that say from the 15th century onwards the key post of India in the Indian Ocean trading network was only partly a function of the midway location of the subcontinent between West Asia on one hand and Southeast and East Asia on the other. Critically more important was the subcontinent's capacity to put on the market a wide range of tradable goods at highly competitive prices. By far the most important of these was textiles of various kinds. These included high value varieties such as the legendary Dhaka muslin and Gujarat silk embroideries the really important component for the Asian market. This is something that Ranabir emphasises. The real component for the Asian market was the coarse cotton varieties manufactured primarily on the Coromondal coast and in Gujarat. There was a large-scale demand for these varieties both on the eastern markets of Indonesia, Malay, Thailand and Burma as well as the markets of Red Sea, Persian Gulf and East Africa. While it is impossible to determine precisely what proportion of total domestic demand for mass consumption of textiles in these societies was by imports from India. The available evidence would seem to point in the direction of this not being altogether insignificant. India's capacity to manufacture these textiles in large quantities and to put them on the market at highly competitive terms made it in some sense an industrial hub of the region surrounded by West Asia on one side and Southeast Asia on the other. It would obviously be wrong to interpolate this kind of scenario backward in time and to say that in the absence of specific information however limited for the relevant period this kind of situation would also have been valid say around the year in 1100 or 1200 CE. But by the same taken it would help to keep in mind that in the absence of evidence suggesting a revolutionary change in the meantime, the possibility of this kind of a scenario being valid for a earlier period as well should not altogether be dismissed out of head. All that I suggest is that when one has very little to go on, as you people do it is useful to keep all possibilities in mind and to see in which direction the available evidence takes one rather than taking an inflexible position. That is precisely what Ranabir has done and I congratulate him for that.

3

The Arab Seagoing Traditions in the Indian Ocean (8th-15th Century AD)

John Carswell

The Arab interest in the Indian Ocean and beyond dates from pre-Islamic times and lasted until the arrival of the Portuguese in the fifteenth century. This paper outlines briefly the topics, before discussing each of them in greater detail. The information comes from two different sources, from my surveys round the coasts of India and Sri Lanka and in the Maldive Islands and subsequent excavations, and secondly from an invitation to edit a new edition of George Hourani's *Arab Seafaring* for Princeton University Press in 1995. Whilst the former provided a great deal of first hand information, the editing of Hourani's peerless or rather timeless text made us think about the Indian Ocean on a larger scale. Additionally, focus on Central Asian trade rather than the maritime routes between East and West, has in turn added another important and contrasting element. Whilst the new edition of *Arab Seafaring* was mainly concerned with supplementing Hourani's pioneer work with material from archaeological sources by scholars working in and around the Indian Ocean. These endeavors were supplemented by the publication of a lengthy article in the Encyclopaedia of Islam entitled *milaha* (that is navigation, seamanship, seafaring) by four more scholars, who dealt with the literary sources.

Thus the sources of information are both textual - from Arabic and other accounts referring to their activities, and material from the architectural and archaeological record. The latter has been supplemented in recent years from the submarine excavation of ancient wrecks, a continuing activity which is likely to become of major importance in the future as a source of new information.

Two factors played a major role in the development of maritime activity in the Indian Ocean, the discovery of the monsoons in the first century CE for sailing directly across the ocean, and an increase in trade between India and the Mediterranean, fostered during the Roman period. This accelerated with mercantile activity further east, as far as China. In all of this India played a pivotal role, poised as it was between East and West.

Whilst China itself was aware of new markets for products such as silk and ceramics, it was Arab (and Persian) initiative which led to the establishment of a highly sophisticated trading network between the Far East and the Mediterranean from the earliest days of Islamic expansion. Because of the pattern of the bi-annual monsoons, eastbound shipping from the Red Sea and Persian Gulf was obliged to make a stopover in south India or Sri Lanka, whereas on the return journey by careful timing the voyage could be made directly. For this reason, ports such as Quilon in Kerala and Mantai in Sri Lanka played a crucial role on the outward trip, whereas the Maldives were equally important on the way home as the shipping had to take a more southerly route.

Surveys of the coast of India and Sri Lanka and of Male in the Maldives, in the 70's have confirmed this seasonal variation of east-west shipping. More to the point, the excavation of Mantai on the northwest tip of Sri Lanka between 1980-84 has revealed its central importance as the greatest emporium between China and the Mediterranean, which lasted for over a thousand years until it finally succumbed to the invasion of the Cholas in the eleventh century. The focus then shifted to south India, and such ports as Nagapattinam, with direct Chola involvement in trade with Southeast Asia weakening the Arab monopoly.

But there is little doubt that the Arab traders continued to play an important role, operating within the Islamic world which in the Early-Medieval period stretched from the Atlantic to the Pacific. For this we have the first-hand testimony of the fourteenth-century Arab traveller Ibn-e-Batuta, with specific references to India, Sri Lanka and the Maldives, which he visited on his way from his native Tangiers to China itself. Even today, there is hardly a town on the west coast of India which does not have a mosque and a Muslim community, surviving testimony of the earliest Islamic traders to the east.

It is now generally agreed that the understanding of the

monsoons and their bi-annual fluctuation and their exploitation for a swift, direct passage across the Indian Ocean and beyond took place some time in the first century CE. It is generally thought that a Greek called Hippalus learned the art of sailing directly from Arabia to India, as described in the *Periplus*, but here it must be clarified (and I quote Hourani) that he did not discover the existence of the monsoons, which had been known to the Greeks at least since the fourth century BCE. His contribution was to spot how the outward voyage could utilise the southwest monsoon, and thus make for a much faster if more dangerous trip than the safer practice of coasting along Arabia until catching the north-east monsoon to journey further south. As with all discoveries involving natural phenomena, it must have evolved from the observation of wind and weather patterns over centuries. What is certain is that using the southwest monsoon in this fashion was a distinct improvement on the slower technique of safely hugging the coast.

In this manner the whole west coast of India was linked to Arabia, the Persian Gulf and Red Sea, and East Africa. In India, from Khambat in the north to Cape Comorin in the south there is a string of ports. Khambat lies on the gulf of the same name and was described accurately by Masudi, who visited it in 915 - 916 CE, also mentioning the strong tidal bore that sweeps up the gulf. Al-Idrisi also describes Khambat in the twelfth century, mentioning the estuary and the fortress. Ibn-e-Batuta visited the town in the fourteenth century and talks about the estuary ebbs and flows, leaving the ships stranded on the mud at low tide. The town has a fine mosque built in 1325 CE from fragments of Jain and Hindu temples, with white marble cenotaphs delicately carved with Arabic inscriptions, including the tomb of the Sharif Al-Samarri, whose impressive mansion Ibn-e-Batuta mentions. He records that the majority of the inhabitants were foreign merchants from as far afield as Isfahan, Gilan, Baghdad and Diyarbakir.

Further south is Bharuch, the Barygaza of classical times, lying thirty miles up the river Narmada. The *Periplus* mentions how the Arabs of Muza (probably Mocha) in Yemen sent ships there to carry away copper, ebony and timber of all kinds, and merchants from Cana in south Arabia traded incense with Oman and Barygaza. This is evidence for Arab coastal trade in the pre-Islamic period. Muslims arrived in the second century of the Hijra, and the little mosque there is said to be dated 1037 CE. Its present custodian is a doctor, and his ancestors, physicians to the courts of Akbar, Jahangir and Shah Jahan. Idrisi in the twelfth century describes it as a fine town full of foreign merchants, from as far afield as China. Bharuch today is a vast mound over fifty meters high, covered with the buildings of the modern city. A fine Chinese porcelain blue and white sherd was found on the slopes below of the Jiajing period (1522-1560 CE). It is to be regretted that no excavations of any significance have ever been undertaken here.

In south India, the most important port with contacts both with the Islamic world and the Far East was Quilon (Kulam Mali in Arabic). A roadstead rather than a natural port, like Calicut further north, shipping of any size had to anchor offshore and goods transported back and forth by smaller craft. Ships from the Persian Gulf would use the northeast monsoon to sail directly across the Indian Ocean to Quilon. In 851 CE, a merchant from Siraf stated the voyage took a month. Quilon was said to be the furthest west that Chinese vessels reached, and Ibn-e-Batuta described the Chinese junks he saw at Calicut. Indeed, on one of these he intended to embark for China and had it loaded with all his possessions, but staying on land for Friday prayers, it was wrecked during the night and all his goods lost.

There is hardly a village, let alone town, on the west coast of India which does not have a mosque; and in Bhakthal for instance over 60% of the population is Muslim. This in conjunction with the evidence rehearsed above serves to underline the importance of maritime contacts with the Islamic world in the past. The same is true for Sri Lanka, mentioned also by Arab travelers and where the remnants of the Muslim community are still to be found on the west coast.

As for material remains, Sri Lanka not only has Chinese material from archaeological contexts but also Islamic ceramics as well, pointing to the fact that Sri Lanka and indeed India was the meeting point for long-distance shipping coming both from East and West. It was also the practice for vessels making the whole journey from the Red Sea and the Persian Gulf to rest between two monsoons in south India or Sri Lanka. A modern example of a sailor who did not do so and who was becalmed for weeks off the east coast of India was Tim Severin, who sailed in his reconstructed medieval *Dhow* from Oman to China some years ago. A medieval Arab sailor would have known better.

Discussing the Maldives Islands may seem peripheral in the

context of the Indian subcontinent, but often looking carefully at a tangent, throws light on the main topic. In the fourteenth century, Ibn-e-Batuta traveled from Morocco to China and back. It took him twenty-seven years. He recorded a curious incident in Damascus when he saw a slave-boy in tears, because he had just broken his master's Chinese dish. In the late sixties, a great deal of Chinese porcelain (mostly broken) appeared on the market in Damascus. While cataloging them a curiosity, how it ever got from China to Damascus, sparked off my continuing interest in trade routes. Ibn-e-Batuta had described the unique qualities of Chinese porcelain when he got to China. He also visited the Maldives on the way. Maldivians were converted to Islam in twelfth century. A Maldivian muslim student informed that his family used porcelain at home on Friday nights. This was good enough for me, and in the summer of 1972, I travelled to the Maldives, from Beirut to Kuwait then Bombay and Colombo, and spent a month on Male, the capital island.

Male is only a mile long and half across, with 20,000 inhabitants, so the space is short. That summer the Maldivians had decided to eliminate the graveyards of the thirteen mosques, to make more building land available. Which meant that they had to dig them up. In the process, hundreds of pieces of porcelain appeared, added to those that appeared on the surface of the sandy streets after the rain. We collected everything (the sherds are now in the Ashmolean Museum in Oxford as a study collection) and discovered that it was all Chinese, from the ninth to the nineteenth century.

The significance of this was clear later, when publishing an article about our finds. I have already referred to the westward journey from the Far East presenting different problems. Simply, if you sailed from China by a certain date in the year, you could get all the way back home in one monsoon, but it meant passing far south of India through the Maldives. Both a Portuguese text and an early Chinese account corroborated this. And what we found was entirely Chinese, with no Islamic or other material, unlike India and Sri Lanka, which had a mix of both. What did the merchants want in the Maldives? Again the answer is fairly simple - fresh water. Although the islands are coral reefs barely above the surface of the sea (and it is predicated they may disappear altogether soon with global warming) there is a heavy level of precipitation, and this soaks through and is filtered by the coral, so wells on any of the islands surprisingly produce fresh water.

If the Maldives produced some clear information about the pattern of trade, this was even more so when I turned my attention to Sri Lanka. Initially I had noticed that there was a mixture of both Islamic and Chinese material, and also vast quantities of local earthen ware as well as imported ceramics from India. This led me to thinking about Mortimer Wheeler's classic excavation at Arikamedu in south India after the World War II, when he managed to correlate imported Roman pottery with local material and thus provide a chronology for the whole of Indian archaeology. But Sri Lanka presented (and still presents) an even more intriguing challenge, for here there is not only imported material from the East and West, but also from the Indian mainland, combined with the local pottery. If it were possible to relate all four of these with the dateable architectural contexts in which they can be found (such as the classic sites like Anuradhapura and Pollonaruwa) with the additional evidence from coins, then it could be possible to set up a tight chronology not only for Sri Lanka and India but also for the whole of Asia.

With this heady prospect in mind, the survey in Sri Lanka was started, and the process resulted in identification of at least one minor site where this chronological co-existence occurred. This was at Vankalai, at the northwest tip of the island, and a short but instructive excavation was conducted there in 1974. With both Islamic and Chinese sherds, and lots of local pottery, it was obviously a small but prosperous settlement which may have only lasted fifty years or so, at the beginning of the twelfth century.

A year later on a little island off the Jaffna peninsula quite by chance a hoard of over five hundred Chinese pots, an abandoned cargo of a vessel sailing from China to the western world was found. These were easily dateable, of the Song dynasty, and a marvelous corpus of all the types which were being exported at that time, more or less contemporary with the material found in excavation at Vankalai. After this accidental discovery the Sri Lankan Department of Archaeology, gave the permission to excavate the most important site of all - Mantai.

Mantai (fig. 3.1) lies at the north-west tip of Sri Lanka, on the eastern side of the string of underwater reefs which link the island to India, known as Adam's Bridge. Less well known is the fact that these reefs successfully prevent shipping of any scale sailing between the two, although there are many historical maps which mistakenly show

3.1. Mantai: Aerial view

trade routes using this channel. What Adam's Bridge does provide is easy access to Sri Lanka from south India, to the island's misfortune even to this day.

Mantai was excavated for three seasons, from 1980 until 1984 when the civil war put an abrupt end. The final excavation lasted three months and was supported by numerous organisations, including the Ford Foundation in this city, the National Endowment for the Humanities, the Metropolitan Museum in New York and the Oriental Institute in Chicago, as well as the British Museum in London. It was an eucemenical cast, with staff from the United States, Britain, Germany and India, as well as students from Sri Lanka including Sinhalese, Buddhist and Tamils. In spite of the threatening clouds of discord and civil war, it was a remarkably friendly group, with over a hundred local villagers as workmen.

The site had an earlier, prehistoric past and had been found in the 2nd millennium BCE as a hunting/gathering community with a special emphasis on the sea and its products. This was the nucleus which when abandoned provided the mound on which Mantai proper was founded, some time about 500 BCE. The site was then continuously occupied until the beginning of the eleventh century, when it was devastated by Chola invasions from south India, from which it never recovered.

Its particular importance lies not only in its location but also its function, as the main point of contact between Anuradhapura and the outside world. With trade flowing from all four points of the compass, and an economy obviously based on the development of the northern dry zone through a sophisticated system of reservoir tanks and canals producing an agricultural surplus, it became an emporium unrivalled throughout maritime Asia for 1500 years. This was recognised over a century ago, and various attempts have been made to unravel its past. None of them has so far proved successful, for a variety of reasons. Our attempt to subject this major site to the discipline of modern archaeological techniques and investigation on a multi-disciplinary level was no more successful ultimally than those in the past, but this time victim to the very forces which had annihilated Mantai in the eleventh century —-the conflict between the indigenous population and that of south India.

But even with this caveat, a great deal was accomplished in the three seasons of work. For instance, quantities of Chinese pottery of the Tang dynasty (618-904 CE) was found. This included white stoneware, Tang splashed ware (until then never believed to have been found outside of China), and Changsha glazed earthenware of a type known to have been exported from its production site in south central China throughout Southeast Asia and as far as the Persian Gulf. Tang white ware was found, in such perfect state of preservation that it looked as if it had been made yesterday. In this tiny respect alone, Mantai proved to be a great educational experience. The reverse side of the coin was an incident which occurred one day whilst sorting the day's collection of sherds, carried out mostly by willing students. Quite by chance a fragment of burnished red pottery emerged amongst the thousands of pieces of humble local earthenware. On it was moulded a scene with a monkey on a crocodile's back, with three letters of a Brahmi inscription in one corner (fig. 3.2). It was from the shoulder of a *kundika,* or spouted water-pot, and it was easy to calculate that it was the only surviving piece of a sequence of twelve such scenes. It depicted an episode from the well-known animal fables of the *Jataka,* the origin of the later *Khalila wa Dimna* stories in the Persian and Islamic world. It also turned out to be the oldest known example of an illustration of this particular story by almost a thousand years.

Mantai was extraordinary for two reasons. First, by virtue of being well within the monsoon zone, with both heat and humidity to conspire

3.2. Mantai: Fragment of burnished red pottery with Brahmi inscription

together, almost no organic material of any sort survived on the site — no textiles, wood, manuscripts or any kind of perishable material. This is in complete contrast to Egypt, where the dry climate is conducive to the preservation of organic material, and also, the Gobi desert in Mongolia, which has the same conditions. So, what remain at Mantai are only indestructible artifacts — such as broken pottery. Pottery therefore, *faut de mieux*, plays an exceptional role in interpreting the history of the site.

Thousands of beads of polished stone, carnelian, glass and crystal were found. Broken and also half-finished crystal beads with the boring never complete before they split proved that Mantai was a major manufacturing site for such beads. Furthermore, a specialist study of the technology used to produce them revealed that they had been bored with a double-hafted diamond bit, the first recorded use of industrial diamonds in history. They date from the 9th-10th century CE.

The tiny multicoloured glass beads were also proved to be made on the spot, for we found the unfinished glass tubes from which they are formed, and the same technology is used in India to this day. These beads have been found all the way from Korea to Africa and again we have a link to Mantai as a major industrial site. Another glass bead (or rather half of one) was an extraordinary find, for it is a sophisticated type made from coloured glass canes, and known to have been manufactured in the Rhineland in the 4th-5th century CE. A complete example has been found in Jutland. How such an object got all the way to Sri Lanka at such an early day can only be conjectured.

What else do we have from farther West? Well, for a start there are numerous examples of white-glazed pottery imitating the Tang stoneware already mentioned, which were made in Mesopotamia in the 8th-9th century. What is intriguing is that they should have been economically viable to the extent that they would have been exported in the Indian subcontinent where they would have been in direct competition with the original Chinese prototypes. The same phenomenon occurs in excavated white ware at Anuradhapura, where Chinese and Islamic white ware is found side by side. Another import was Islamic glass, of which great quantities were found. In the medieval period glass was a commodity much in demand in China, which had not native rival product. And besides fragments of glass vessels, large pieces of glass cullet were also discovered, perhaps the base material for the Asian-Pacific glass beads manufactured at Mantai.

Other specifically Islamic material includes lustred glass, and also luster pottery of the early Islamic period, almost certainly manufactured in southern Iraq. Its export westwards is well known, where it became the springboard for the Fatimid luster industry in Egypt and later for luster-wares in Tunisia, Andalusia and Islamic Spain. But it is rare to find it being exported in just the opposite direction. Large quantities of turquoise-glazed earthen ware have also been found at Mantai, and also complete examples from other sites in Sri Lanka. These were also Mesopotamian products, and in this case exported even further afield, for examples have been found in a dateable 9th century tomb in China.

And finally, a 9th century carved Islamic tombstone from Mantai indicates that a Muslim not only lived in Mantai, but also died there. Similar gravestones have been found elsewhere in Sri Lanka and the National Museum in Colombo has a substantial collection of such objects. Islamic and Chinese coins have also been found in quantity from sites in Sri Lanka.

The present paper so far concentrated on the material evidence, there is of course the legacy of Arab methods of shipbuilding and seafaring, which are also closely intermingled with the indigenous tradition in India, Sri Lanka and the Maldives. From pre-Islamic times the Arabs have been the great innovators, the adventurers whose initiative spread far beyond that of the Indians and even the Chinese.

They established a monopoly, which lasted for fifteen hundred years. In simple terms, they knew more, whether it was seamanship, navigation or simply greater experience of the oceans and their vicissitudes.

It is one of history's great ironies that it was the knowledge of an Arab, Ahmed Ibn-e-Majid Al-Najdi, that revealed the secrets of Indian Ocean to the Portuguese in the fifteenth century, thus paving the way for the conquest and domination of the East by the West in the centuries to come. It is obvious that the Europeans would have discovered the passage to India anyhow sooner or later, just as Magellan rounded America and sailed across the Pacific. In our own times, we know to our cost perhaps that whoever has the superior technology can rule the world, but we should acknowledge that the scientific knowledge and skill of Arab mariners once linked the civilised world, from the Atlantic to the Pacific.

Comment

B. Arunachalam

When I accepted the invitation to be a discussant for this paper, I was thinking that he is going to talk something on actual navigation and when I got the paper I thought it was an opportunity for me to learn something more of what I didn't know.

Prof John Carswell has highlighted a couple of points which I will first mention. He says that India is in a very significant location for all East West trade between areas to the west of India and areas to the east of India. And to that extent south India has derived certain benefits. As examples he mentioned Quilon on the west coast and on the east coast he mentioned Mantai which is on the north-western coast of Sri Lanka just close to Talai Mannar which provides a link with Anuradhapura which was the capital during the late Chola times.

He has also mentioned that ceramics provide a very important set of evidences specially in the tropical countries where other perishable products do not remain. Sri Lanka and the Malabar coast were the meeting points for the long distance shipping both from the West and East and in this the actual progress of Arab navigational route development from the 8th to the 15th century has to be scrutinised and examined.

Carswell also mentions that the Arab vessels on their outward voyages towards the east, into the Bay of Bengal used to halt in one of the ports of south India or Sri Lanka while on the return journey they could sail directly back to Arabia using the sea and wind conditions.

And in this he also mentions that on the return voyage Maldives constituted an important port to replenish the water supply.

He goes into great detail of the importance of ceramics and he has based his entire work on his field experience. And I think he has visited 28 to 29 sites on this Sri Lankan coast and about 25 on the Tamil Nadu coast. He did his work in the 70's and this work has been further followed up and continued later on by the Japanese scholar Karasimha and by the Thanjavur Department of Archaeology. Now in this total evolution of the Arab trade routes from the 8th to 15th century number of interesting facets come out. Some have been answered and some will probably have to be probed into deeper in days to come. On the Arabian front there has been a shift in the port sites, which have been engaged in the Chinese and the eastern trade. It is progressively shifted to the Persian coast on to Hormuz then into the Red Sea ports but that is only in the Arab side and ultimately by the 15th century on to the east coast of Africa. And what is important to notice is that all the ceramics evidence found in this section is all on the coastal side and another point of importance to note is that till the 14th century no Chinese *Junk* ever went west of the west coast of India, which means that all the ceramics product which reached the Arab coast had been taken in Arab vessels.

Now coming to the importance of Mantai the question is why was Mantai chosen. Why was it so important? My personal feeling is that in the first phase of Arab navigation routes extending to the Arabian sea they reached upto the east coast of India and Sri Lanka and at that time the trade was not only with the ports further east but also with the Indian requirement. Arab horses were extremely significant and the port of Mantai is known as Parumdurai in Tamil literature, which was importing Arab horses in the 11th-12th century CE. So, this was one set of exchange points and the east coast ports have been known since earlier times to have trade links with ports of Malay and Indonesia. There was much profit coming from Indonesia and Malay or the Chinese products used to be exchanged on the east coast ports or one or two on the west coast ports. These were prominent exchange point locations. But as the Arabs developed the trade routes further into the Bay of Bengal, which may be, 10th-11th century CE that a new set of exchange points became important. These new exchange points are on the Malayan coast, Malacca and some on the

Sumatran coast. In this regard one has to analyse that navigational routes which were followed by the Arabs. The location of Mantai which is on 9^0 latitude means that the detour south of Kanniyakumari and for the sea going vessel to further to the eastern part of the Bay of Bengal ports it will have to come back and then follow south of the Sri Lanka coast. A number of references exist both in the Arab voyage accounts and later accounts as late as the beginning of the 19th century. The navigators were informed not to hug the Sri Lanka coast as it was notorious for its lightening and enormous number of navigational hazards and therefore it was necessary to keep at a good distance from the south Sri Lanka coast.

And more important reason was if you travel further south of Sri Lanka you are entering into a belt where the wind and current circulation will be extremely favourable to take the square-rigged vessel since we are talking about the 12th and the 13th century. And these vessels could be taken to the southern coast of Sumatra very easily. Infact there are references which state that they could in twelve days reach any port on the west Sumatra coast from south of Sri Lanka.

So to me it appears that the Chinese trade was developed in three phases. The first phase was sailing through the Arabian Sea to reach the east coast which was a well set route which the Arabs knew as the Persians had been traveling in that route and it has been very convenient. The second phase is this extension to the eastern section of Bay of Bengal upto Malacca or Sumatra coast. The third phase of development would be further extension from Java/Sumatra to the Chinese ports.

There is one more aspect, which we will have to consider. That is the Chinese trade has always been vacillating that is, open and shut.

4

The Relative Maritime Technologies at the Time of the European Encounter (15th-16th Century AD)

K. Raja Menon

India is surrounded by the seas on three sides and has a long maritime history but it is rather surprising that the Indian history taught in school hardly has any references to the sea as though the oceans did not exist. There are references of about 15 critical battles which were fought on Indian soil between the battle of Plassey and the last battle in the Sikh wars. It was only while researching maritime strategy almost forty years later that I came across some detailed accounts of the decisive battles fought on the seas in the 17th and 18th century. Indeed, the first of the great maritime historians, Alfred Thayer Mahan wrote a book on the Anglo French battles off Madras in the late 18th century to illustrate how control of the seas was vital to launching operations on land (Mahan 1908). The Battle of Nagapattanam on 6th July 1782 between the French Admiral Baillie de Suffren and British Vice-Admiral Sir Edward Hughes with 11 ships on either side was probably the biggest sea battle (fig. 4.1) to decide the fate of India, however no Indian ships participated in the battle. Mahan wrote his work in 1905 after which there have been of number of authoritative books on Indian history. Sadly, none of these refer to the contest at sea having been won or lost before operations on land could be undertaken. The absence of this narrative is partly understandable, for at sea, there were no battles the Indians waged to compare with the intensity and decisiveness of battles like Plassey, Wandiwash, Seringapatam, Assaye or Sobraon. True there was the storming of Gheria, the fortress at Janjira, and there were the skirmishes between the Portuguese and the Samudri of Kozhikode and Mangalore, but the decisive battles at

4.1. The Battle of Nagapattanam

sea, off the Indian coast were fought between the British fleet and the French, allied to Tipu Sultan.

This is curious because economic historians have commented quite successfully that up to the early 17th century, India produced about 25% of the world's GDP with China accounting for 30%. It is curious because, how did a country that could command 25% of the world's wealth not have access to a matching level of seapower (Kennedy 1988: 188-191).

In the first part of this paper a relative comparison is made of the maritime technologies available to the European and the Indian mariners, and the second part analyse the causes of the mismatch between the two technologies. The first part is naturally factual. In assessing the levels of the maritime technology available to a nation, most analysts would prefer to look at the subject under some specific headings such as shipbuilding size and modernity, propulsion, maritime infrastructure, such as dockyards and ports, weapon technology, navigation and hydrography, systems to support the maritime endeavour—research, finance, medicine, law, insurance and material.

Shipbuilding

Upto the 13th century, pictures of ships from the West, the Middle East and the Indian Ocean show that their size and sail outfit looked fairly similar. For instance one of Christopher Columbus' ships had lateen-rigged sails while all heavily built Western sailing-ships had a mizzen mast that was rigged with a lateen sail. The tonnage of these ships varied from 70 to about 200 tons. However, from the late 14th century onwards one finds that ships in the west began to grow in size until by the middle of the 16th century, capital ships of close to 1000 tons were not uncommon. It is of course accepted that the sailing ship was never the product of one particular country or region. Influences from all over the world often went into the evolution of ship design. From the western point of analysis the best work of analysing sailing-craft in the Indian Ocean remains that of Admiral Paris (1843; Rieth 2004: 105-115; also see pp. 187-210). Of his many works, it is the work on the sailing vessels of the non-European world that will remain one of the world's classics. Assembled between 1826 and 1840, the authentic drawings of Indian Ocean crafts indicate little change from the 14th century to the present time.

It has often been argued that Indian sailing craft did not need to be any bigger than they were, owing to the requirement to avoid large-scale risk of loss of bigger cargoes. This may be true, but it is inconceivable that the ability to build larger ships would not have been attempted by merchant guilds, princes or navies. If ships were getting through to their destination, it stood to reason that a bigger ship made better sense, that would be more able to defend itself and weather the elements better. Coming back to the influence of other traditions on a particular style of shipbuilding, up to 1650, English naval vessels of smaller size existed that were virtual copies of Algerian designs (Bowen 1669: 289). This is interesting because the Arab *Dhow* is a virtual replica of the Algerian *Xebec.* There are in fact many who hold the view that the Europeans got the lateen rigged sail from the Arabs.

So, the question is why, if the Indian and Arab *Dhows* are so similar to the Algerian *Xebec,* didn't Indian shipbuilding adopt the design techniques of western shipbuilding as the latter progressed faster? It has been accepted that the English master shipwright in Queen Elizabeth's reign used Venetian designs (fig. 4.2) and drafting methods in his *Fragments of Ancient English Shipwrighting* (1586), the earliest acknowledged English treatise on shipbuilding.

As ship designers attempted to give the customer what he wanted, the laws of ship design became clearer. The effect of length on speed came to be understood. However designers could not achieve greater lengths until construction methods became more precise. For

4.2. A Ventian Galleas at the Battle of Lepanto. Note the lateen-rigged masts with long yards made by fishing two timbers together, as is the practice in the Indian Ocean (*Source: John Fincham*)

4.3. A scale model of the French warship *l' Ocean* on the basis of which the warship was actually built in 1775 (*Source : MNM, Paris*)

instance, when the American frigate *Constitution* burst upon the naval scene and outfought and outsailed bigger vessels, it wasn't quite realised that the length to breath ratio of the *Constitution* was about 6:1. Until then the cruder construction methods attempted to ensure safety by both using unnecessarily heavy timber and increasing the breadth. Achieving high length to breadth ratio within a reasonable thickness of timber, without producing a vessel that sagged or hogged was the design problem. It is inconceivable that these problems can be tackled without drawings (Chapelle 1967: 29).

Production drawings require that the displacement be known and the formulae for finding out the displacement was defined by Hardingham in the early 18th century. But even prior to this, in the middle of the 16th century, drawings for ships had been produced, although there is some controversy that in the 18th century, ships were produced from models (fig. 4.3), and not from drawings.

The earliest record of marine architectural drawing in England goes back to 1586 (Fragments). From the drawings, lofting drawings were produced for every third or fourth transverse frame, called the mould frames. These were joined to the keel and battens run fore and aft to show that the ship could be planked. Naturally the drawing could be used to produce copies of the same ship in other shipyards (Baker 1954: 262-277; Dean 1670).

Why wasn't this process repeated in India? It is not sure, but fairly certain, that the traditional Indian shipbuilder used no drawings (fig. 4.4-4.6). This didn't mean that the Indian ship design was deficient. In fact the speed-length ratio of an Indian *Dhow* (fig. 4.7) is allegedly very good. This ratio is obtained by dividing the maximum speed achieved by the vessel by the square root of the length.

This index should be as high as possible and not lower than 4:1. By a process of refining over time there is little doubt that the underwater hull form of the sailing *Dhow* is very efficient and weatherly, when loaded.

This latter quality describes the vessel's ability to sail upwind without drifting to leeward. The process of drawing the design would still not accurately establish the thickness of planking required to prevent the vessel sagging and hogging.

However experiences of previous vessels would show how much water the vessel would ship in (Chapelle 1967: 26). It was common

4.4. Mandvi: Traditional shipbuilding in Gujarat

4.5. Mandvi: Naturally shaped wood of poor quality used as ribs

4.6. Mandvi: 14 inch nails and the heaviness of the planking relative to the ribs

4.7. Model of a two-mast lateen-rigged *Dhow*, with a small mizzen

practice when handing over a watch in a sailing ship to indicate how much water was in the bilges and whether the watch had to be called to pump out the bilges.

It is interesting to see how a traditional boatbuilding yard in Beypore has got into the business of making transverse frames accurately by what looks like the obvious process of lofting (fig. 4.8-4.11). This yard, although a traditional boat-yard has adopted modern building techniques, something that Indian yards could have done in the 17th and 18th centuries, but didn't for reasons that will be examined later. Here are also some shots of the older method of boatbuilding where very poor quality wood is being used in such a key area as transverse frames. This may not be accepted technical practice but the effects of mere poverty and lack of finance, or simply the non-availability of good teak.

It is a fact noticed by the Indian Maritime History Society that much of the wood being used in traditional boat-yards today is

4.8. Beypore: Modern Indian shipyard

4.9. Beypore: Handmade blocks for mainstay

4.10. Beypore: Mould frames produced by lofting drawings (*Source: MHS, Mumbai*)

4.11. Beypore: Mould frames produced by lofting drawings (*Source: MHS, Mumbai*)

Malaysian wood. The European method, needless to say enabled a country or a company to establish the best design and finally a master shipwright so that the best design could then be replicated in all the other yards. In this way, the French developed the famous model (MNM: 24) which became the standard design of battleship not only in France, but in Britain, where the 74 gun battleship (fig. 4.12) with a displacement of 2550 tons, soon became Nelson's favourite fighting platform.

In France, it is believed that accurate 1:24 and 1:48 models (fig. 4.13) were built and put up for approval to the king before the design was executed. France perhaps took some early action to establish state support to shipbuilding with the setting up of a school of naval architecture in Paris in 1741 and an Academie de Marine in 1752. The only school of naval architecture was guided by the Inspector General, a scientist of 'many talents' (MNM: 33).

Wood Processing and Wood Working

Building a navy is obviously a national pastime. This can be understood only when it is made clear that a 74 gun ship would consume anything from 3000 to 3400 fully grown Oak trees (Dodd and

4.12. A scale model of a 74 gun ship of the late 18th century (*Source: MNM, Paris*)

4.13. A 1 : 48 model used for construction in a French shipyard with the side opened up to show the arrangement in the hold (*Source: MNM, Paris*)

Moore 1984: 13). A 48 gun frigate could take upto 2500 fully grown Oak trees. Many of the trees used in shipbuilding had to be permitted to grow to 120 years so that the heart of the trunk was used for the keel, frames and the stem and stern. Single pieces two feet thick were required for these key areas.

In the case of Indian shipbuilding many kinds of timber was considered acceptable, although, the Malabar yards stuck mostly to teak. In both Western and Indian shipbuilding tradition, naturally curved wood was mostly used for the knee supports. The shapes which were required in English shipbuilding, for which naturally shaped timbers called compass timber was used, were grown specially by tying down the branches of oak trees to grow them squatter. A ship's side was made of four to six inch timber with frames two feet apart so that in most places the timber thickness was 18 inches deep to stop round shot. No single country in the West was able to meet the requirements of their own shipbuilding.

Masts for English, French, Danish and Dutch ships came from as far away as the Siberian forests. Dazing was the chief export port for most timber. The countries which provided the most timber were treated as strategic assets and maintaining access to them became one of the chief war aims.

In the Napoleonic wars, a blockade was set up to prevent French access to Baltic timber (Dodd and Moore 1984: 15). The lesson is apt for India. Although there were strong shipbuilding yards in Malabar, Tuticorin and Saurashtra, the great teak forests of Malabar would have been barely enough to meet the requirements of Beypore and Tuticorin. Oak was simply not available in large parts of England and France and those who owned large estates put down legacy forests of Oak to leave in their will to their grand children a hundred years later. In Malabar, teak over 100 years old had disappeared by the early 20th century and the last solitary Malabar teak tree, almost 400 years old with a diameter of almost four feet is now a national monument in the Valparai forest range.

Arunachalam (2002) has conducted extensive research into the woods used in Indian shipbuilding. In the south India, builders preferred *Calophyulym inophylum* and *Terminalia tomentosa* for the keel and *Babul* and *Khair* were preferred for the stem and stem-posts and ribs. None of these trees grew to the size of the Oak, either in the straight or in the curve. So Indian construction often used joined wood in the keel to get sufficient rigidity and the natural curves used would never have supported four inch thick planks. It is a moot question whether Indian shipbuilding was crippled by the lack of access to a sufficiently large and strong stock of timber. Certainly in 1780, it was estimated that there were 15,000 merchant and warships in England which if built at one go would have required wood which could not have been grown in an area the size of the whole of England (Dodd and Moore 1984: 14-16). When it comes to processing enormous quantities of wood of this size, it could have been achieved in an adequate time frame only by mechanical means. Normally horse powered reciprocating saws or saws powered by falling water were used.

Propulsion

It would not have been possible to think of designing large vessels without the means of hoisting much larger sails. This problem was solved in the West by the use of compound masts, where straight spars were fitted one on top of the other to produce mast heights of over 180 feet. Since some of the 100 gun ships with a displacement of 5200 tons could produce almost 10 knots of speed in a good wind, it can be calculated that they were generating about 4000 horsepower.

In India, mast timber of the kind used by the West was available in the Himalaya, but there is no record of any Himalayan timber being sought for shipbuilding.

The Indian designer, partly solved the problem of creating a larger sail by hosting a single sail on a yard which was so large that they were often two pieces of timber fished together. Large areas of canvas, made of flax, required miles of cordage and hundreds of blocks. If the blocks for a 74 gun ship had to be made by hand, the block making would have taken as long as the construction of the ship, because 922 blocks were required. The great inventor and engineer Isambard BruneI made a block making machine which could mass produce blocks and the *lignum vitae* sheaves which were difficult to cut by hand.

In the case of Indian ships it is difficult to tell whether the lack of a desire to build larger ships crippled the search for a large sail area or was it the other way round? But suffice it to say that while the Indian sailing ship did go through a process of evolution in some areas such as the method of steering, method of fastening the planks and the methods of navigation, the sailing rig remained unchanged for a thousand years.

Navigation

The primary skill in navigation is to master coasting by day. The eastern and western schools of navigation used pretty much the same procedures for coasting. While the Mediterranean navigators relied a little more on the compass and the distance run, the Indians relied less on the magnetic compass. Measuring speed by paying out a float on a rope marked with knots and timing the number of knots against a 30 second sand glass, or a quick flowing sand glass was a common technique. But Nicolo Conti who traveled in the East during the early 15th century correctly identified the chief difference in the method of ocean navigation of the Indians and in the West.

In India, said Conti rightly, boats are steered by the pole-star (Taylor 1954: 123-125). This fact had actually been noticed earlier by Marco Polo who had been entrusted with the escort of a Chinese princess being sent to the Persian court by the sea route. Marco watched the captain of his ship make no comment about 'losing' the pole-star in the region of Sumatra, and being overjoyed at finding it again in the latitude of Kanniyakumari.

Apart from using the pole-star as a directional aid, Indian navigators had a series of tables giving the pole-star altitude of various ports. The altitude was measured by a *Kamal* which may not have given exact measurements of altitude but this did not matter, because the central string had knots on it to indicate various altitudes of the star at various ports, and whatever errors existed would appear each time and therefore be cancelled out. Marco Polo and Conti both commented on the use of the *Kamal* (but not in so many words) on their returrl to Europe, they also confirmed that the Indians had and used the compass but relied more on star navigation.

Long before Vasco da Gama came to India, Indian pilots navigated regularly to Aden, Mombasa, Madagascar and the East Indies. It could be said that up to the end of the fifteenth century the Indian and Arab skills in ocean navigation were perhaps ahead of the West (Taylor 1954: ch. VI). This was soon made up for in the Age of Discovery. Partly this is explained by the absence of the transfer of scientific information from European scientists to their own mariners. For it is also reported that Chaucer the poet, for instance could tell the time from his astrolabe or determine the latitude by the upper and lower transit of a star in the same period when the European mariner could not. The European equivalent of the *Kamal,* the cross staff was in fairly wide use in Europe by 1550, but the astrolabe was obviously more popular for measuring the sun's altitude. The difficulties of European navigation in finding clear skies and their reliance on dead reckoning on charts led to the earlier development of charts, a benefit that seems to have passed the Indian mariner by. The establishment of the Royal observatory and the longitude prize of 20,000 pounds sterling for inventing a chronometer that could determine the longitude to half a degree or 30 miles led to the famous Harrison Chronometers produced in 1736 (NMM: 15). In the mean while the octant had been developed in 1730. The chronometer, the octant, the sextant and competent hydrography took European navigation decades forwards of the Indian style, although the Indian style had held on up to the 15th century.

Arunachalam has done some commendable work on translating the traditional systems of ocean navigation, all of which were oral traditions. For instance the knowledge of the latitudes of Indian ocean destinations, the compass bearings to those destinations, which could be translated to star headings to be used at night were in fact taking Indian navigation to a chartless state, where all the knowledge remained in the heads of the pilot (Arunachalam 2002: 216-232). Whether this was a good or a bad thing is not for us to judge except to say that the Indian style led mariners away from the use of charts - and that had deleterious consequences for Indian navigation in the Age of Discovery when the charts of the world were being drawn from all the hydrographic data available. It is certain that the use of accurate time to calculate a complete geographical position from a Nautical Almanac was a process that the Indian navigator never mastered.

Hydrography

Indian tradition of navigation as route oriented and detailed instructions were handed down from *Muallim* to *Muallim* whereas the Europeans went in for perfecting the art of hydrography. The latter method opened up the art of navigation to everyone and removed the restricted oral tradition from practical use. At the same time the oral traditions of the West were codified in what is today called the Pilots or sailing directions. When looking at say the sailing directions for entering Aden, the western school and the eastern school give remarkably similar advice on the safe passage from Socotra to Aden (MNM: 10-11). But the tradition of chart making advanced by leaps and bounds from the discovery of the Americas onwards. One of the earliest charts of the western Indian Ocean was produced between 1515 and 1519 by Lopo Homem and is called the Miller Atlas (fig. 4.14).

Lopo Homem was the master of nautical charts in Lisbon in 1517 and he produced this chart by compiling all the existing navigational data into one chart. It can be seen that barely 20 years after the arrival of Vasco da Gama in India, the hydrography of the entire western Indian Ocean including the west coast of India had been accurately surveyed. Two more charts indicate the speed at which hydrographic information was disseminated.

The Sebastao Lopes chart of the Mediterranean, produced in 1555, was also compiled in Libson (fig. 4.15).

Except for a few inaccuracies around the Island of Crete, the chart is almost perfect. Another chart, the Pieter Goos chart produced in Amsterdam was compiled from all the navigational charts then available (fig. 4.16). What is truly amazing is that while the Miller Atlas contained no accurate information on the area east of India in 1520. By 1660, the data on the Goos chart shows that all the coast-line had been surveyed up to Shanghai, Kyushu, New Guinea and Tasmania. What the western method of navigation did was to bring the world closer for all mariners who could read a chart; whereas the Indian pilot

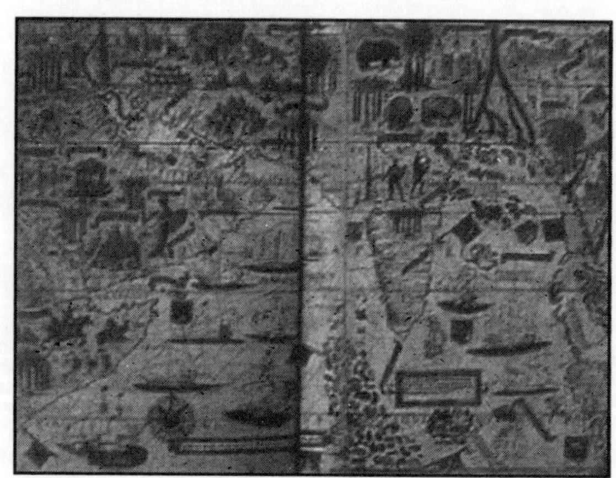

4.14. The Miller Atlas by Lopo Homem, c. 1515 CE (*Source: MNM, Paris*)

4.15. The Mediterranean chart of Sebastao Lopes, c. 1555 CE (*Source: NMM, Greenwich*)

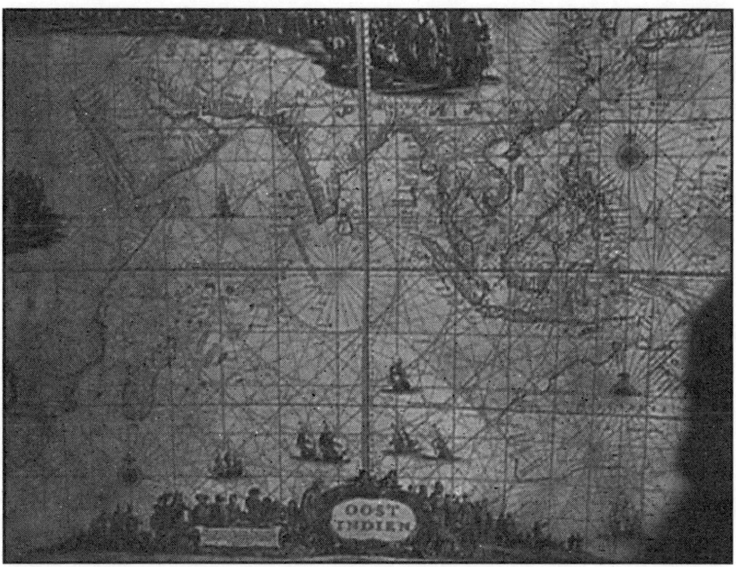

4.16. The Chart of the East Indies by Peter Goos, c. 1660 CE
(Source: NMM, Greenwich)

remained a specialist with knowledge of a particular route. His knowledge of that route was unmatched and his knowledge included the state of weather, colour of the water, fish and bird life, the texture of the sky and depth of water to be expected. But specialist route navigators never became universal navigators. The famous *Mallam ni pothi* (NM), the Gujarati text of a sailing direction for Peninsular India indicates that the understanding of the uses of a hydrographic chart had come to the local sailor but perhaps 300 years late to matter to the general safety and well being of Indian navigation.

Weapons

The effect of superior weapons on an inferior system has traditionally been one of the primary causes of decisive defeats and revolutionary changes in governance. When one looks at the growth of British power in India, one cannot fail to notice that the East India Company extended its influence, primarily from Calcutta, Madras and Bombay, all of them port cities. In each of these port cities, massive fortresses were built, all of which had one unique feature. Instead of the main defences of the ports facing the sea, as coastal forts generally do, these forts had their main battlements facing the land, indicating the low priority to threats from the sea. How did this maritime superiority arise?

So far we have looked at ships as sailing platforms. We need to now look at sailing ships as weapon platforms. There is no doubt that the 17th and 18th centuries were centuries of the sea. The true sailing ship was capable of remaining at sea for long periods. Victuals were generally taken on aboard for six months, water for three to four months particularly after iron ranks were fitted in the late 18th century. The growth of colonial power and the enrichment of Europe was made possible by the true sailing ship which transferred the center of power of the 'civilised world' from India and China to Europe. The superiority of European weaponry from the 16th century onwards rested on developments in metallurgy that took place during the renaissance. Indian metallurgy had rightly established its fame, its chief advertisement being the Damascus sword, the technology for which came from India.

By the early 18th century cannon in Europe was identified by the weight of shot fired. Guns on board ships began to be standardised, just as ships began to be standardised. In 1716, the Carronade was invented, designed to throw a heavy ball a short distance or a large number of small balls to sweep the deck of the enemy in a close encounter. The weights of a 32 pound gun and a 32 pound carronade tells its own story. The gun weighing 3 tons needed a crew of 10-12 men, of whom 4 were earmarked for the boarding party. The carronade weighted 17 hundredweight and had a crew of 4 men. The huge number of guns required by a navy forced the European powers to shift to cast iron guns in 1700, except for the larger 42 pounders where brass was used. These guns were the lowest deck guns in the 1st rate ships.

There is no record of any special naval gun being cast for ships in India. Quite often land cannon were embarked in gun carriages in Indian warships. It would appear that the heaviest cannon carried by Maratha warships were nine pounders. A casual British survey of Indian cannon making was made at various times and we are told of the result of their survey in Irvine's book on the Mughal army (Irvine 1994: 118). There is hardly a ship that comes to India that does not sell them cannon and small arms; the most of the gun powder they make themselves. They cast shot in abundance, but there is no prince that casts cannon but the king of Travancore. The cannon and military stores are smuggled into the country. This quotation fairly well answers the query in many Indian minds as to why when Kerala had an old and illustrious bell metal foundry process, they could not make cannon. The answer is that they did, but it would appear that cannon in India were made by extremely diverse sources. Duperron a French officer has stated that the cannon of Siraj-ud-dowlah was made by casting molten brass around an iron cylinder. There is nothing backward about this technology if it could have been standardised and in fact shows that somewhere in India some gunsmith had worked out for himself the art of combining the advantages of cast iron and brass in one composite design.

In general the cannon industry in India was diverted from making what was needed by poor higher direction. Talking of Indian cannon, Cambridge, an unknown writer quoted by Irvine says 'they give them pompous sounding names.... And some have pieces which carry a ball of seventy pounds. When we march around them with our light field pieces, and make it necessary for them to move their heavy pieces, they get into utter confusion and are easily destroyed.' A survey of Indian cannon can be made as a large number of pieces are still visible in various Indian museums and on historic sites. It would appear that some of the best of the early pieces were made in the Deccani kingdoms of Bijapur, Ahmednagar and Golkonda (fig. 4.17-4.19). This can be ascribed to the presence of Persian and Turkish gun makers who arrived in central India through the sea route in the 16th century.

When it came to personal weapons, all European accounts speak of the high level of skill of the individual soldier or sailor, but when massed together they seem to have been led by people with no knowledge of their profession. The famous painting of Maratha *Gurabs* and *Gallivats* attacking a frigate has much to tell. It is doubtful whether the 9 pounder cannon carried by the Indian ships would have had any effect on the hull of a European warship. The cannon could have been used with chain or bar shot to cripple the rigging of a man-of-war, but this could at most be a defensive measure to delay pursuit. Looking at the cannon and matchlocks made in India in the 17th and 18th century, there are some pieces which are equal to the best made anywhere, but when a good piece was made, there doesn't seem to have been any effort to locate the competent manufacturer and give him larger orders. Despite the absence of the industrial revolution in Europe in the 18th century it has been estimated that using better engineering techniques, almost two million muskets were made in France during the Napoleonic wars while Britain made a million and a half during the same period (EB: 547).

4.17. *Malik-e-maidan* - made by a Turkish designer in Ahmednagar in c. 1548 CE

4.18. *Zafarbaksh* probably the best Mughal piece cast in Aurangzeb's time, c. 1674 CE, 12.25 tons—32-40 lbs shot (*Source: Banerjee*)

4.19. A cannon cast during Adil Shah's reign, Bijapur, late 16th century, fired a shot over 32 lbs (*Source: Banerjee*)

Systems

As ships became larger and the crews increased in size, the length of the voyage began to test the systems that went into running a large warship, as well as running a navy. A first rate battleship could carry as many as 1200 men and often stayed away from its home port for as long as three to four years. Ships sometimes returned to their home port having lost about 30% of their crew to disease. Ships were sent to the Atlantic and then diverted to the Indian Ocean, something they could not do unless they carried the charts for their new destination, along with the relevant nautical tables.

The task of maintaining the health of the crew fell upon the lonely surgeon on board who would send foraging parties to lonely Islands or strange coasts to collect 'ascorbutic' plants, or plants that had vitamin C, to ward off scurvy. Ships were often dismasted in battle or in a storm and had to have access to dockyards to replace lost sailing gear. On a ship away for three years, living in close proximity to each other, discipline became a serious problem. Captains had the authority of life and death over men and the punishment often had to be brutal. Surgeons performed intricate operations including abdominal surgery and the repair of cracked skulls with silver plate. Mail reached ships half way across the globe and ships could deposit their mail in any port with a government representative. Deserters and mutineers were hunted down in any part of the globe, no matter what the effort cost. All this was part of a globe girdling system in which the ship itself was only one complex part.

Causes of the Decline in Indian Maritime Technologies

In absolute terms Indian maritime technology never declined. In relative terms, while technology in other parts of the world progressed, Indian technology remained rooted in the 10th century. We need to examine the causes. Undoubtedly no single cause could have created such catastrophic handicaps, but some of them may have just come close to it.

Engineering Drawing

Perhaps the most intriguing one is the failure of indigenous Indian ship-builders to understand or use engineering drawings. Engineering drawing or model building began to be used in Europe from the early sixteenth century and the benefits of producing lofting drawings have been explained. But once drawings appeared in Europe, ships design began to increase in size - indeed the profession of ship designer began to make his presence felt beyond that of the master shipwright who carried a plan in his head. Engineering drawing had existed in some form with the Greeks and Romans, but the Renaissance was about discovering many things that the Greeks and Romans knew. Perhaps engineering drawing was carried furthest by Leonardo da Vinci who left behind detailed drawings of various futuristic inventions including aeroplanes. So why didn't engineering drawing come to India in the same way that knowledge flowed from one continent to another?

It is known that the Greeks influenced early Indian astronomy. Many astronomic Indian treatises were translated by the order of the early caliphate, and the Europeans got much of their astronomy from the Arabs. Brahmagupta's *Brahma Siddhantha,* along with the medical treatise, the *Charaka,* and the Indian folk story the *Panchatantra* were translated into Arabic by the order of Al-Khwarizm. Al Beruni also expressed his astonishment with the philosophy of the *Bhagavad Gita.* All this occurred upto the end of the 12th century. Did something catastrophic occur between the 12th and 15th century that created a barrier between India and the West?

In economic terms, it would appear so, as the transit trade between India and Europe fell drastically between the 12th and 15th century. Indian shippers virtually disappeared in the western Indian Ocean, frightened by rampant piracy in Socotra and Aden giving rise to the expression the *Bab-el Mandap* or gateway of sorrow for the entrance to the Red Sea. Some historians have tried to explain this and the explanations raise even more queries. One is that the Mongol conquest of the Middle East, followed by massive outbreaks of plague devastated present day Iraq, Syria, Jordan and Lebanon. The resultant fall in economic interest crippled the great international Jewish and Muslim trade financiers of Cairo.

Macpherson believes that Indian Ocean trade as a whole remained vibrant and dynamic and that states and regimes had little effect on the free conduct of this trade. But the author tends to differ on this. States, or governments made most of their money, not in trading but in taxing that trade. Quite often the state would be unable to fulfill its share of the deal in suppressing piracy or in the case of Aden, Dahlak, and Socotra, there were government officials hand in glove with the pirates who would prey on one nationality or another. When the Portuguese took over the taxation of Malacca, they allied themselves with the *Klings* (or Tamil Muslim community) to shut out the Gujarati, the *Moplahs* and the Jews. But there are only some of the explanations on how India became partially isolated from the beneficial developments of the renaissance, particularly structural drawings. It is clear that when one is building ships of upto 200 or 300 tons, a master ship-wright can build it from plans in his head, but when ships grow to 1000 or 2000 tons with three sets of compound masts it is impossible to proceed without drawings.

Weapon Technology

The same process repeated in metallurgy, cannon making, and gunpowder technology. While in the first ten centuries we have fine examples of Indian metallurgy in the form of pillars and the Damascus sword, the speed at which Europe left India behind in weapon making needs further analysis. Gun powder technology which originated in China probably went to Europe through the Arabs. But the Europeans soon realised that after sieving, coarse grained powder was ideal for cannon, medium grain for muskets and fine grain powder for pilots and priming. In the late 18th century Benjamin Robins, an Englishman, invented the ballistic pendulum to measure muzzle velocity and from then on the state intervened in establishing norms and standards for ordnance. These processes simply bypassed Indian makers of gunpowder.

When it comes to cannon making European cannon of the late 14th century roughly resembles Indian cannon of the 17th century. Both had huge barrels and attempted to fire stone shot upto 400 lbs. Early attempts to fire cast iron shot of that weight caused the cannon to explode because iron shot weighs two and a half times its stone equivalent of the same volume. Using cast iron shot of that size virtually blocked the cannon. Early cast bronze cannon technology was developed from sculpture and bell metal foundry work.

South Indian bronze suggested that Indian bronze casting technology for sculpture was very high. Cannon making only required that the brittleness of bell metal be removed by using different proportion of copper and tin, as well as to remove the air cavities in casting, which required much higher temperature. French sources tell that to cast a 24 pounder cannon it took the fire of 400 trees to get the right cast. It becomes clear as to why the state of Travancore where bell metal casting is a famous industry was the only one casting bronze cannon in the 18th century according to an English commentator.

Europe also made some other significant improvements. In about 1740, boring machines were used in Europe to bore the barrel. Until then the cannon were cast hollow. Boring produced more accurate weapons. All that the industrial revolution did was to replace the horse with steam as the prime mover of the boring machine. So, while it is commonly heard that India missed the industrial revolution, it is not as well understood that India missed the scientific developments of the renaissance too. It is difficult to conclude, one way or another, that missing the renaissance occurred owing to physical barriers between India and Europe, or if physical barriers did not exist between these continents, would Indian society have accepted and adopted European scientific developments? After all, in 1903, public violence against inoculation against the plague caused one and a half million deaths in India, which was repeated three years running. Would inherent societal factors have prevented ideas from other continents from being accepted? Probably so, as we shall see.

Verbalising Knowledge

If large structures such as ships of 1000 to 2000 tons cannot be built without drawings how did the Indians build some large structures anyway. It is well known that the advantages of a drawing is that, if successful it can be copied in many shipyards. The Indians had fairly standardised temple designs. How was this done? It was witnessed recently when the author saw a new temple being marked off in NOIDA by a Tantri from Kerala. He marked off the floor plan on the basis of certain verses he knew. The process that described is the conversion of scientific information into verse. This process has societal and technological implications beyond the process itself. Narasimha has done some pioneering work in the use and conversion of large numbers into phonetics and verse. It is not being suggested that this process existed for building ships in India, but whatever process existed for building a large structure like the *Brihadeeswara* temple was not replicated for shipbuilding in India. The process described is a phenomenon by itself and a subject into which my research on shipbuilding has not taken me. What really happens when knowledge is verbalised, is that its possession and transfer is restricted to a very few. This side-effect may be intentional or otherwise. But from the time of the *Vedas*, which were verbalised, for centuries we see a tradition in India to verbalise all kinds of knowledge.

The Effects of Religion

Where and when did the concept of *Kalapani* come in, that is to say, the loss of whatever caste a man possessed, by going overseas beyond territorial India. Once he did this, he had to be re-ordained into his original caste through religious rites. If such rules ever were codified they should be found in the *Dharma Shastras* of either *Manu* or *Yajnavalkya.* One of the great ceremonies described in the *Manu Smriti* is the one on *Shraddha* or the old vedic ceremony of honouring one's deceased ancestors or *Pitris,* performed daily, monthly, at funerals and at notable occasions. Here there is a strong and large paragraph on persons to be excluded from a *Shraddha* and the list is "sellers of meat, a public servant of the Raja, a feeder of cattle, a seller of the moon plant, a navigator of the ocean, an oil man, a tamer of elephants and horses, a maker of bows and arrows, one who teaches the use of weapons, a breeder of sporting dogs, a falconer, and one who removes dead bodies for pay". This list is the one where the forbidden professions are described. There are three other lists — one of persons excluded on moral grounds and of physical appearance. But if one looks carefully at the list of people excluded on account of their profession it can clearly be seen that this is nothing but a Brahmanical attempt to dominate the *Kshatriyas.* While the attempt to isolate the ocean navigator is what concerns us, there is little doubt that this law has had a crippling effect on the ability of this society to defend itself.

What practical effect would this have had? In India, the upper castes never went to sea. Some outstanding navigators produced in India were either Muslims or from the lower Hindu castes. But even this achievement does not produce any degree of maritime power. To command a fighting ship, or a fleet, one is looking for qualities of military and weapons knowledge, of man management and leadership, of navigation and seamanship and of social standing to represent one's profession at the court of the king. In Europe we often hear of men of high social status assuming command of fleets without adequate professional knowledge. In India we see a different picture. The navigator is fairly competent, but has little military knowledge. His social standing is wretched, and he can command no presence in the court. The court does not recognise the need to couple navigational skills with military command, which if acceded to would have automatically increased social standing. We see the same problem in the British Royal Navy which in the 17th and 18th centuries had a sailing master attached to each Captain of a warship. One exception to all this, is the case of Shivaji and Kanhoji Angre where Shivaji created senior naval officers out of seagoing personnel.

This discussion could be conducted at a theoretical level and the arguments may be inconclusive. But on another statistical level the evidence is fairly conclusive. The British had conducted detailed surveys after 1857 to assess the employability and numbers available for recruitment into the armed forces. They did not bother then about equal quotas for states on a population basis, as is done now. The result was that in 1947 a great majority of seamen in the old Royal Indian Navy came from just two areas, the Ratnagiri and the Makran coast. All this shouldn't really be too much of a revelation, because even in Akbar's court, while there was a strong established hierarchy of military command down to commander of a hundred horse, there was no hierarchy of naval command centrally or regionally. The 1901 census is revealing on the state of the all India schedule castes. It lists the major fishing and sea going communities in the coastal states as Andhra — Zalari; Bengal — Bagdi, Dewar, Jalia, Kyot and Mallah; Kerala — Kanakan, Vaalon; Maharashtra — Koli; Orissa — Bagheti, Gokha; Tamil Nadu — Bharator also known as Paravan and Vaalon.

BIBLIOGRAPHY

ARUNACHALAM, B. (2002) *Heritage of Indian Sea Navigation*, Maritime History Society, Mumbai.

ARUNACHALAM, B. (2002) Ship Building in the North Konkan 1550-1750 AD, In: *Essays in Maritime History*, Vol. II, Maritime History Society, Mumbai.

BAKER, William A. (1954) Early Seventeenth Century Design, *American Neptune*, Vol. 14, No. 4.

BALASUBRAMANIAM, R. (2002) *Delhi Iron Pillar*, Aryan Books, Delhi.

BOWEN, Frank (1669) *The Sea: Its History and Romance*, Halton & Truscot Smith, London.

CHAPELLE, Howard (1967) *The Search for Speed Under Sail, 1700-1855*, Norton & Co., New York.

DEAN, Antony (1670) *Doctrine of Naval Architecture and Tables of Dimensions, Material, Furniture and Equipment*, Original Manuscript in the Magdalene College Library.

DODD, James and MOORE, James (1984) *Building the Wooden Fighting Ship*, Hutchinson, London.

EB - Encyclopaedia Brittanica, The Technology of War, Vol. 29.

IRVINE, William (1994) *The Army of the Indian Moghuls*, Low Price Publications, Delhi.

KENNEDY, Paul (1988) *The Rise and Fall of the Great Powers*, Harper & Collins, London.

MAHAN, A.T. (1908) *The Influence of Seapower on the French Revolution and French Empire*, Sampson, Low, Marston & Co., London.

MNM - Tresors d' Oceans, *Musee National de la Marine*, Paris.

NM - *Mullum ni pothi*, Original Gujarati Document, National Museum, New Delhi.

NMM - The Collections, *The National Maritime Museum*, Greenwich, London.

PÂRIS, F.-E. (1843) *Essai sur la construction anavale des peuples extra-européens*, 2 Vol., Paris.

RIETH, Eric (2004) Traditional Boats and Ships of India Through Ethnographical Studies of Admiral Paris (1806-1893), In: Alok Tripathi (ed.) *Proceeding of the International Seminar on Marine Archaeology*, Organising Committee of the ISMA, New Delhi.

TAYLOR, E.G.R. (1954) *The Haven Finding Art*, Hollis and Carter, London.

Comment

R. Narasimha

Admiral Raja Menon has presented a comprehensive survey of the history of maritime technologies around the time of what he calls the European encounter. The paper is full of insights and ideas, and I would like to begin by considering two episodes in the technological history of India that reinforces his points.

The first episode relates to the way that the Wadias in Bombay built excellent ships for the East India Company - acknowledged by the British to be better than what they could have built in the England at that time (early 19th century). Among the well known ships that were made in the Bombay dockyards of the Wadias were *HMS Minden* and *HMS Trincomalee.* Indeed one of the Wadias, Ardeshir Cursetji, became the first Indian to be elected to the Royal Society in 1841. It is interesting that all those excellent ships built in Bombay were made to British designs, highlighting the fact that even while the technology of building ships was available in India; however, the science behind designing it was not. The next Indian to be elected to the Royal Society was S. Ramanujam, the mathematician and that came 77 years later (1918). He was in many ways from a totally different world from that of Cursetji - reinforcing once again in the cultural separateness of the engineers and the craftsmen from the scientists and the mathematicians in India.

The second episode, reinforcing the same point, concerns the rockets that were built by Tipu Sultan in Mysore in the second half of the 18th century. Their range of about 1000 yards was at that time twice as long as what the best European rockets could achieve. Although they were not decisive in winning battles for their masters, the rockets terrified the British troops exposed to them. There is for example the well known story of the 'Sultanpettah incident' involving Col Arthur Wellesley, who was attacked by a hail of rockets and musket-fire as he was on a reconnaissance sortie the night before the British assault on the Srirangapattna fort in the 4th Anglo-Mysore war of 1799. The attack clearly scared Wellesley out of his wits, and his hasty disordered withdrawal might well have attracted a court-martial if it was not for his influential brother, the Governor General in Calcutta. Even after defeating Napolean and becoming a great national hero as the Duke of Wellington, he used to talk about the incident in an attempt to erase what some saw as a black mark on his otherwise distinguished career. But the interesting point about this story is that a few year after 1800, William Congreve in Britain was making a whole wide variety of rockets far superior to those that his countrymen had encountered in India. And the reason for this was that Congreve was bringing to bear on the problem a whole scientific armoury — from Newton's laws to chemical science and production technology - that were totally outside the ken of Tipu's rocket makers. The Congreve rockets were used with great effect in attacks on Boulogne during the Napoleonic wars and in Washington during war of 1812.

The main reason for the higher performance of the Mysore rockets appears to have been metallurgical quality of the steel casing that they used. In 1790, India was still exporting iron and steel to Britain for it was of better quality. But once again European science ensured that Indian technology was surpassed very fast. By the middle of the 19th century the traditional iron and steel industry of India had been virtually wiped out.

The fascinating question raised by Adm. Menon regarding maritime technology is therefore actually valid for the whole science and technology system of Indians at that time. In the early 19th century Raja Rammohan Roy bemoaned the state of 'the mechanical arts' in India, and went to England in 1830 because 'he longed to see the country of Lockes, of the Bacons, of the Newton's, of the Hampdons and of the Watts'. Note the conspicuous absence in this list of the empire builders like Clive and Hastings. Roy appears to have realised that behind the

British merchants, writers and war fighters he saw in Bengal was a world of new ideas and thinkers in science, technology and philosophy.

This question has acquired a new intellectual dimension following the work of Joseph Needham, a distinguished British biochemist who devoted the latter half of his life to making a monumental survey of science and technology in ancient China. As his project took definitive shape Needham said 'with the appearance on the scene of intensive studies of mathematics, science, technology and medicine in the great non-European civilisation, debate is likely to sharpen, for the failure of China and India to give rise to distinctively modern science while being ahead of Europe for fourteen previous centuries is going to take some explaining'.

There have been many answers attempted in India about this question and I have argued elsewhere that such explanations as the idealistic philosophy of Hindus, their other-worldly nature, the lack of protestant ethic etc. are really not convincing if only for two reasons. First, Indian philosophy is too diverse to lend itself to such simple characterisation and secondly, as the Needham quote above admits, Indian science and technology were ahead of the west during the long centuries when these 'explanations' must still have been operative. The issue is of course very complex and has to do with the culture, history, sociology, economy and geography of India. But let me highlight what I consider to be the two major components of the answer as I see it.

The first part is recognition that the formal knowledge system of India had a deep, rational and not unwarranted distrust of the Greek axiomatic method. The Indian view, as expressed by thinkers of various religions and philosophical persuasions, was that a formal deductive system was inadequate and useless. This view was most eloquently expressed by Nilakantha (1502 CE), the great Kerala mathematician — 'Astronomy is primarily a discipline of observation, (numerical) experimentation and computations, and deserves to be revised periodically (formal deductive logic) is of little substance, endless and often not decisive'.

To over simplify matters, we could say that the Indian ideal was the empirical, rulemaking brevity of Panini, rather than the parsimonious, model making, deductive system of Euclid. The rational Indian objection to Greek axoimatism of the Euclidean type was basically that there were no self-evident truths that one could begin from, and whatever might be proposed as axioms would themselves have to be result of an appropriate inference process. Indeed these Indian objections to axiomatism and to the Platonic philosophy that truth can be derived from sheer logic without need for observation, found an echo in the contempt of Francis Bacon, the patron saint of modern science, to Greek learning. These differences between Greek and Indian views are mostly clearly seen in their approaches to astronomy. The Indian view is best understood as reflecting an attitude of what I have called 'computational positivism', namely the belief that all that matters is agreement between observation and computation: everything else is irrelevant. This model free approach led in fact to astronomical predictions that were not surpassed in accuracy even a hundred years after Newton's Principia (1687). For example in 1790 the famous British mathematician John Play fair concluded, in a review of the *Astronomy of the Brahmins* 'Of such antiquity, therefore, must we suppose the origin of this astronomy, unless we can believe, that all the coincidences which have been enumerated, are but effect of chance: or what indeed were still more wonderful, that some years ago there had arisen a Newton among the Brahmins, to discover that universal principle which connects, not only the most distant regions of space, but the most remote periods of duration; and a Lagrange, to trace though the immensity of both, its most subtle and complicated operations'.

Newton's system however showed that it was possible to create physico-mathematical models that could be extraordinarily accurate, invoking only three laws (incidentally also called axioms in the *Principia)* and the principle of gravitation. We can see the revolution of modern science as a part of European miracle was triggered by (among other things) technological imports from China and mathematical inventions from India, both transmitted by creative West Asian intermediaries. But Newton combined these with the ancient Greek penchant for model making in an enterprise whose result were epoch-making. It must be understood that this extraordinarily success of Newton's was unexpected and still remains a mystery, as has been so eloquently articulated by such modern scientists as Diarc, Winger and Einstein. The fair success of Ptolemy's model (2nd century CE) realised to be entirely false since Newton shows why it is not reasonable to expect the scientific enterprise may be philosophically unstable. That it has actually turned out to be robust in the last 300 years continues to be a source of fascinating debate in the world of science.

The second part of explanation has to do with the relative prosperity of India till about the middle of the 18th century. As Adm. Menon points out, India accounted for 25% of the GNP of the world at that time: scholars like Andre Gunder Frank would put it even higher. Globally, therefore, India was in a very comfortable economic position. This relative prosperity was possible in part because Indians were fortunate to live in the most productive topics, where land in itself (blessed by plenty of sun and rain) was an asset of great value. This surely accounts for the generally favourable view that Indians took of nature as basically bountiful, in contrast to the view of environmentally deprived northern Europe, where Francis Bacon preached that nature and her creation have to be enslaved and even tortured. Indians had no great economic incentive to conquer their neighbourhood, which was largely desert, high mountains or dense jungle. On the other hand, India itself was always an object of desire from west of the Indus; it was a sub-continental oasis for the inhabitants of a warm, dry West Asia, and a sub-continental hot house for those from Europe.

Indians did not need to develop technology for offensive purpose of the kind that western countries have traditionally done for 1000 years. Here perhaps lie the roots of Indian accommodationism and (as Needham called it) the western psychology of dominance, which gained from an aggressive pursuit of modern science and technology.

In conclusion let me say that the questions raised by Admiral Menon in his paper are profound, and part of much larger central issue that faces India even today, although in contrast to the situation 300 years ago, we are now the people suffering from deprivation. A new dynamic that assimilates a western model making culture with our ancient creativity in computation and craftsmanship is what is needed if we are not to remain part of the Needham question.

BIBLIOGRAPHY

NARASIMHA, R. (1985) *Rockets in Mysore and Britain, 1750-1850 AD, PD DU 8503,* National Aeronautical Laboratory, Bangalore.

NARASIMHA, R. (1999) Rocketing from Galaxy Bazaar, *Nature*, Millennium Essay, 400.

NARASIMHA, R. (in press) The Indian Half of Needham's Question: Some Thoughts on Axioms Models, Algorithms and Computational Positivism, *Interdisciplinary Science Reviews*.

5

Archaeological Investigation of Princes Royal off Bangaram Island - Introduction and Underwater Operations

A.K. Patnaik

Not a professional archaeologist though the author had short excursions into this field off and on in professional diving career. Truly, such adventures have been incidental to my ability to operate underwater rather than with an agenda of a classical archaeological investigation. Therefore, the instant investigative work is my first formal work in marine archaeology since during this endeavour we had specialists from the field from the Archaeological Survey of India as collaborators in the excavation.

The Navy happens to be the protector of maritime interests in all its facets. Whereas priorities of focus get determined due to compelling demands, the issue of marine archaeology has been a late starter in India in many ways. Investigations off the coast of India have been few and far between and it would not be too far off the mark if it is said that the excavation report that is going to be presented is perhaps the first serious indigenous endeavour undertaken by the Archaeological Survey of India and Indian Navy in a professional and coordinated manner.

This paper would touch upon the background, brief on facts of the general excavation area, record of earlier dives on shipwreck *Princes Royal,* planning for the excavation and the conduct of underwater operations.

Background

The mystery of the deep and the vestiges of human history that lay beneath the abysmal depths of the oceans have always beckoned mankind to devise new ways to explore them and, where possible, connect them by time and space to the forgotten cultures, technologies and trade practices of the bygone eras. By far, the ships are known to be the noblest artefacts of man's quest to conquer the mighty oceans and to that extent, they represent the historic realities of these periods. It is a known fact that a number of ships that ventured into the oceans either for trade or otherwise, were wrecked in that process and today they lie beneath the very surface of the medium they chose to conquer, some of them in a well preserved state. These shipwrecks are valuable 'Time Capsules' the study of which help to unfold the maritime history of the nations whose flags they flew. A thorough and systematic study of these wrecks further reveal a wealth of scientific, cultural and traditional data of the periods to which they belonged.

Closer home, the Indian Ocean has been one of the oldest basins that was crisscrossed by brave mariners. This Ocean has also been a silent witness to a number of ships that went down to the 'Davy Jones Locker'. A large number of wrecks stand testimony to this fact, which, though recorded in the archives, has not been excavated till to date. The wreck of *Princes Royal* is one among these wrecks that was studied and excavated jointly by the Archaeological Survey of India and Indian Navy between November-December 2002 off the Bangaram Island **[15]** in Union Territory of Lakshadweep.

Earlier Excavation Efforts

Before dwelling on the excavation proper few words about the Lakshadweep Islands and the history of dives on the wreck of *Princes Royal.* Lakshadweep is an archipelago consisting of 12 atolls, 3 reefs and 5 submerged banks, irregularly spread over an area of nearly 77,000 sq. km. in the Arabian Sea between latitudes 8^0 and 12^0 30′ North and longitude 71^0 and 74^0 East. These islands with a total area of only 32 sq. km. have a lagoon area of about 4,200 sq. km., 20,000 sq. km. of territorial waters and about 4,00,00 sq. km. of Exclusive Economic Zone. Of the total 36 islands only 10 are inhabited (*Gazetteer* 1977). They lie astride the trade route between Africa, Arabia and the Malabar Coast and act as landmarks for ships sailing in the Arabian Sea. Ships have played an important role in the history and culture of these islands, as these were the only means of their survival and contact with the outside world. A number of ships were wrecked around these islands, which is probably due to the fact that these water were uncharted and the subsurface reefs were not visible to the mariners. However, these waters were systematically surveyed by the Indian Navy in the early 70s and charts for safe navigation are now available to all mariners. Of the many shipwrecks in the area, only a few have been recorded (Tripathi 2004: 5) in the archives while many more still lie buried unrecorded.

The Bangaram Island is a small, uninhabited Island (fig. 5.1). Like most other islands in the Lakshadweep group this tiny coral island lies north to south. General pattern and disposition of all these islands is almost identical. They are boat shaped with stormy beaches on the eastern side and lagoon on the western side.

On 25 December 1990, a group of sport divers chanced upon a shipwreck that lay on the outer slope of the coral reef off Bangaram atoll and in their enthusiasm they lifted some artifacts lying exposed on the surface. When the Union Territory Administration came to know of this discovery, they confiscated these articles (Tripathi 1991: app. I-II) and requested the Archaeological Survey India to investigate the site and assess its archaeological significance. As a consequence, the Archaeological Survey of India carried out a systematic investigation on this shipwreck, in April 1991 (Tripathi 1994: 125-128). Sections of the wreck were found lying between depths of 9 to 40 m. with a variety of artefacts indicating that the site had the potential of a detailed

5.1. Bangaram Island, Lakshadweep

archaeological excavation. In April 1992, a diving team from the Naval Marine Archaeology Club, Mumbai visited the site and carried out diving at the site. Later, in March 1995, Indian Naval divers from the Southern Naval Command carried out diving on the site and *Indian Naval Ship Sujata* using the triangulation method fixed the exact position of the wreck. No sooner was this discovery known, the site became a popular dive spot among touring divers which may have accounted for the pilferage of few artefacts by default. During one such dive a bronze bell with inscription of the name of the ship, and the year of commission was also recovered.

The Archaeological Survey of India, which is the custodian of the cultural heritage of the nation including all historic wrecks in the Indian waters, approved the excavation of the wreck during the field season 2001-2002. It was also around this time that the first joint pilot project of an archaeological investigation between the Navy and the ASI was being mooted. Therefore, logically, the site off the Bangaram Islands got the nod from both quarters. Consequent to this decision, a joint investigation of the site to ascertain the scope of work was carried out in May 2002. This team sailed onboard *Indian Naval Ship Jamuna* **[16]** a survey ship that is well equipped with the latest survey equipments like DGPS, side scan sonar, echo-sounder, magnetometer, current meter, gravimeter, satellite navigation system, etc. A team of divers from the Navy's Diving School at Kochi also formed part of this team.

Upon arrival at the site, the team carried out a systematic study of the upper two terraces of the wreck site and collected few samples of archaeological significance without disturbing the site. These included pots and sherds of brown glazed ware, pieces of copper sheet and iron objects. Based on the results of the preliminary survey that were very encouraging, and the recommendation of the survey team, it was decided to undertake a full-fledged excavation, in November-December 2002.

Planning for the Archaeological Investigation

As mentioned earlier, this excavation being perhaps the first such indigenous endeavour, called for a high level of coordination and planning between various participating formations. The site itself being 230 nm. from the Indian mainland and the islands themselves commercially underdeveloped posed considerable problems of logistics.

Repairs and replacements for every bit of equipment had to come from the mainland. All these, and many more other associated factors put together made this a complex and challenging task. To overcome these challenges a series of meetings were conducted in New Delhi where various contingencies were discussed and necessary instructions were passed on to the participating formations. Extensive planning efforts were put in to ensure that adequate provisioning for consumables, equipment and accessories was carried out and wherever required, fabrication of associated gears was undertaken.

Having completed the basic planning and coordination, it was now time to work out the modalities for the excavation. As the wreck lay embedded in a coral reef carrying out excavation with minimal damage to the ambient ecosystem was one of the prime considerations. Moreover, the excavation work itself involved various intricacies, which was an altogether new subject to most of the participants. Therefore, in order to execute the task in hand systematically, it was decided to carry out excavation in two phases. The first phase, which was conducted in November 2002, involved confirmatory survey of site, mapping the extent of the wreck by laying of grids and marking of artefacts. The second phase, in December 2002, involved excavation of the site and recovery of the artefacts for archaeological analysis.

Since the excavation involved prolonged and extended diving, it was decided to utilise the naval dive support ship *INS Nireekshak* **[17]**. This ship is fitted with contemporary diving systems and equipment that includes re-compression chambers to cater to any complication arising due to decompression, submersible dive-bell, cage for diving, underwater communication system and videography and photography equipment. This ship is also capable of supporting six saturation divers to a diving depth of 200 metres. In addition, the ship carries two Rigid Hull Inflatable Boats (fig. 5.2) and inflatable crafts each that aid in conduct of diving operations **[18]** away from the ship and in supporting such operations subsequently. A total of 4 officers and 25 sailors all of whom were qualified saturation divers with a vast experience in diving took part in this joint effort. Apart from this, the ship has a qualified underwater marine medicine specialist and a 4-bedded medical inspection room with adequate facilities to cater to field medical emergency. Mobilisation of such a comprehensive diving effort is indicative of the fact that nothing was left to chance once the team arrived at the excavation site.

5.2. *Princes Royal*: Rigid Hull Inflatable boat used in diving operations

Conduct of the Archaeological Investigation

Having completed the initial planning and coordination at all levels for the ensuing task, the team assembled at Kochi, a major port in south-western India and embarked the Navy's Dive Support Vessel, *INS Nireekslak.* The ship sailed for Bangaram Island along with naval divers and participants from ASI. Few personnel from the Reef Watch Marine Conservation (RWMC) which is a non-governmental organisation also joined us in this maiden endeavour to study the biological effects of shipwrecks on the surrounding flora and fauna. During the long passage from Kochi to the excavation site, the ship became a hub of various activities (fig. 5.3). These included preparing and testing of all diving equipment, onboard dive support systems, laying out of gears for mapping the site, functional tests on airlifts and water-dredge and conduct of instructions to familiarise the participants on marine archaeology techniques and procedure. Apart from these, extensive briefings on the various aspects of underwater excavation, covering documentation, use of low pressure equipment for controlled probing and retrieval, packaging and preservation of artefacts, etc. were discussed.

In the initial stages of the excavation, it was found that the weather was unpredictable with the sea state changing frequently. It also rained, though intermittently, and the currents in and around the site were

5.3. *Princes Royal*: Preparation for the excavation during the passage to the wreck site

variable. The underwater visibility was conducive to diving in the early hours of the day and gradually decreased as the day progressed. Since the wreck was lodged right on top of the reef with the surf becoming heavy during the rising tide, it was decided to anchor the dive support ship at the specified anchorage which was at a distance from the site for safety reasons.

In view of these considerations, an elaborate dive plan was devised to support diving and excavation activities. The day usually started with loading of the hired motorised boats (fig. 5.4) with heavy equipment like compressors, storage bottles, high-pressure hoses, etc., well before sunrise. They were dispatched to the site so as to arrive there by the time the faster boats fetched up. The inflatable boats which had considerable speed advantage followed up thereafter with the diving personnel and their equipment which included personal diving sets, superlite helmets, photography and videography equipment, and the underwater communication gear.

Since diving operations were conducted in depths up to 45 metres, where the likelihood of a decompression illness could not be ruled out, one of these fast boats was earmarked to standby to ferry any such affected diver to the ship where elaborate facilities for treatment were kept ready.

5.4. *Princes Royal*: Wooden boat to support diving and excavation activities

5.5. *Princes Royal*: Diving on site

With the first light, diving operations were commenced to take advantage of the calm surface conditions and better underwater visibility earlier in the day. As the participating divers were new to this task, few initial dives were conducted for training and orientation in underwater archaeology work. Divers were quick to learn the various techniques and procedures and the dives for excavation work progressed smoothly.

Being the first systematic excavation a lot of emphasis was laid on correctness of the procedure and excavation techniques. Meticulous care was taken to ensure that each dive was accurately recorded so that no diver went into the regimes of repetitive dive or combined dives because of the wide variation in depth in the area of excavation. Divers were rotated around systematically so that each of them were employed both at the shallow as well as deeper end of the wreck site alternately.

Most of the dives were carried out in buddy pairs (fig. 5.5) so that while one diver did the excavation, the other was busy in marking and recording the artefacts as well as photography and videography. To learn from the proceedings of the dives and in order to get all participants into one grid, on completion of the days work, a combined debrief was carried out with the help of video recordings of the day's proceedings.

Having established these cardinal business rules, the excavation proceeded systematically and smoothly. As part of the excavation plan, an initial acoustic survey of the site was carried out using side scan sonar and echo-sounder from a boat (fig. 5.6). However, the results of

5.6. *Princes Royal*: Conducting acoustic survey

this survey using the side scan sonar were not encouraging, as the survey did not accentuate the presence of the wreck, probably because the entire hull of the ship seemed to have become flush with the seabed and the artefacts strewn around in the general area were covered with layers of encrustation and marine growth that made it almost a part of the coral reef. The echo-sounding around the site indicated well-marked submerged terraces at 7-12 metres, 15 metres, 21-36 metres and 51-61 metres after which there was a sudden increase in depth to a few hundred metres.

The GPS coordinates of the site as well as the position by triangulation which were available to the team from previous dives on the wreck came in very handy as with the first dive itself, the divers were able to locate the exact position of the wreck. Once the site was physically located, two datum's were established by the divers, one at 37 metre, which was the position of the ship's anchor, and another at 9 metre, which was the position of the ship's cannons. Once this was done, a 12 mm. line was laid in east-west direction to a distance of 100 metre (fig. 5.7). A second line running in north-south direction was laid perpendicular to the first line. Both these reference lines were marked at intervals of three metre with a small float and at nine metre with a large float. The bigger floats were inscribed with distance and direction from the datum to facilitate easy identification and marking

5.7. *Princes Royal*: Laying of grids for excavation

of the artefacts that were located in the site. With the help of these grid lines, the entire site was divided in three metre square grids by laying 8 mm. lines running in east-west and north-south directions. Each grid was given a specific number and the topside display board indicated the exact grid reference from where each of the artefacts had been located. With these grids as reference, concurrent diving operations were progressed at the shallow and deeper ends grid by grid for a systematic investigation.

During the excavation a classical 40 metre dive lasted for approximately 20 minutes without any decompression stops. However, decompression stops were provided in few cases where the diver needed to stay for longer durations in order to complete his task. The underwater communication and video facility enabled a continuous two-way talk between the diver and the surface attendant and the operation could be directed very well. The pneumatic lift bags were extremely useful in lifting heavy items underwater and as a matter of fact one of the cannons was lifted for preservation using this facility.

The airlift was deployed where large areas of terraced stretch were required to be de-silted. Similarly, water dredge was used wherever light probing was required which were lying embedded. The surface team continuously updated the grid map of the area with information from each dive. Whereas the position of the anchor and the cannons as they lay on the seabed gave some indication of the fore and aft portion of the ship, the location of artifacts such as pottery led us to identify the potential areas of other artifacts which was very useful in progressing the excavation. As the excavation progressed with each day the divers gained in experience and confidence and were able to make considerable contribution towards the closing stages. In fact the excavation proceeded very smoothly during the second phase in December where briefing sessions (fig. 5.8) became shorter and shorter due to easy comprehension of the task at hand by all members. The vast experience of the divers in diving in adverse conditions of high currents and low visibility was a definite advantage as they went about their tasks meticulously as per briefing received on surface.

The divers were allocated specific grids for searching in a progressive manner so that all artefacts were photographed (fig. 5.9) and videographed *in situ* before retrieval. Accurate measurements of

5.8. *Princes Royal*: Briefing in progress

5.9. *Princes Royal*: *In situ* documentation

the objects were taken in relation to reference lines. Care was also taken to ensure that the objects were correctly marked (fig. 5.10) and recorded. Each object was allotted a specific reference number and the location of the objects found was documented.

The paper would not be complete without a mention of the rich

5.10. *Princes Royal*: Marking of objects

and myriad marine life encountered by the divers. The benevolent benediction of Lord Varuna turns the graveyard of ships in her belly into a veritable habitat for a rich marine life. The divers had the opportunity of cohabiting these waters for over a month with a diverse array of marine life. The RWMC is providing a yeoman service by monitoring and creating public awareness for preservation of marine eco-system in India in all its diversity. This joint venture also provided an opportunity to RWMC to share the experience of the underwater environment around the wreck of the *Princes Royal*.

In the end, I would like to very briefly touch upon the broad findings on which the analysis has been progressed. The dimension of the anchor **[24]** and the presence of four large cannons **[25]** each measuring 2.2 metre in length indicate that the ship was a large vessel. The copper sheathing on the hull indicates that the ship was built in the 18th century as prior to this copper was not used for the protection of wood (Tripathi 2001: 112-116) which was predominantly used for construction of the ship's hull. The inscription of date on the ship's bell **[30]** corroborates this finding. The recovery of porcelain and brown glazed ware, which may have been the main cargo of the ill-fated ship, suggests that the ship could have been heading from Southeast Asia. Next paper by Alok Tripathi gives a detailed account about the excavation and identification of this historic wreck.

BIBLIOGRAPHY

Gazetteer of India - Union Territory of Lakshadweep (1977) Coimbatore.

TRIPATHI, Alok (1991) *A Report on Bangaram Shipwreck and Proposal for Shipwreck Exploration and Excavation in Lakshadweep*, Archaeological Survey of India, Unpublished Report, Khajuraho.

TRIPATHI, Alok (1994) A Historic Shipwreck in Lakshdweep Waters, In: S.R. Rao (ed.) *The Role of Universities and Research Institutes in Marine Archaeology*, Society for Marine Archaeology, Dona-Paula.

TRIPATHI, Alok (2001) Identification of Marks and Dating of Historic Shipwreck: Marine Archaeological Investigation in the Arabian Sea, In: A.V. Narasimha Murthy, K.M. Suresh, K.P. Poonacha and K.R. Basavaraj (ed.) *Hemakuta - Recent Researches in Archaeology and Museology*, Bharatiya Kala Prakashan, Delhi.

TRIPATHI, Alok (2004) *Excavation of Princes Royal,* An Interim Report, Organising Committee of the International Seminar on Marine Archaeology, New Delhi.

Comment

M.C. Joshi

I am happy to learn about the achievements made with the cooperation of the Indian Navy and Archaeological Survey of India. I am still more happy that a real shipwreck excavation has been carried out, which was my one time dream.

Furthermore, Lakshadweep is also an area of my interest. I initiated archaeological exploration here in the mid 80s, keeping in view further developments. While coming to Lakshadweep I do not know from where should I begin - from modernity or antiquity? The Arabian Sea has been mentioned in classical Sanskrit literature as *Ratnakar* and Indian Ocean as *Mahodadhi.* And so far as Lakshadweep Islands are concerned from about the 11th-12th century, the area was known as Lakshadweep and it included Maldives also. It is something like *sahastra* multiple islands and initially they were perhaps being exploited for their coir. Even Portuguese tried to capture some of the northern islands for the same purpose and the Lakshadweep inhabitants poisoned them and thereafter no attempts were made. Well its possible because I saw there the earliest historical vestiges. There were found two Buddha heads, made of coral and those pertain to the Karnataka idiom. So, it is possible that sometime at least the northern islands were under the control of Indian Kings. It is also clear if we compare what is there in terms of Buddhism. What was found in Lakshadweep and what is in the Minicoy and other Maldives Islands. In one place Buddhism is of the Sinhalese brand, Sri Lankan brand, but at least two of these heads suggested that it is closer to the western Indian idiom and seems more connected with Maharashtra and Karnataka.

The local inhabitants were certainly under the Cheras, rulers of Kerala and that is why Perumal Cheraman legend is also associated because we have many kings of that name and one of them is also depicted in one of the Thanjavur murals. There is a tradition among the Muslims that he got himself converted to Islam. The local Muslims maintain a tradition that they were converted by the great grand son of Abu Bakar, the second Khalifa. How far it is correct because I had the occasion to see the earliest Quran Sharif there but it was written on paper, of course the script was Kufic that paper was not being used in earlier holy books, they were written on skin. They also showed me certain cloth which also appeared to be of a later date. Survey of the graves there revealed that the dates were not earlier than the 17th century. Similarly Koya is supposed to be the corruption of the word *Khaja*. So, far as the tradition of the Lakhsadweep Islands are concerned they have more to do with Islam. And even earlier Lotika Varadarajan pointed out to several technical terms not one of them is either of Malyalam origin nor of Sanskrit origin. So the only tradition they were maintaining is the Khorrum tradition.

15. *Princes Royal*: Buoy showing location of shipwreck off Bangaram Island

16. *INS Jamuna*: Survey ship

17. *INS Nireekshak*: Dive support ship

18. *Princes Royal*: Divers getting ready for dive

19. *Princes Royal*: Divers laying lines

20. *Princes Royal*: Diver transporting debries from the trench

21. *Princes Royal*: Pottery and copper objects exposed in excavation

22. *Princes Royal*: Divers retrieving a brown glazed jar

23. *Princes Royal*: Brown glazed jar of small size

24. *Princes Royal*: Iron anchor

25. *Princes Royal*: Cannons

26. *Princes Royal*: Diver excavating a cannon

27. *Princes Royal*: Cannon retrieved with lifting balloons

28. *Princes Royal*: Ballast stones after partial excavation

29. *Princes Royal*: Wooden planks of the ship

30. *Princes Royal*: Bronze bell inscribed with PRINCES ROYAL 1792

6

Archaeological Investigation of Princes Royal - Excavation and Identification

Alok Tripathi

Indian Ocean is one of the oldest basins, crisscrossed by great maritime people in different kind of boats and ships since antiquity. Many of these ancient ships did not reach to their destination and some of them are still lying buried on the ocean floor. Systematic study of ancient shipwrecks reveals a great deal of information about the past societies, who built and used them. Ships are said to be the noblest artefacts of man's multifarious skills. They are not only moving machines but also the historic reality of any age. Ship may be seen as a detached, floating, isolated culture with its own rules and hierarchy, a replication of the parent state, perhaps an exaggerated version of it (Hicks 2001: 164). In 2002 Underwater Archaeology Wing (UAW) of the Archaeological Survey of India (ASI) in collaboration with the Indian Navy (IN) carried out first systematic excavation of an ancient shipwreck in the country, off Bangaram Island in Lakshadweep.

Lakshadweep is the tiniest Union Territory with a land mass of mere 32 sq. km. It is an archipelago consisting of 12 atolls, 3 reefs and 5 submerged banks, irregularly scattered over nearly 77,000 sq. km area in the Arabian Sea between 8^0 and $12^0$30′ North latitude and 7^0 and 74^0 East longitude (*Gazetteer* 1977).

The first direct reference to these islands is found in the Vayalur Pillar inscription of Pallava king Narasimhavarman II (680-720 CE). It mentions *Dvipalaksham* under Pallava kingdom (Krishna Shastri 1925-26: 152). An inscription in the Rajarajeswara temple at Thanjavur (c. 1018-1019 CE) also mentions the conquest of these islands by Rajarajendra Chola (Venkayya 1913: 82). The history of Lakshadweep from 13th century is well recorded.

Ships have played an important role in the history and culture of Lakshadweep. These coral Islands, lying directly on the trade route between Africa, Arabia and Malabar coast, were the landmark for the ships sailing in the Arabian Sea. Ships and boats were necessary for the people living on these isolated islands separated by deep sea from the mainland. They were the only means of their survival and their contact with the outside world (Tripathi 1993: 92-98). Since antiquity many ships have wrecked around these dangerous reefs (Tripathi 2004 :5). Being far away from the main land they were not disturbed and some are still preserved on the seabed.

An ancient shipwreck was noticed on the outer slope of coral reef off Bangaram atoll **[15]**. It was a chance discovery on 25th December 1990 when some sport divers noticed some objects lying on the reef. In April 1991[1], the Archaeological Survey of India investigated the site to access its archaeological importance. The wreck with abundance and diversity of objects of archaeological significance was recognised as the deepest submarine archaeological site in the country (Tripathi 1991; 1994: 125-128; 1996: 111-117; 1997: 781-787). Initial investigation by the author suggested that the site needs to be studied in detail (Tripathi 1991). However, the wreck was identified for archaeological excavation but it could not take place for over a decade.

Meanwhile, the site became a tourist attraction and obviously suffered some loss. During one such dive a bronze bell was recovered from the site. This bell was inscribed with the name and date of the ship.

With the establishment of Underwater Archaeology Wing in the Archaeological Survey of India this long pending project was taken up afresh. On the recommendation of the Standing Committee of the Central Advisory Board of Archaeology the Director-General, ASI approved the excavation of this shipwreck in the field season of 2001-2002.

In May 2002, a joint team of the ASI and the IN carried out preliminary investigations at the site. Indian Navy's survey ship, *INS Jamuna* **[16]** was deployed for survey and investigation. The archaeologist of ASI and divers of IN conducted about 20 dives. The reef was surveyed and upper two terraces of the wreck site (fig. 6.1) were systematically studied. This investigation provided vital information about the extent of the site, spread of antiquities and

6.1. *Princes Royal*: Remote sensing survey

condition of the wrecked ship (2002a; 2002b: 72-80), which was needed for proper planning of the excavation.

The excavation conducted in November-December 2002 was planned carefully. *INS Nireekshak* **[17]**, fitted with state-of-the-art diving facilities, was deployed for conducting the first shipwreck and the deepest underwater excavation in India. A wide range of diving equipments **[18]** such as SCUBA, devitor and Superlite with two-way communication systems were used for conducting excavation at different depths. The site spread on three terraces at depth varying from 9 to 61 m. All the artefacts lying exposed on the reef were identified, marked and numbered with the help of plastic tags (fig. 6.2). Divers explored the whole area and tied these tags to every antiquity. Later on position of these objects was fixed. A group of two divers took precise measurements to fix the position of each object. The location of these objects was later drawn on the measured drawing. The accuracy of the drawing was further confirmed with the help of the vertical recording of the site.

The entire slope of middle terrace, between 20-36 metres, is strewn with variety of artefacts (fig. 6.3). The antiquities lying on the steep slope include copper vessel, grinding stone, parts of the ship, objects of iron, pieces of copper sheet, wooden planks, etc. Due to exposure on the surface the wood has been deteriorated and at places the planks are very thin. The marks of nail and the nails are still attached in metallic form or rusted form. Most of these planks lie almost north-south.

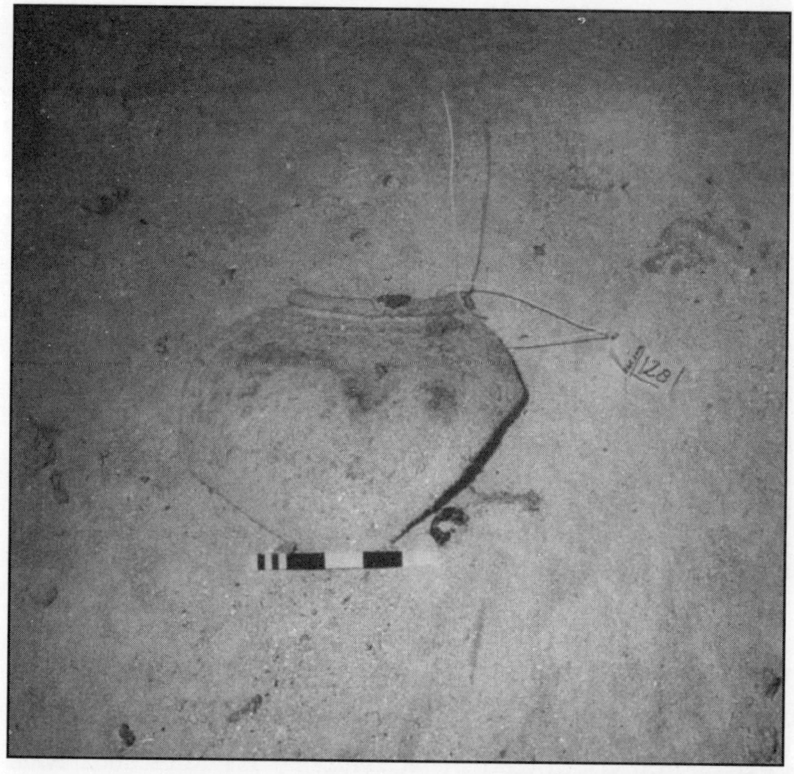

6.2. *Princes Royal*: Marking of exposed objects

6.3. *Princes Royal*: Artefacts lying on middle terrace

Area on upper two terraces was divided into two sectors-northern and southern sectors. After fixing the reference points a 12 mm. thick line, running north to south, was laid **[19]** to a distance of 100m. Another line, running east to west was laid perpendicular to it. These lines were marked with small float at every three-meter and with a bigger float at every nine meter. Three meter square grids were laid (fig. 6.4) with the help of thin lines running east – west and north – south. Each grid was given a specific number for proper documentation.

Excavation

Excavation was conducted both manually as well as mechanically. In the areas where small fragments and artefacts were found layers of sand and dead corals were removed manually. Manual excavation was also adopted near the cannons so that the position of these cannons is not disturbed.

A low-pressure airlift was also used in southern sector to carry out excavation in few grids where there was a thick deposition of sand and dead corals over the artefacts. The airlift with long flexible pipes on either end was fixed on a high area far away from the site. The inlet pipe was kept longer than the outlet pipe to transport the excavated sand and corals to a safer distance so that they do not disturb

6.4. *Princes Royal*: Grids laid for excavations

the visibility at the site. The outlet of the airlift was kept in an area, after thorough study, so that the sand deposited there did not affect the submarine conditions.

One grid in southern sector was identified for initial training of the naval divers who were working for the first time on an archaeological site. At the deeper end, in northern sector, two grids were taken up for excavation at a depth of 35 meter.

In phase I excavation was conducted in southern sector. Entire site in southern sector is covered with a layer of dead corals. Divers manually lifted these dead corals lying on the surface, and transported them in perforated plastic troughs **[20]** to a fixed point. A cavity near a huge coral growth, to the east, outside the marked area was selected to dump the dead corals. The corals lifted from the grids are dumped at one place in order to avoid any damage to marine environment as well as to fill the excavated trenches at the end of the excavation. This area was not only outside the working area but also close to the trenches where the excavation was to be carried out. To the east of it was a huge and high ridge running north-south which prevented the spread of these corals as well as merged the dump with the ridge which otherwise would have given an ugly look in the photographs. It saved

6.5. *Princes Royal*: Underwater video camera used for documentation

time and energy as every moment on the ocean floor is precious and need to be fully utilised.

After removal of uppermost layer of dead corals, cannons and other artefacts became clearly visible. Excavation near the cannons revealed small sherds of brown glazed ware **[21]**, pieces of green glass bottles, copper nails, lead and iron objects.

In Phase II, excavation in northern sector was taken up. The area selected was on the lower edge of the terrace. A grid was prepared by using perforated iron angle. The grid was assembled on the boat and lowered near the mooring point. Later a group of divers lifted the grid and carried it to the site and fixed it as per the site plan. The whole operation was documented (fig. 6.5) and guided from the surface (fig. 6.6) with wireless communication attached with diving masks.

The divers had gained good experience and performed very well. The excavation was conducted systematically with thorough documentation. Large number of bricks, pieces of brown glazed ware, thick storage jars, pieces of wood, corroded and cemented parts of the ship, copper fry-pan, a sherd of porcelain, sherds of pottery with embossed design, etc. were retrieved from these trenches. Two complete brown glazed pots **[22]**, one green glazed sherd, one bowl were also retrieved from the trench.

6.6. *Princes Royal*: Constant monitoring of underwater operations on the surface

Pottery

A variety of pottery found on the wreck site includes small, medium and big size jars of brown glazed ware, blue on white porcelain, thin potsherds with embossed designs, green glazed potsherd, etc.

Brown Glazed Jars

A deposit of pots with thick dark brown glaze on outer and inner surface forms the main part of the assemblage. The shapes included jars, bowls and dishes. The jars varied in dimensions but generally fall into three categories — small jars without handles and medium and big jars with four handles. The upper portion of big storage jars is also decorated with embossed designs, generally a line of dots, wavy lines, etc. Few of these containers were intact (fig. 6.7). Pottery retrieved from the lower layers was largely free from marine encrustations **[23]**.

The ship was carrying a large quantity of these jars as they were found spread everywhere on the site — from the top most to the lower most terrace (fig. 6.8). Large number of pottery was also found in the excavation. The sherds retrieved in the southern sector were of very small size whereas those found in the northern sector were bigger in size. The contents of the jars could not be found because their mouths were open. Glazed inner surface of these vessel suggests that they might have been used for carrying or keeping liquids or food items.

6.7. *Princes Royal*: Brown glazed jar

6.8. *Princes Royal*: Pottery lying on lower most terrace

Porcelain

Another important antiquity retrieved from the wreck site was small sherd of blue and white Chinaware (fig. 6.9). Many European ships of trade often carried huge quantity of blue and white Chinaware. So far only two sherds are found but there is possibility of finding more pots lying buried beneath sand and thick layer of corals. On the basis of the typology one of the sherd found earlier was dated to 16th century (Muhammed 1994: 181), but now with the finding of ships bell and other studies (Tripathi 2001: 112-116) we know the date of the wreck.

Anchor

The anchor of the ship **[24]** is the largest object found on the site. It was measured precisely. The anchor is lying up side down on the lower end of the middle terrace. Its upper part, arms, is at a depth of 35 metre where as the lower part, ring, is in 40 metre depth. The shank of the anchor is 4.40 metre long. The ring measures 70 cm. in diameter, each arm is 2 metre long and flukes are 45 cm. wide.

Cannons

Four cannons, around 2.2 metre long, are the most prominent objects in southern sector. These cannon have heavy encrustations.

6.9. *Princes Royal*: Sherd of blue and white Chinaware

6.10. *Princes Royal*: Copper vessel

Three cannons **[25]** are lying in a group roughly forming a triangle. These are the most eye-catching objects on the site. It was decided not to disturb these cannons and preserve them on the site. The fourth canon, lying isolated to the east of these cannons, was selected for lifting. This cannon was almost half buried in the dead coral and sand.

The area around fourth cannon was excavated manually **[26]**. Pick-axe, chisel and hammer, knifes, etc. were used for removing hard deposits around cannon. Detailed measurements were taken and the drawings were prepared. On 4th December 2002, the Navy Day, the cannon was lifted with the help of two lifting balloons of 1 ton and 500 kg. capacity **[27]**. A boat towed the cannon, while it hanging with the balloons. After a journey of about 10 nm. it reached the ship where it was lifted with the help of ships crane and kept on the quarterdeck of the ship.

Iron Objects

A large number of iron objects were found on the site, earlier, but the function and nature of many of these iron objects remained obscure as they were found highly oxidised and in small fragments. In some cases, only encrustations remained and the objects disappeared completely. A large group of heavily rusted and encrusted iron objects was noticed to the south-east of the trench in northern sector. The group was too big for lifting. Being out of the trenches it was not disturbed but documented *in situ*. Some small fragments of iron, which were retrieved during the excavation, were stored in water so that the environment of these objects is not changed suddenly.

Copper Objects

A good number of objects of copper were found during the excavation. These were copper vessels, ships fastenings and pieces of thin sheet.

Ship carried vessels of copper onboard. A complete vessel of copper, around 52 cm. high, was found earlier (fig. 6.10). The upper half of the similar pot was found close to the iron anchor. A pan of copper was retrieved during excavation from northern sector (fig. 6.11).

Several heavy, thick and round through bolts with clinch ring were noticed on the slope close to keel (fig. 6.12). Copper fastenings were developed in 1783 to use below the water surface to eliminate the problem of galvanic action. These bolts were 135 cm. long and were probably fitted in the keel. Earlier also similar copper bolts and other smaller bolts of similar thickness were found. They were about 35 cm. long with outspread head and might have been used in the frames of the ship. A large number of square headed flat pointed nails were also recovered earlier (fig. 6.13). These nails were used for fitting the planks to the frames.

While excavation a large number of small nails were also found. These small nails were used for fixing the copper sheet on outer surface of the hull. Most of the nails retrieved from the lower layers were in good state of preservation with very sharp pointed end.

Thin sheet of copper had covered the outer surface of hull to protect it from wood borers. In 1781, first time, a frigate of the British

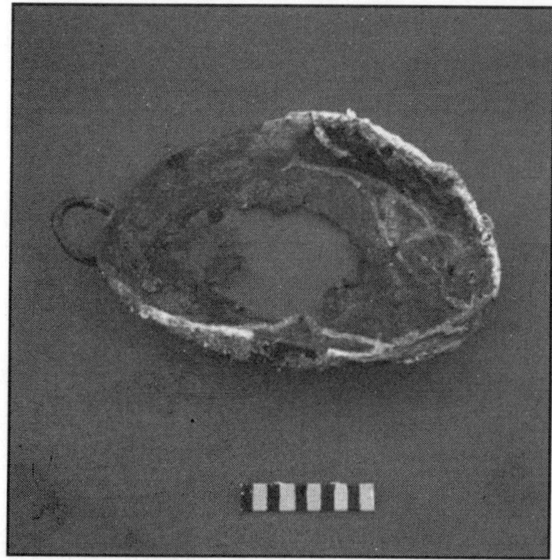

6.11. *Princes Royal*: A pan of copper

6.12. *Princes Royal*: Thick and round through bolt

6.13. *Princes Royal*: Copper fastening

Navy was "coppered". Soon copper sheathing was adopted by most of the European Navies and many merchant ships were also coppered. Small, torn and fragile pieces of protective copper sheet of the ship was noticed all over the site.

Lead Objects

Some objects made of lead were also found in excavation. Exact nature and function of these objects could not be determined as they were fragmentary and are covered with marine incrustations.

Stone Objects

Rotary querns were the main objects of stone. These querns were of two types (fig. 6.14 – 6.15). Besides, some other smaller objects and slabs were also found.

6.14. *Princes Royal*: Rotary quern

6.15. *Princes Royal*: Rotary quern

Ballast Stones

Heaps of ballast stones **[28]** were noticed in southern sector to the south-west of the cannons. The surface of these stones was covered with calcareous deposit due to which they appeared like dead lumps of coral and it was very difficult to identify them. During the excavation deposits over these stones were removed, before documentation.

Bricks

A large number of bricks were found spread in a large area to the east of anchor, close to lower edge of middle terrace. Most of these bricks were intact (fig. 6.16) and covered with marine encrustations. Broadly they are of three different sizes. Cemented material with these bricks suggests that these bricks were used on the ship. Perhaps these bricks might have been used for lining in the galley of the ship to protect the wooden ship from fire. A copper pan found below these bricks also supports this assumption.

Glass

Number of fragments of green glass was found on the site. Many small fragments were also excavated from the lowest layer in southern sector along with small sherds of brown glazed pottery. Study of these fragments reveled that they are from broken bottles. Bases of these bottles were quite thick and heavy. Bottles were the only glass objects noticed on the site.

6.16. *Princes Royal*: Brick

Ship's Timber

Excavation did not intend to expose hull of the wrecked ship. With the study conducted so far it appears that the hull of the ship has been largely damaged by marine woodborers. Long copper through bolts lying exposed on the bottom, outer planks covered with copper sheet, and absence of frames of the ship also support this assumption. Among the remains of the hull mostly wooden planks forming the shell of the ship were found spread on the site. Main concentration of these planks was noticed to the south of the anchor on the slope. Most of the planks **[29]** lie north-south, roughly parallel to the keel of the ship (fig. 6.17). Some of the wooden planks measured up to 325 cm. in length (fig. 6.18). The thickness of the planks was not much and at times two planks were used together to provide required thickness. Some of the planks, which were examined by the author earlier, were well preserved and free from woodborers. Many planks lying exposed on the surface had deteriorated considerably and were very thin. Some of the small wooden pieces found during the excavation were also badly damaged by woodborers.

6.17. *Princes Royal*: Keel of the ship

6.18. *Princes Royal*: Wooden planks

Bell

Recovery of the bronze bell **[30]** by amateur divers, was an important find. This 49 cm. high bell is inscribed with "PRINCES ROYAL 1792". Its circumference is 83 cm. at the top and 147 cm. at the bottom and it weights about 100 kg. The bell is in good condition except one of the upper rope hole, which has broken. Since it was found prior to the excavation, care was taken to relate it with the wreck. Our study indicated that the bell belongs to the wrecked ship.

Date and Identification

Since prior to it no shipwreck was excavated in India, the real problem was the date and identification of the ship. Wooden planks and ropes examined on the site suggested it was a sailing ship of historic period. Cannons indicated a date after the arrival of Europeans. Other evidence, which could help to date the wreck were pottery, construction technique of the ship, fastenings and also copper sheathing.

Before the recovery of the bronze bell, the only evidence that could be used to solve this mystery was a small piece of a copper sheet bearing two stamps (fig. 6.19) on it (Tripathi 2002b: 72-80). Here it was sure that the sheet belonged to the wrecked ship. The sheathing could have been done either at the time of her construction or during some later repair when the ship was operational. Stamps bear four letters 'PMCo' in a circle, could tell about her manufacturer and subsequently the date of the ship and country it belonged to. Attention was then focussed on these marks.

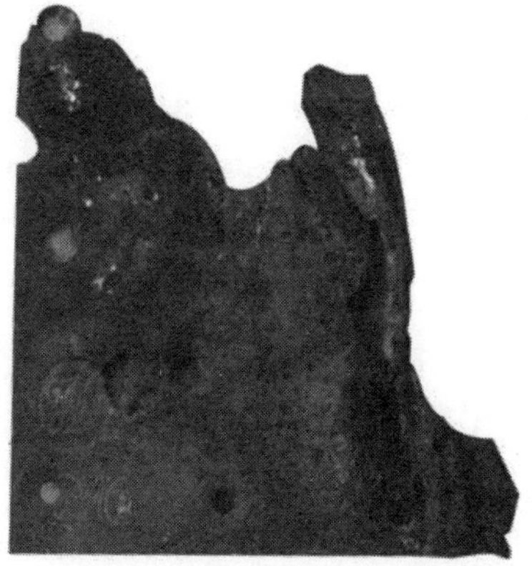

6.19. *Princes Royal*: Piece of a copper sheet bearing two stamps

History of marks in Europe, especially in Britain, is well recorded for the last few hundred years but no information could be found about the marks on material like copper sheet. These letters in stamp were compared with known letters in the stamp of British silver marks. Letters of stamp closely resembled various date letters used by the British silver smiths in the 18th century which suggested that the marks on copper sheet could be dated to 18th century (Taylor 1968: 31-45).

Since the marks in question were on copper and not on silver their resemblance with silver date marks could not be taken as final. However, it helped in narrowing down the area of search. Further researches in European marks, shipping and copper trade, economic and transport history proved helpful and information collected from various sources led to the identification of the stamp and helped to date the wreck with some certainty (Tripathi 2001: 112-116).

Destruction of the hull of the wooden ships by woodborers was a problem, which was faced by seafarers since the beginning. It forced them to invent and use a variety of protective methods. In 1761 British Navy for the first time used copper to cover one of its 32-gun frigates - *Alarm.* Soon it became popular and by the end of 18th century copper sheathing was adopted by most of the European navies. Many merchant ships were also coppered (Harris 1996: 550; Rees 1971: 72-86).

Lloyds register gives very useful information about copper sheathing of the ships. Use of copper sheet to cover the hull of the ship caused galvanic action and resulted in fast deterioration of its metal fittings. To eliminate this problem copper bolts were developed in 1783 to use below water surface. Finding of copper, bolts and nails once again suggested it to be a ship of 18th century (Tripathi 2001: 112-116).

Though these studies helped to date the ship but the marks were needed to be identified. Further studies in economic history, copper industry and patent records provided information about 'Parys Mine Company' at Holywell, one of the major companies with sales to French, Dutch and Spanish navies. It is also recorded that the firm's mark was stamped on 'the improved copper sheet'.[2] Comparison of the stamp on copper sheet with the copper tokens issued by Parys Mine Company left no doubt that the sheet was manufactured by this company.[3]

With the help of the stamps found on the copper sheet and copper fastenings of the ship it could be concluded with reasonable certainty that the ship is datable to the end of 18th century which goes well with the date on the bronze bell. The recovery of the bronze bell from the wreck site had provided an important evidence. Now there was no doubt that the bell belonged to the ill-fated ship *Princes Royal.*

With this identification, the search for the history of *Princes Royal* began. Now the main task was to find the details of the ship. Search of archival records was carried out, a large number of archaeologists interested in this field were contacted and web sites were browsed. The studies so far have revealed that several ships bearing similar name were built during last few centuries. Not only that but more than one ship bearing the name *Princess Royal* was plying in the last quarter of the eighteenth century. In India too, a 14 gun Grab named *Princess Royal* was constructed by Wadia's for Honourable Company's service in 1768 in Bombay dock (Wadia 1955: 331).

A database on the VOC ships revealed that there had been a VOC vessel with the name *Prinses Royaal.* It had made 15 voyages successfully and had been laid up and sold in 1666.

Right from the beginning it seemed most probable that the wrecked ship might be a British one. Thus search in Lloyd's register seemed useful. There are 15 vessels listed as *Princess Royal* in the Lloyd's Register.[4] There are fourteen ships in the main part of the register whereas one ship in the supplement pages at the end of the register. These ships were constructed between 1763 and 1791. None of these are shown as built in 1792. Three of them were constructed in 1791 and nine of them were sheathed (Table 6.1).

There is only one ship named *Princes Royal.* This ship recorded from 1793 to 1795 editions of Lloyd's Register is very interesting (Table 6.2). List of the lost ships of British East India Company also mention a ship *Princes Royal.*[5] There is not much information about this 805-ton ship except its tonnage.

If the date on the bell is the year in which the vessel was built then none of these can be identified with the wrecked ship. The last mentioned vessel was constructed in 1786 about 6 years before the date inscribed on the bell. However, it may be possible that an earlier built ship was refitted in 1792 when the bell was also inscribed. Its voyage to China points to the wrecked ship as it was also carrying porcelain and Southeast Asian brown glazed ware. In the 1796 edition,

Table 6.1

Construction		*Tonnage*	*Rigging*	*Sheathing*	*Port of*	*Voyage*
Year	*Place*				*Survey*	*(Sailed for)*
1763	London	320 tons	Ship, 2 decks	Sheathed in 1786	London	Memel, Constant trader
1764	Yarmouth	123 tons	Brig, single deck with beams	Sheathed in 1782	London	Rotterdam, Constant trader
1768	London	120 tons	Brig, 2 decks	Sheathed in 1785	London	Trieste
1778	Cowes	150 tons	Brig, single deck with beams	-	Petersburg	Leith
1781	Newcastle some repairs in 1889	350 tons	Ship, 2 decks	Sheathed in 1789	London	Smyrna, Constant trader
1786	Wales	170 tons	Ship, 2 decks	Sheathed in 1789	London	Oporto
1787	Berwick	101 tons	Brig, single deck	-	London	Nantz
1787	Neath	185 tons	Ship, 2 decks	-	Bristol	Oporto, Constant trader
1788	Berwick	102 tons	Brig, 2 decks	-	London	Charentes
1788	Poole	152 tons	Ship, 2 decks	Sheathed in copper in 1970	London	Gibraltar
1789	Yarmouth	174 tons	Ship, 2 decks	-	Yarmouth	Petersburg
1790	Liverpool	405 tons	Ship, 2 decks	Sheathed in copper	Liverpool	Africa, Constant trader
1791	Cork	142 tons	Brig, single deck with beams	Sheathed	Cork	Dominica
1791	Dartmouth	108 tons	Brig, 2 decks	-	Dartmouth	Jersey
1791	New Brunswick	246 tons	Ship, 2 decks	Sheathed in 1792	London	New Brunswick

Table 6.2

Construction		*Tonnage*	*Rigging*	*Sheathing*	*Port of*	*Voyage*
Year	*Place*				*survey*	*(Sailed for)*
1786	London	805 tons	Ship	-		St. Helena and China on April 5, 1793

the vessel is no longer listed. It seems it was lost in 1795-96 during her voyage from China to the Persian Gulf.

This excavation did not aim to excavate or salvage the wrecked ship. Rather it was to test the skills of our knowledge and available knowledge by applying them to different depths for planning development of the discipline in the country. Excavation conducted from 9 m. to 61 m. depths with accepted procedure demonstrated that the underwater archaeologists and naval divers can undertake operations of any complexity following internationally accepted procedures. Much of the parts of this shipwreck are lying in the limit of conventional compressed air diving. Detailed study of the archaeological evidence excavated from the wreck site corroborated with historic documents will shed welcome light on the maritime history and trade in Indian Ocean.

NOTES

1. In some of the publications (Tripathi 2006a :43; 2006b :131) date of first investigation was given May 1991, by mistake.
2. P.R.O.—AND/106/3408, 3.11.1791 and 106/2636, 1.1.1791.
3. http://www.mining-memorabilia.co.uk/Token.htm
4. http://www.reach/~sc001198/ShipsP1.htm
5. http://www.eicships.info

BIBLIOGRAPHY

Gazetteer of India — Union Territory of Lakshadweep (1977) Coimbatore.

Harris, J.R. (1996) Copper and Shipping in the Eighteenth Century, *Economic History Review*, Vol. 19.

Hicks, R.D. (2001) What is a Maritime Museum, *Museum Management and Curatorship*, 19.2.

Krishna Sastri, H. (1925-26) The Vayalur Pillar Inscription of Rajasimha II, *Epigraphia Indica*, XVIII, Government of India Central Publication Branch, Bombay.

Muhammed, K.K. (1994) Chinese Pottery - A Study in Decorative Motifs and Inscriptions, In: S.R. Rao (ed.) *The Role of Universities and Research Institutes in Marine Archaeology,* Society for Marine Archaeology, Dona-Paula.

Rees, G. (1971) Copper Sheathing - An Example of Technological Diffusion in the English Merchant Fleet, *Journal of Transport History.*

Taylor, M.M. (ed.) (1968) *A Dictionary of Marks,* New York.

Tripathi, Alok (1991) *A Report on Bangaram Shipwreck and Proposal for Shipwreck Exploration and Excavation in Lakshadweep,* Archaeological Survey of India, Unpublished Report, Khajuraho.

Tripathi, Alok (1993) Traditional Boats of Lakshadweep, *Journal of Marine Archaeology,* Vol. 4, Dona-Paula.

Tripathi, Alok (1994) A Historic Shipwreck in Lakshadweep Waters, In: S.R. Rao (ed.) *The Role of Universities and Research Institutes in Marine Archaeology,* Society for Marine Archaeology, Dona-Paula.

Tripathi, Alok (1996) An European Shipwreck off Bangaram Island (India) in Arabian Sea, In: G. Kuppuram and K. Kumudamani (ed.) *Marine Archaeology - The Global Perspectives,* Sundeep Prakashan, Delhi.

Tripathi, Alok (1997) Submarine Archaeology: In search of Unknown Past of Lakshadweep Islands, In: Jagat Pati Joshi (ed.) *Facets of Indian Civilization - Recent Perspectives,* Aryan Books International, Delhi.

Tripathi, Alok (2001) Identification of Marks and Dating of Historic Shipwreck: Marine Archaeological Investigation in the Arabian Sea, In: A.V. Narasimha Murthy, K.M. Suresh, K.P. Poonacha and K.R. Basavaraj (ed.) *Hemakuta - Recent Researches in Archaeology and Museology,* Bharatiya Kala Prakashan, Delhi.

Tripathi, Alok (2002a) *Excavation of Shipwreck off Bangaram Island - Preliminary Investigations with Indian Navy,* Underwater Archaeology Wing, Archaeological Survey of India, Unpublished Report, New Delhi.

Tripathi, Alok (2002b) An Eighteenth Century Shipwreck Off Bangaram Island, Lakshadweep, *History Today, No. 3,* The Journal of the Indian History and Culture Society, New Delhi.

Tripathi, Alok (2004a) *Excavation of Princes Royal,* An Interim Report, Organising Committee of the International Seminar on Marine Archaeology, New Delhi.

TRIPATHI, Alok (2004b) *Excavation of Princes Royal off Bangaram Island in the Arabian Sea, Journal of Indian Ocean Archaeology*, No. 1, New Delhi.

TRIPATHI, Alok (2004c) Princes Royal - Excavation of Ancient Shipwreck in the Arabian Sea, *Current Science*, Vol. 86, No. 9, Indian Academy of Science, Bangalore.

TRIPATHI, Alok (2006a) Recent Underwater Excavation in India, In: A.S. Gaur and K.H. Vora (ed.) *Glimpses of Marine Archaeology in India*, Society for Marine Archaeology, Dona-Paula.

TRIPATHI, Alok (2006b) Excavation of a Historic Shipwreck, In: P. Chenna Reddy (ed.) *Mahasenasiri: Riches of Indian Archaeological and Cultural Studies*, Sharada Publishing House, Delhi.

VENKAYYA, V. (1913) *South Indian Inscriptions*, Vol. II, pt. IV, Government Press, Madras.

WADIA, R.A. (1955) *The Bombay Dockyard and the Wadia Master Builders*, Bombay.

Comment

M.C. Joshi

When I was the Director-General of the Archaeological Survey of India, I picked up Alok Tripathi for training to be sent abroad because most of the young archaeologists were reluctant to join this particular branch of archaeology and today he has proved his mettle, and that gives me great happiness.

Why there was a taboo of using iron or copper? The tradition could be located in *Yuktikalpataru,* a 11th century text, which says certain boats to keep themselves in shape should not use any kind of metal particularly iron because of the magnetic rocks in the sea. How far this tradition is based on religion only the mariners can tell.

Finding porcelain on shipwreck is very interesting. Among the ports the earliest one sherd have been located in Kanchipuram. The Hindus only used it for external purposes and not internally but Muslims did use it and that is why even imitation of Ming pottery was manufactured in certain parts of Central Asia. It was then imported through Europeans. What about porcelain pieces with Devnagari inscription? I was surprised to see two things from Japanese excavation about 14th-15th century porcelain with Devnagari inscription, whether they were made to order by the Sultan of Gujarat. Another interesting object, two bones in the Kyoto museum of Japan, which also contain Devnagri inscription of a slightly earlier date — may be of the 11th or 12th century. But I could not solve the problem and this continues to be an enigma and this needs to be explored because amongst the

Hindus, porcelain was taboo. So this area is very interesting particularly the western coast because as far as I know geologically there is a narrow channel.

Lakshadweep in the popular parlance was being used as the graveyard of the ships that is why the larger ships always avoided it and the Roman ships they actually went the western part of Lakshadweep and then descended at Mantai and if they had to come to India to the eastern coast they took the full round of Sri Lanka and came over there. But those in the western ports wherever we see the Roman material they must have followed the old route from the Persian Gulf then along the Gujarat coast they may have entered Sopara very close to present day Mumbai. There might be more shipwrecks, at least the Portuguese, French and the English records show and there are earlier traditions because most of the Indian ships were plying along the coast. There may be cases of shipwreck that need to be explored. I have many more things to say but I would stop here to provide more opportunities to the audience to have discussions.

7

Traditional Boats and Ships of India Through Ethnographical Studies of Admiral Paris (1806-1893)

Eric Rieth

Francois Edmond Paris (fig. 7.1) was born on the 2nd of March 1806 and died in 1893 at the age of 87 after a long professional life of seamen, scientist and curator of museum.

From the multiple facets of his existence, it is that of founder of nautical ethnography, which we will retain here.[1] Indeed, Admiral Paris, Curator of the naval museum from 1871 to his death, established the methodological and scientific bases of the nautical ethnography by studying the European and 'Extra-European' boats and ships as cultural testimony and matters of history (Rieth 1992; 1993; 1999: 6-20; 2001a: 231-244; 2001b: 663-674; 2002: 67-70).

7.1. Admiral Francois Edmond Paris (MNM)

In his ambitious and innovative project to give an account of all the great nautical traditions, Paris granted a significant place to the nautical spaces of the Indian Ocean. The coast of India, taken as revealing example of his vast investigation on the scale of the oceans, occupies, within the nautical spaces studied by Paris, a major position. They are some characteristics of the exceptional documentation collected by this seaman and scientist along the shore from Pakistan, in the west and Bangladesh in the east, which we would like to present. The paper examines the conditions under which this documentation was collected, describes its contents and underlines some significant aspects of perception that Paris had of the Indian nautical traditions.

The Collection of Documentation

It is primarily at the time of his participation in three campaigns of circumnavigation that Paris gathered testimony on the boats and ships of India.

The first campaign, organised in the tradition of the great scientific expeditions of the 18th century, those of Cook, Bougainville or Laperouse, was directed, from 1826 to 1829, by Dumont d' Urville on board the ships *Astrolabe*, ex *Coquille*. The essential mission entrusted to Dumont d' Urville consisted in undertaking an exploration, in particular, of the Strait of Magellan, Solomon Islands, coasts of New Zealand and New Guinea, archipels of Gilbert, Marshall and Carolines, coasts of Philippines and Borneo. The goals of the expedition around the world were to draw hydrographic charts, to observe and study the human and natural peculiarities of these territories, in great part unknown. The goals were also, to locate the remains of the two ships of Laperouse which were tragically wrecked in Vanikoro, in 1788.

The campaign of Dumont d' Urville was a great success (Dumont 1830-1833). In the field of hydrography, for example, forty-two charts were drawn. In addition, thanks to the information provided by English Captain Peter Dillon, the shipwrecks of the two ships of Laperouse were located on the reefs of Vanikoro, Soloman Islands.

Hardly re-entered France, the young midshipman Paris embarked aboard the *Favorite*, a frigate placed under the command of Captain Laplace to carry out, a second campaign around the world from 1829 to 1832. The purely scientific goals of the first voyage were transformed into politico-commercial objectives during this second expedition. It was a question 'of showing the French flag' to affirm the economic and cultural presence of France in zones where the influence of Great Britain became, year-by-year, more and more. Since the treaty of Paris in 1763, the French influence was limited to the narrow space of five territories (Chandernagor, Karikal, Mahe, Pondicherry and Yanaan) whereas the influence of the East India Company extended in India. However, the politico-commercial perspective was the main goal of the expedition of the Favorite, but the scientific and hydrographic aims were also not entirely forgotten.

On his return to Toulon, Laplace had fully fulfilled his mission. Four thick books describe the results of the expedition (Laplace 1833-1835). Charts as, for example, those of the Seychelles Islands and of the shore of the Cochin chine, testify amply to the success of the hydrographic part of the expedition.

The third and the last campaign of circumnavigation in which Paris took part were that of *Artemise*, also ordered by Laplace from 1837 to 1840. The objectives of this expedition were similar to those defined at the time of the *Favorite* as the instructions given to Laplace by the Minister for the French navy mean it clearly: *'La mission que vous allez remplir a essentiellement pour but de rendre au commerce francais tous les services qui pourront dependre de vous, soit en recueillant dans les etablissements etrangers que vous visiterez des informations utiles aux speculations des armateurs et des negociants de nos ports, soit en offrant a leurs navires la protection que partout les batiments de l'Etat sont appeles a leur donner'*. (The purpose of the mission that you will fill is primarily to render to the French trade all the services which will be able to depend on you, either collecting in the foreign settlements which you will visit useful informations for the speculations of the shipowners and the merchants of our harbours, either offering to their ships protection that the French navy have to give them everywhere).

The results obtained by Laplace and his crew answered waiting of the minister for the French navy (Laplace 1841-1854), on the other hand this expedition, was tragic for Paris. At the time of the port of call of the Artemise in Pondicherry, in June 1838, Paris, envoyed in mission at Porto Novo had to observe the operations of a factory (it was the politico-commercial dimension of the expedition), where he met with an accident. The only solution offered by the surgery of the time was the amputation. This tragic accident blocked the maritime career of Paris.

During these three campaigns around the world, the official function of Paris within the staff was to control hydrographic charts and, possibly, to correct the old charts. He had also to draw new charts.[2] One can suppose that his talent as draughtsman, already recognised, as his scientific competence, already greatly estimated, were not foreign with this function of hydrography at one time when to draw up a chart required nautical and scientific but also artistic competence.

The drawing of the charts made it possible for Paris to leave the ship to embark on board a launch in order to undertake soundings or, still, to unload to take bearings. In both the cases, it was a marvelous means, to approach the boats and sailors of these remote shores. It is certain, however, that Paris at that time of his work as hydrographer, was to remain completely absorbed in his task. In all probabilities, it was during the port of call that he could devote to his non-official activity, that consistent, in particular to observe, study and draw the Extra-European boats and ships.

Thus during the campaign of the *Favorite*, Paris as he underlines himself: *'a continue a relever le plan de tous les navires et pirogues'* (continued to record the plan of all the ships and dugouts) (Paris 1863).

Of his mission along the coast of India in 1830, three original plans are preserved at the Musee National de la Marine (MNM). These include a *catimaron* (fig. 7.2 - 7.3) of the coast of Golconde (MNM, J9n/8192), another of a 'shoe-boat' of Coringui, coast of Golconde (MNM, B9n/8150), and the plan of a 'basket-boat' of False Divi Point, Coast of Golconde (MNM, B9n/8029). With these records are added five water-

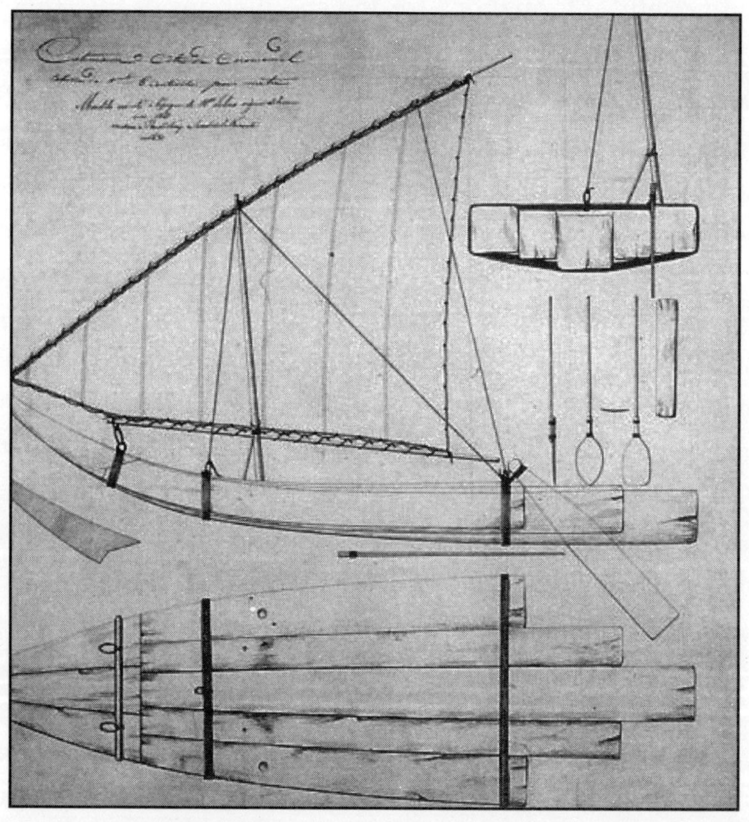

7.2. Plan of a *catimaron* (MNM J9n/8192)

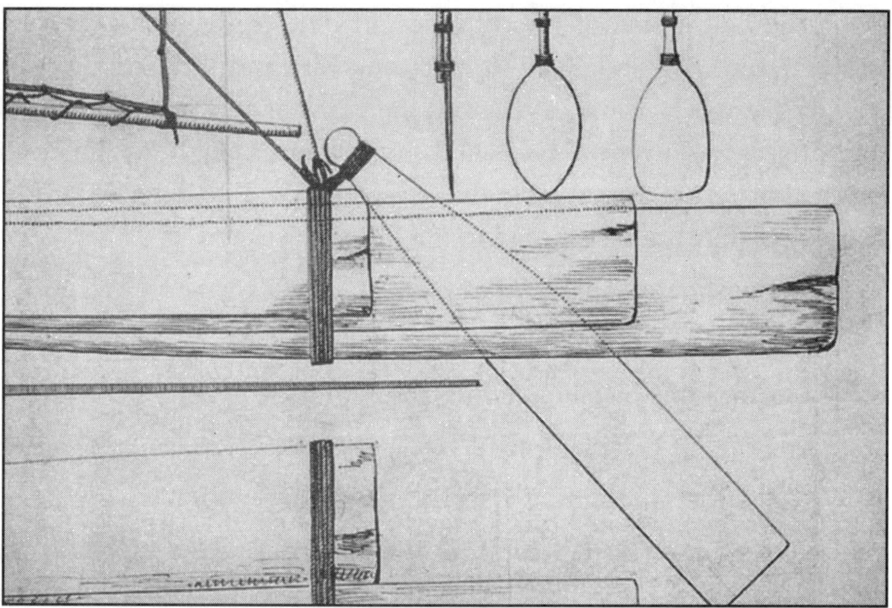

7.3. Details of a *catimaron* showing the aft part of the logs with a lee-board used as a rudder (MNM J9n/8192)

coloured plans drawn at the same time and gathered in an album called the 'Favorite album'.[3]

All of the plans of boats and ships of India collected by Paris between 1826 and 1840, some were not recorded by himself. Two authors of plans are identified. The first is, according to Paris, a certain J. Edye, in reality John Eyde, who was master shipwright of the naval yard of Trincomalee, and author of an excellent study, illustrated with original plans of traditional Indian boats (Eyde 1834).

The second is Sir William Seppings, chief shipbuilder of the East India Company, and according to Paris, skillful engineer, *habile ingenieur*. Two plans of boats recorded by Seppings, a boat of river Ganges, and a *dak* of the same river,[4] were published by Paris.

It should be noted that, in a preoccupation with an intellectual honesty, Paris always mentions the name of his information or the origin of his sources.

Let's now examine the contents of this collection and the way in which this one was worked by Paris by always attaching us to the only nautical space of India.

The Documentation on Boats and Ships of India

Three periods can be distinguished which correspond, moreover, with three stages of the professional and scientific activity of Paris.

The first is that of the collection of the documentation carried out between 1826 and 1840. In fourteen years and three expeditions around the world, Paris established the plans of almost two hundred and fifty Extra-European boats and ships. Concerning India, nineteen original plans are preserved in the collection of the National Maritime Museum. They are distributed in the following way:

— Coast of Malabar: eleven plans

— Coast of Coromandel: three plans (fig. 7.4)

— Coast of Golconde : two plans

— Coast of Bengal : two plans

— River and mouth of Ganges: one plan

Four of these plans were drawn in 1830, during the expedition of the Favorite, four in 1837 and ten in 1838 at the time of the expedition of the Artemise. One plan, which is not dated, was probably drawn in 1838.

With the strict technical documentation, only drawn and annotated, is added, that also original, gathered in the *Favorite* album. This documentation comprises the architectural records, drawing and water-colour drawings (fig. 7.5 - 7.7) of three boats of the coast of India (Paris 1829-1832: pl. 8-9, 11, 14-15). These architectural records are supplemented by six water-coloured drawings (Paris 1829-32: pl. 2-3, 5, 7, 10, 12) and two — pen and inks drawings (Paris 1829-32: pl. 4, 6)

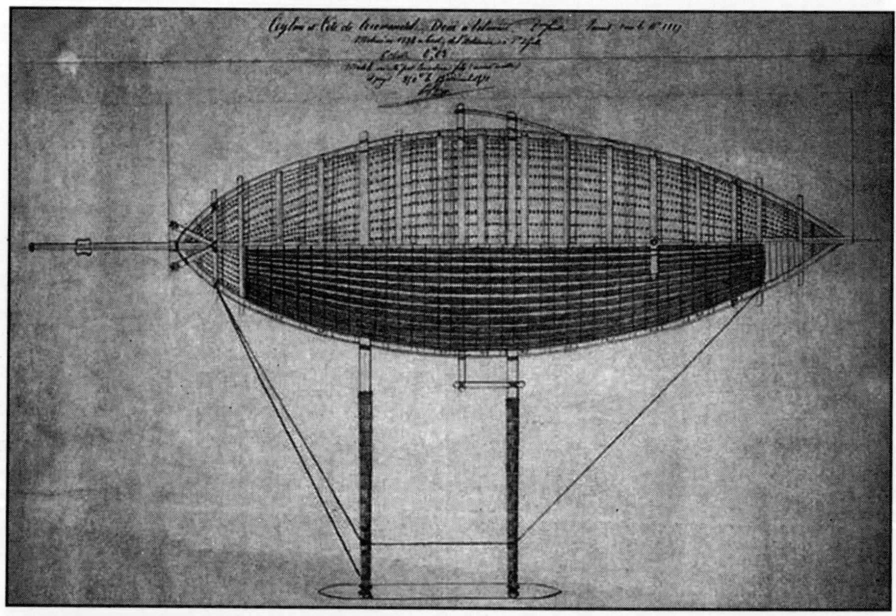

7.4. Plan of a *doni* outfitted with an outrigger (MNM J9n/8225)

showing the boats and the sailing ships in their context of use (fig. 7.8 - 7.11). Beyond their aesthetic interest, these documents are very informative from the point of view of the functional knowledge of these Indian boats and ships.

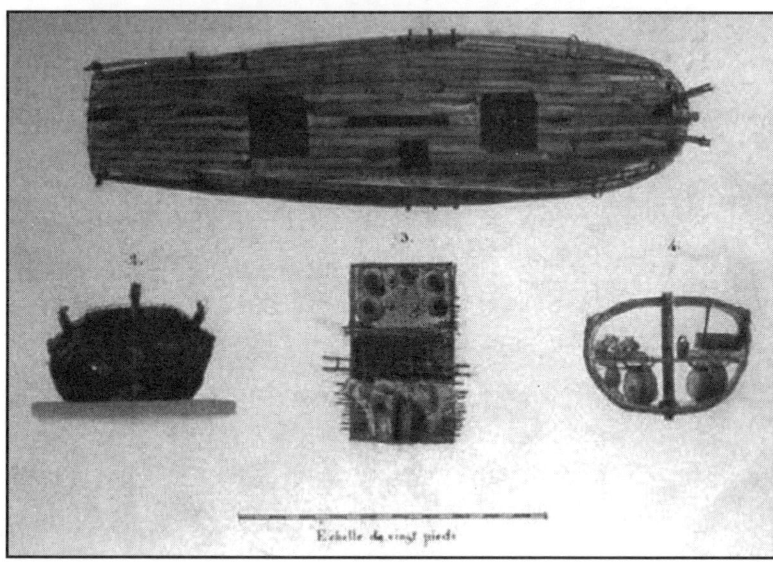

7.5. Plan of a coastal ship. The uncommon record of the kitchen and berth for the crew is very interesting for the knowledge of conditions of life coast of Golconde, mouth of the Godavari river (MNM, manuscript B178, *Favorite* album pl. 8)

7.6. Plan of sailing-ship with lugsail rounded by Paris at False Divi Point, coast of Golconde (MNM, manuscript B178, *Favorite* album pl. 9)

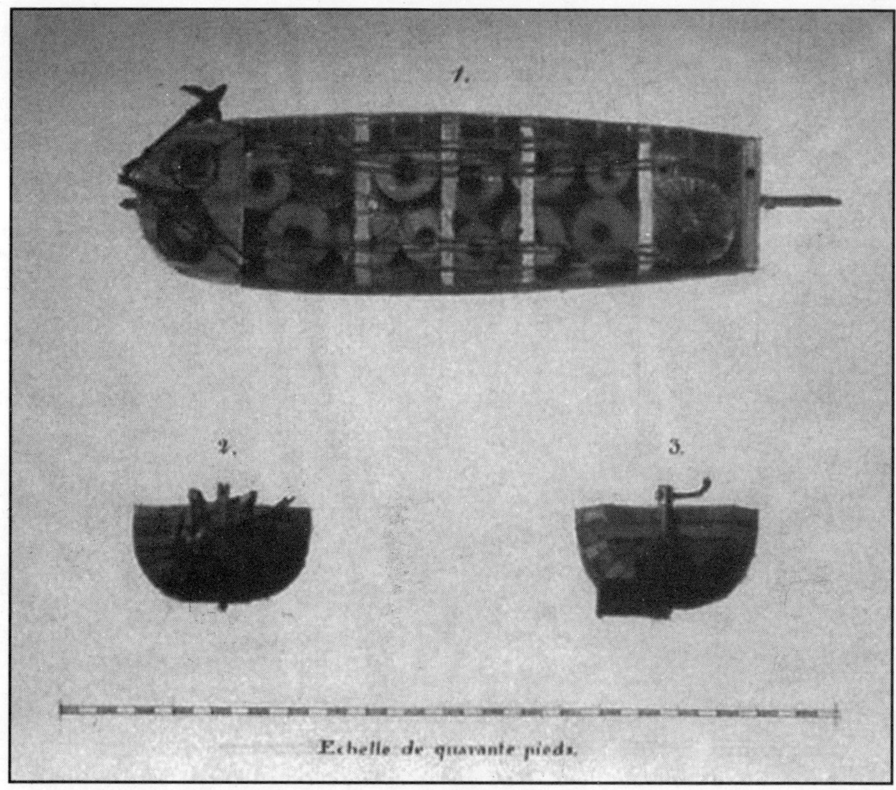

7.7. Plan of a sailing tanker for fresh water carried in 14 jars, coast of Golconde, mouth of the Godavari river
(MNM, manuscript B178, *Favorite* album pl. 12)

The second stage relates to working of this documentation of fieldwork. It falls under a very short lapse of time: between 1840 and 1843. To conclude this immense task, Paris benefited from his assignment in the capital of France at the navy board of plans and charts *(Depot des cartes et plans de la marine)* to assist Laplace in the preparation of the official publication of result of the expedition. In three years, Paris ensured the elaboration of his graphic documentation (plans and drawings), wrote a manuscript and supervised the composition of his *Essai sur la construction navale des peuples extra-europeens* (Essay on the Shipbuilding of the Extra-European Peoples). This work in two volumes, one of the text (one hundred and sixty pages) the other of pictures (one hundred and thirty pages), published in 1843 under the auspices of the ministry of the French navy, truly represents, in our opinion, the book founder of the nautical ethnography.

For the first time, indeed, four basic and always current concepts are clearly formulated: the choice to regard boats and nautical traditions

7.8. The sailing tanker with a fore lugsail and an aft sprit sail (MNM, manuscript B178, *Favorite* album pl. 12)

7.9. A *doni* (MNM, manuscript B178, *Favorite* album pl. 4)

as authentic subjects of history; the study of boats and ships as technical systems implying the taking into account of the socio-economic context of use as the natural environment; the collection of documentation during fieldwork and in particular by taking architectural records (forms and structure); finally, the urgency to preserve the memory of the traditional boats and ships of the Extra-European countries in particular, most fragile and most threatened of disappearance. Paris notes in the last lines of the introduction of his *Essai* .... *'Le temps approche*

7.10. *Doni* of Karakal pictured at Pondicherry. The rigging of the aft shows European influence gaff sail, top mast table, shrouds with deadeyes (MNM, manuscript B178, *Favorite* album pl. 3)

7.11. A sailing-ship rigged with lateen sails, Pondicherry (MNM, manuscript B178, *Favorite* album pl. 5)

ou il deviendra difficile de retrouver les pirogues des naturels, et elles auront le sort de leurs habitations et de leurs divinites, deja bien alterees, dont la plupart n'existent plus que dans les dessins des voyageurs'. (The time approaches where it will become difficult to find the dugouts of the natives, and they will have the fate of their dwellings and their divinities, already well deteriorated, whose majority do not exist any more but in the drawings of the travellers).

For India, Paris has preserved the memory of thirty-four boats and sailing-ships in form of architectural plans (Paris 1843: pl. 10, 12, 14-17, 20-21, 26-27, 29, 31-33).

With this documentation, fourteen lithographs are grafted locating the boats within their daily framework of use (Paris 1843: pl. 11, 13, 18, 22, 25, 25 bis, 28, 34-39). This corpus of graphic and iconographic data is prolonged by a dense and precise text, rich in technical but also historical informations (Paris 1843: 16-44).

Lastly, the third and last stage of the treatment of this documentation takes seat between 1871 and 1893, the period during which Paris directed the naval museum at the Louvre. It is during this period, Paris used his talent[5] and collected models of European and Extra-European traditional boats and ships. A part of this collection was financed, in fact, with his personal money.[6] If there are other museographic collection of models of traditional exotic boats in Belgium, Germany, England, Italy or Netherlands, rare are the collection quantitatively so important for which are known in a so precise way the plan dated from which the model was built, the technical and historical context of the boat, the name of the model maker, the price of the model or the date of its building. It is as an historical source that the Paris collection is exceptional. There are thirty-six models concerning the coast of India, in this collection. All the types of boats and ships recorded in *Essai sur la construction navale des peuples extra-europeans* (Essay on the Shipbuilding of the Extra-European Peoples) are present and distributed in the following way:

— Bombay - one model

— Coast of Malabar - thirteen models

— Gulf of Mannar - one model

— Coast of Coromandel - six models (fig. 7.12 - 7.14)

— Coast of Golconde - two models

— Coast of Bengal - four models

— River and mouth of Ganges - nine models

Essentially, these models, on a scale of 1 : 50, 1 : 25, 1 : 16.6 (this third scale being majority), were built between 1872, second year of function of Paris to the Director of the naval museum. Some very rare models, such as *catimaron* (fig. 7.15-7.16) of the Coromandel coast (MNM 9EX8, 1846), were manufactured in the year 1846-1848, whereas

7.12. Model of a *chelinque* or *masula* (MNM 9EX 6): General view

Fig. 7.13. Model of a *chelinque* (MNM 9EX 6): Details of stem

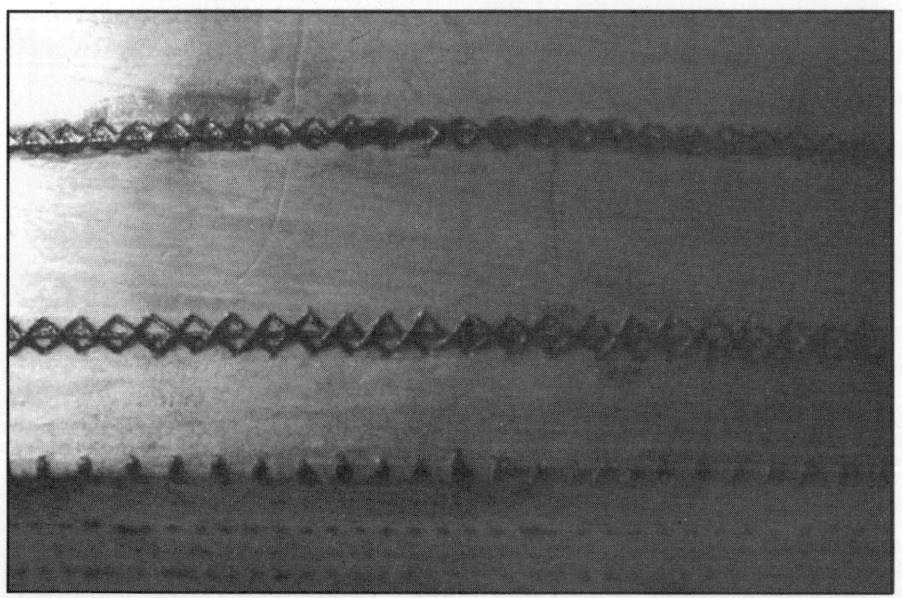

7.14. Model of a *chelinque* (MNM 9EX 6): Details of sewn joints

the naval museum was directed by the naval engineers Lebas, curator from 1836 to 1852.

These models of boats and ships, built with care and respect of the details on the basis of a rigorous documentation, constitute an extraordinary testimony from the point of view of the history of the nautical techniques. It is through a vision in three dimensions, which

7.15. Model of a *catimaron* (MNM 9EX 8): General view

7.16. Model of a *catimaron* (MNM 9EX 8): Details of logs and mast step

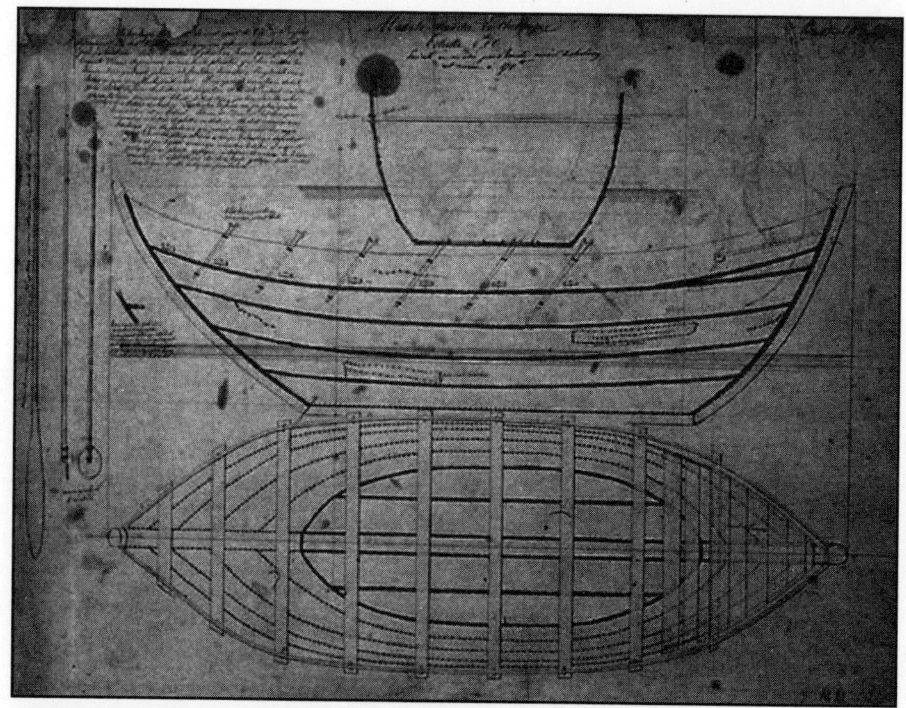

7.17. Plan of *chelinque* or *masula* (MNM J9n/8190)

one perceives best, for example, the shape of a hull, the structure of an outrigger or the different elements of standing rigging. But these models have another value which concerns the cultural memory of the shipbuilders and sailors of these Extra-European boats the Paris wanted to collect, study and preserve. How did he understand and analyse this memory?

The Indian Nautical Tradition According to Paris

The first aspect taken in account by Paris to study these exotic boats and ships, with a such unusual architecture for an European, is of geographical nature. As seamen, he is particularly sensitive to the safe, or dangerous, nature of the shore.

To cite an example of safe nature of the shore he mentions the possibility of finding a good anchorage in Bombay,[7] *'seul port de la cote de Malabar ou il soit possible de mouiller pendant la mousson du Sud-Est'*, (only harbour of the Malabar coast where it is possible to lie at anchor during the monsoon of the south-east).

He also cites an example of dangerous nature of the shore mentioning the impossibility of unloading as along the coast of Coromandel where, *pendant la mousson du Nord-Est...., the cote est*

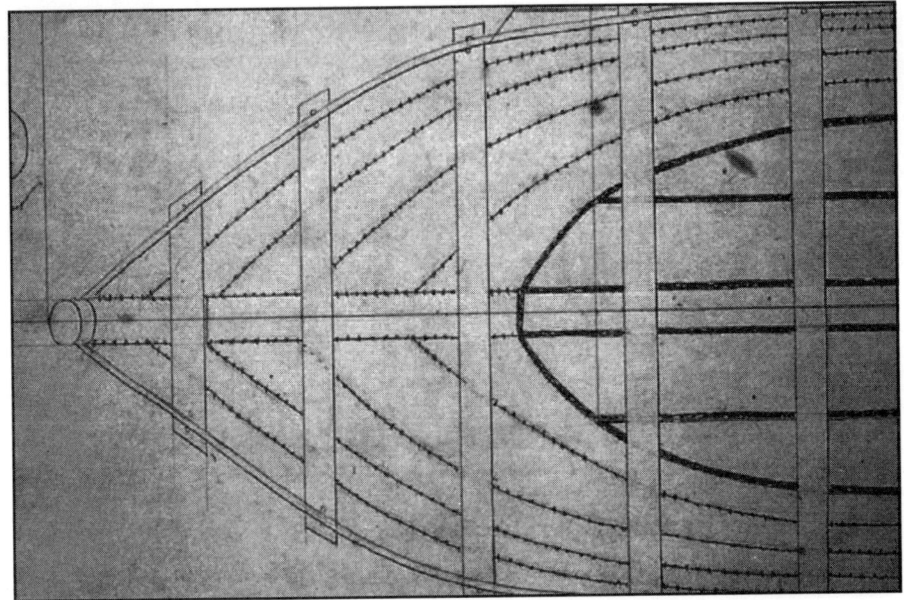

7.18. Detail view of the fore part of the hull of a *chelinque* (MNM J9n/8190)

impraticable et n'offre point de refuge: la mer, poussee par des vents violents, vient s'y derouler en longues lames d'ecume, sans aucune interruption depuis le Bengale jusqu'a Ceylan (during the monsoon of the north-east...., the coast is impracticable and does not offer any refuge: the sea, pushed by violent winds, makes long waves without interruption from Bengal to Ceylon).

Through his presence of a favorable or unfavorable coast with the development of a nautical activity, appears, in filigree, the close connection linking, within the framework of a coastal navigation architecture from morphology, structure, propulsion, direction, operation...., in a natural environment. The example of the boats of the Coromandel coast in general, and that of the *chelingue* or *masula* in particular (fig. 7.17-7.20), are very significant of these relations.

Non-decked and flat-bottomed boat, with the high sides reinforced transversely by a series of crossbeams, with sewn planks taking support on no frame, the *chelinque* (or *masula*) has a silhouette not very harmonious but perfectly adapted to its main function: that to sail on the heavy surf that beats on all beaches. Paris notes: '*ces formes lui sont en quelque sorte indispensable car il faut qu'elle soit legere pour etre aisement soulevee et pouvoir monter sur la crete des lames les plus vives: haute, pour l'eau y embarque moins; plate, afin de ne pas chavirer lorsque la mer que se retire et l'abandonne sur la plage; enfin tres elastique pour ne pas etre brisee*

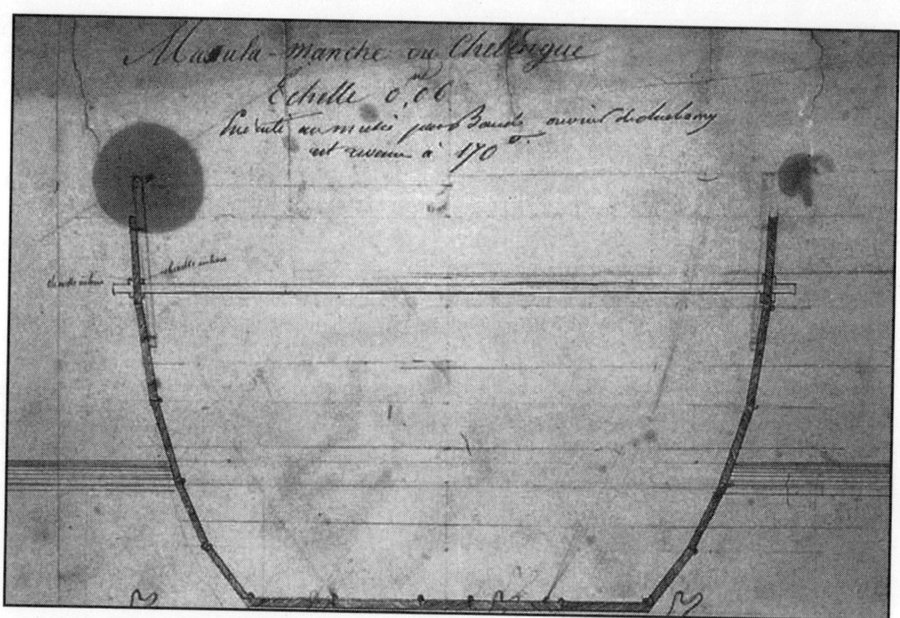

7.19. Master frame of a *chelinque* (MNM J9n/8190)

7.20. Water-colour drawing of *chelinque* on a beach (MNM, manuscript B178, *Favorite* album pl. 7)

par les chocs violents qu'elle eprouve et dont un seul demembrerait un canot europeen'. (These forms of hull are to some extent essential because it is necessary that the *chelingue* is light to be easily raised and to be able to sail on the top of the most tall waves; high, so that water embarks there less; flat-bottomed in order not to capsize when the sea is withdrawn from the beach; finally, very elastic not to be broken by the violent impacts one that it tests and whose only one would dismember an European boat).

Still, this family of boat is used, more or less modified, but without rupture with the description given by Paris more than one century ago (Kentley 1985: 303-317; 2003: 120-160).

This determinism of natural order is not the only dimension considered by Paris in his analysis of the nautical traditions of India. The second is of religious and socio-economic order, quite as outstanding as the first. In the east of the Cape Comorin, southern land of India, the coast of Coromandel provides of it, according to Paris, an excellent illustration. He writes: *'Une autre cause s'est jointe a ces difficultes naturelles, et peut-etre son influence a-t-elle ete plus grande encore: c'est la religion qui differe de celle de la cote de Malabar, ou l'islamisme permet aux classes aisees de s'adonner a la navigation, tandis que dans la partie orientale de l'Inde, le culte de Brahma.... defend a ses sectateurs de s'exposer sur la mer'.* (another cause joined these natural difficulties, and perhaps its influence was larger still: it is the religion, which differs from that of the Malabar coast, where Islamism allows the easy classes to be devoted to navigation, while in the eastern part of India, the worship of Brahma... defends with its followers to be exposed on the sea) (Paris 1943: 26-27).

According to Paris, Hindu religious system would have resulted in giving up with the poorest classes of the Indian society the nautical activities. He notes: *'Les seuls marins sont les parias, rebuts de la population et objets du mepris general'* (the only sailors are the parias, rejects of the population and objects of the general contempt).

The boats of these men, fishermen and sailors, would be with the image of their social status and precariousness of living condition: simple rafts in form of *catimaron* and modest coastal ships, *faibles et mal construits* (weak and badly built) underlines Paris.

If this determination of religious and socio-economic nature coming to superimpose itself on that of natural order would require, with any doubt, to be moderated and corrected, it is any case revealing of a deeply new vision of the time: that of a nautical architecture resulting, from a whole influence, positive or negative according to cases, related to the environment and the society.

To measure the influence of such scientific position on the contemporary research undertaken in the field of nautical ethnography, it is enough to point out what in 1972 Francois Beaudouin wrote in the introduction of his ethnographical monograph devoted to the boat of Berck (north of France), marking the revival of the studies on the traditional boats and ships in France. *'Nous nous proposons... de montrer que l'etude du bateau et des techniques nautiques constitu le meilleur moyen d'acces a la connaissance de l'homme de l'eau: nous nous efforcerons pour cela de mettre en evidence la facon dont le bateau est determine dans ses formes, sa structure, ses dimensions memes par un grand nombre de facteurs geographiques, historiques, techno-economiques ...a l'inverse, nous essaierons de montrer de quelle facon il peut temoigner de l'action de ces multiples facteurs et ainsi constituer un document de grande valeur pour l'ethnologue'* (We propose.... to show that the study of the boat and the nautical techniques constitute the best means of access to the knowledge of the man of the water: we will endeavour for that to highlight the way in which the boat is influenced in its forms, its structure and its dimensions by a great number of geographical, historical, techno-economical factors.... contrary, we will try to show how the boat can testify to the action of these multiple factors and thus constitute a document of great value for the ethnologist (Beaudouin 1970: 1).

Expressed in another manner, we find through these words the intellectual process of Paris illustrated, since 1843, in his Essay on the shipbuilding of the Extra-European people. This one takes support on a methodical recording of the whole of the specific technical data to each type of boat and ships, from the more general, such methods of shipbuilding, to the most particular, such nature of the sewn joints.

Concerning the methods of shipbuilding Paris describes, for example that the building of the *Patamars* (Paris 1943: 17-20), ships which he regards as best of the Malabar coast and even from India, begins with the installation on the axial structure (keel, stem, stern), of the master frame, only moulded frame. Then, it is the planking which is built. As Paris writes *'la forme generale du navire (est) donnee par la courbure qu'ils (les bordages) affectent en s'appliquant sur l'etrave, le maitre-couple et sur l'etambot ou la traverse de l'arriere (lorsque que le 'patamar' est dote d'un arriere a tableau)'* (the general shape of the ship (is) given by the curve which the strakes affect while applying on the stem, the master-frame and the stern or the wing transom (when the *patamar* has a flat transom stern).

From this method of building mainly 'shell first', only the master-frame is used as mould, the frames, notes Paris *'au lieu d'etre a la base de la construction ne (font) que suivre la courbure que le hasard a donne'* (instead of being at the base of the building, do nothing but follow the curve that the chance gave).

Concerning the use of sewn joints that one finds on many parts of the Indian coast, Paris draws up an extremely rich and complete catalogue of it. He underlines, the nature of the sewn joints of the *Patamar* is different from that of the outrigger dugout off Goa (Paris 1943: 22, pl. 15, fig. 1). The first is composed of crossed lashings, with driving in wooden pegs in the holes intended for the passage of lashing, right from the start of the building and insertion of an internal wedge tightening of these lashings. The second sewn joint is made up of a series of square lashings put in notches and tightening a coir wadding on the internal and external faces of the joints. The sewn joint of the *chelingue* or *masula* of the Coramandel coast are also different from the two others. They are composed of crossed lashing tightening a coir wadding laid out outside the sides planking and inside the bottom planking. Wooden pegs are driven in the holes used for passage of the lashing after a certain time of use of the boat (Paris 1843: 36, pl. 37, fig. 1-2).

This attention with the technical features, which constitute as many 'cultural fingerprints', is not limited to the only architectural aspects. It also relates to the functional characteristics of the boats as, for example, the mode of rowing on the dugout, without outrigger, of Goa (Paris 1843: 22). Paris notes that the sailors row: *'a la maniere espagnole en se dressant sur le banc precedent et se laissant ensuite retomber de tout leurs poids'* (with the Spanish manner while being drawn up on the preceding thwart and then letting themselves fall down of all their weight).

One could, in reality, multiply the examples comparable with those we have just evoked briefly. All would highlight the scientific interest of the documentation collected by Paris in general, and on the part devoted to the Indian boats and ships in particular.

Conclusion

In conclusion of this series of remarks, two aspects appear to us to have to be underlined.

We remain perfectly conscious of the incomplete nature of the documentation recorded by Paris. As hydrographer in mission, Paris collected data at the time of the port of the call whose places were not chosen according to ethnographical aims. In spite of this unfavorable context of ethnographical fieldwork, it seems to us that this documentation-manuscripts, plans, drawings, models, books...

represent an inventory representative of the Extra-European nautical traditions of the first decades of the 19th century.

With this intrinsic maritime importance of the Paris collection preserved at the National Maritime Museum, another dimension of the ethnographical work of Paris is superimposed: that which made it possible, for the first time, 'to metamorphose' the image of the exotic boats and ships by making them pass from the status of 'objects of curiosity' to that of 'subject of history'. From this point of view, the ethnographical studies of Admiral Paris can be interpreted, accordingly to the subject of this seminar as scientific "Gateway to the Continents", and, in particular, between India and France.

NOTES

1. This paper is based on an earlier paper published in French under the title of 'La collection de maquettes du fonds amiral Paris (1806-1893) au Musee national de la Marine : l' example des bateaux de l' Inde et du Sri Lanka', *Revue Francaise d'Histoire d'outre-Mer*, t. 88, n 332-333, 2001; 231-244.
2. During the expedition of the *Astrolabe*, for example, he made the charts of Tongatabou (Tonga Islands), haven of Carteret (New Zealand), one part of the Moluques and Manado (Celebes Islands).
3. MNM, manuscript B178, pl. 8, 9, 11, 14, 15. The plans X (*Patamar*) and XII (*Doni*), for example, recorded by Eyde (1834: 1-14) were published by Paris (1843, pl. 10, fig. 1-3, pl. 20, fig. 7-9).
4. According to the notice of the model (MNM 11 EX5), the plans of this boat were given to Paris in 1838 during the port of call of the *Artemise* in Calcutta.
5. The models of India (and Sri Lanka) were built, under the direction of Paris, as curator of the naval museum by the model makers Frederic Baude, Jean Couadoux (father), Francois Couadoux (son), Eugene Gueret, Charles Hamelin, Emile Noyon, Auguste Penaud.
6. About one half of the collection of models was financed by Paris.
7. The geographical position given by Paris (1843: 17) is a mistake. In reality Bombay is in the north of the coast of Malabar.

BIBLIOGRAPHY

BEAUDOUIN, F. (1970) *Le bateau de Berck*, Paris.

DUMONT d'Urville, J. (1830-1833) *Voyage de la corvette l'Astrolabe pendant les années 1826-1829*, 5 Vol., Paris.

EYDE, J. (1834) Description of the Various Classes of Vessels Constructed and Employed by the Natives of the Coast of Coromandel, Malabar, and the Island of Ceylon for their Coasting Navigation, *Journal of the Royal Asiatic Society*, Vol. 1.

KENTLEY, E. (1985) Some Aspects of the Masula Surf Boats, In: S. McGrail and E. Kentley (ed.) *Sewn Plank Boats*, British Archaeological Reports, International Series, 276.

KENTLEY, E. (2003) The Masula - A Sewn Plank Surf Boat of India's Eastern Coast, In: S. McGrail (ed.) *Boats of South Asia*, London.

LAPLACE, C. (1833-1835) *Voyage autour du monde par les mers de l'Inde et de Chine exécuté sur la corvette de l'état la Favorite pendant les années 1830, 1831, 1832, sous le commandement de M. Laplace, capitaine de frégate*, 4 Vol., Paris.

LAPLACE, C. (1841-1854) *Campagnes de circumnavigation de la frégate l'Artémise pendant les années 1837-1840 sous le commandement de M. Laplace*, 6 Vol., Paris.

PÂRIS, F.-E. (1829-1832) *Voyage de la corvette la Favorite en 1830, 1831, 1832, Collection de bateaux, Dessins d'après nature par E. Pâris, lieut. de vais.*, MNM, manuscript B 178.

PÂRIS, F.-E. (1843) *Essai sur la construction anavale des peuples extra-européens*, 2 Vol., Paris.

PÂRIS, F.-E. (1863) *Notice sur les travaux de M. le Contre-Amiral Pâris*, Paris.

RIETH, E. (1992) *Le voyage de la Favorite*, Paris.

RIETH, E. (1993) *Voiliers et pirogues du monde au début du XIXe siècle, Essai sur la construction navale des peuples extra-européens de l'amiral Pâris (1843)*, Paris.

RIETH, E. (1999) François-Edmond Pâris : homme de mer, de science et de musée, Introduction, *Souvenirs de marine conservés*, Vol. 1 (New ed.), Douarnenez.

RIETH, E. (2001a) La collection de maquettes du fonds amiral Pâris (1806-1893) au Musée national de la Marine: l'exemple des bateaux de l'Inde et du Sri Lanka, *Revue Française d'Histoire d'Outre-Mer*, 2001, t. 88, n. 332-333.

Rieth, E. (2001b) Observer, dessiner, décrire, comparer, analyser: unenouvelle méthode d'étude de l'architecture navale selon l'amiral Pâris (1806-1893), In: J.P. Brun and Ph. Jockey (ed.) *Techniques et sociétés en Méditerranée, Hommage à Marie-Claire Amouretti*, Paris.

Rieth, E. (2002) A propos d'un bateau-citerne du delta du fleuve Godavari (Andhra Pradesh, Inde) dessiné par F.E. Pâris (1806-1893), Note d'architecture navale comparée, In: L. Rivet and M. Sciallano (ed.) *Vivre, produire et échanger: reflets méditerranéens, Mélanges offerts à Bernard Liou*, Montagnac.

Comment

Jean Marie Lafont

It is not an easy task to comment upon a lecture by Dr Rieth on Admiral Paris' collection of drawings, since he is the specialist of that collection which he published in a magisterial volume and studied in several publications. Starting from Admiral Paris' collection, I shall briefly mention about a complementary topic, the modernisation of the Indian navies and some aspects of Indo-French naval cooperation in the 18th century, mainly through documents kept in French archives.

Let's start from Paris' drawing of "Touranne, La corvette de S.M. l'Empereur de Cochinchine Tao-Faung (Le Phenix), En appareillage". Paris' comment on this drawing deals with the military cooperation between France and Cochin China in early 19th century, and mentions the fact that this corvette was built by the local people after the Government of Cochin China purchased and dismantled a French man-of-war in order to understand its structure and military apparatus. At the end of his short description Paris adds, *'Nous avons vu a Touranne deux corvettes a batteries couvertes'* (A rare and interesting example of transfer of naval technologies).

The question is whether something like that was ever attempted by some Indian naval powers in the 18th century, before British paramountacy put the whole of India under one specific system of development, the colonial one, with its well known priorities and forbidden fields of activities. In other words, did the Indian powers, the coastal ones at least, try to develop and modernise the blue water navy in the same way they modernised their land forces in the fields of infantry, artillery and fortifications all along the 18th century?

I do not know of technical (documented) evidence of such an attempt, but it is clear that by the 1750's, when the English East India Company embarked upon its long course of mastering the Indian Ocean in order to master the Indian continent, they had to reckon with three major naval powers: the Angrias, the Sidis and Hyder Ali.

The Angria fleet was not a small, despicable blue water navy, as subsequent histories try to make us believe, but a powerful fleet which required the firing power of the whole English navy in the Indian Ocean to destroy it. In 1756, the combined fleets of the Madras and Bombay Presidencies, both Royal Navy and Company's ships, fourteen man-of-war all, together under the command of Admiral Watson, bombed for 24 hours the citadel and town of Vijaydurg, or Gheria, on the Malabar coast. Then a marine unit, 1800 strong under the command of Robert Clive, landed and stormed the harbour, and burnt down the Angria fleet which consisted of "Three three-masted ships carrying twenty guns each; nine two-masted ships carrying twelve to sixteen guns; thirteen gallivats carrying six to ten guns; thirty others unclassed; two on the stocks, one of them pierced for forty guns". It is this 40 gun ship under construction in 1756 in Vijaydurg which seems to me a significant example of an indigenous Indian navy adapting itself to the new challenges in the Indian Ocean.

Mangalore was the main trading and military harbour of Hyder Ali and Tipu Sultan, although Ali Raja of Cannanore also had a fleet — fourteen ships in 1772 carrying together 60 guns which sailed upto Moka, Muscat and the Coromandel. Ali Raja was under the nominal authority of Hyder. In 1768 a Bombay detachment captured Mangalore for a short while, but the British garrison surrendered the same year to Hyder Ali and was taken as prisoners. In 1781 Admiral Hughes carried out intensive destruction in the port, and in 1783 General Mathew at the head of a Bombay army stormed the city again and occupied it for one year. In 1784, the English garrison under Colonel Campbell surrendered to Tipu assisted by his French troops and a French contingent. It is difficult to evaluate the strength of Hyder and Tipu's navy, and we shall come back later on their development. For the time being, let's say that according to Portuguese sources, who closely followed the growing military power of Hyder and Tipu, the Mysore fleet was 30 ships strong in 1765, and 80 three-masted ships of 20 to 40 guns in 1778. Whatever the correctness of these figures, these man-of-war pierced for up to forty guns also bring the question of the modernisation of an Indian blue water navy in the 18th century, long before the British Raj.

The Marathas and Mysore were the two great Indian powers to seriously oppose British expansion from 1750 to 1803. After the annihilation of the Angria's fleet in 1756, after the destruction of the Sidi's in 1759, after the repeated threats or attempts against Mangalore since 1768, the Indian naval powers were looking for some external support to help them keep a kind of security at sea for their vessels. In terms of naval infrastructure, the French at that time were second only to the English in sea-power. Considering the level of military cooperation developing between these, same Indian states and France in the fields of infantry, artillery and fortifications, it is hardly surprising that something also evolved in naval cooperation. And in terms of Realpolitik, as V.G. Hatalkar observed, it was in their interest to play this card also.

As for the French, the disastrous war of 1756-1763 (culminating in the Treaty of Paris) had shown them that English power in Europe rested on two trans-oceanic supports: North America for sailors, wood and harbours, and India for money. Hence, the decision taken by Choiseul in the late 1760's that the next war, whenever it came, should be fought in India and North America.

At a seminar in the Asiatic Society of Calcutta, in 1999, I dealt at some length with the comparative history of French interventions in America and in India in 1776-1785, which included the cruise of Admiral-Bailli de Suffren in the Indian Ocean and his capture of Trincomalee. To have a French fleet permanently stationed on the Malabar coast, and therefore not obliged to sail back to Mauritius during the monsoon, was a *sine qua non* condition both for the French and the Indians to put a check to the Royal Navy's hegemony in these areas. Negotiations were taking place in the late 1760's on this possibility which would relieve the French from the obligation of capturing Bombay as a preliminary step for any operation in the Indian Ocean. And the two Indian powers with whom negotiations were taking place were the Marathas and Hyder Ali.

Negotiations in Pune came to an apex in 1776, when the French

Government through Saint-Lubin offered a 'Traite de commerce' to the Peshwa. Out of four articles in the preliminary draft, two of them dwelt with the navy.

Maratha vessels would have free access into all French ports, and vice-versa. Maratha vessels would be under French protection at all times in all ports of that government, and vice-versa. Then followed the Minister's instructions to Saint-Lubin concerning the margin of negotiations: in exchange for Maratha protection of the French possessions in India, the King would authorise his shipowners to deal in arms and ammunitions with the Marathas. Saint-Lubin was to demand grant or use of Maratha ports not only for trade, but also for shipbuilding, 'He was to collect exact information regarding the price of wood and labour and prepare an estimate of the cost of vessels of all classes to be constructed in the Maratha workshops. He should attach a statement of the naval stocks and of the number of workers that would have to be sent from Europe for the construction of these vessels, as well as of the number of vessels that could be constructed in a certain given time...' . A French diplomatic mission (consulate) could be set up in Chaul, or Gheria, or Rajapur which should be acknowledged by the Maratha government.

The treaty was signed in Pune in 1777 and sent to France for ratification. Concerning shipbuilding, the ship *Sartine* which had brought Saint-Lubin to Chaul was repaired in that port by Indian workers, an excellent way to test Indian know-how in handling a European man-of-war. Saint-Lubin wrote to Versailles, 'Maratha shipbuilders are much better workmen than their French counterparts, they are slower no doubt, but their planking is so perfect that a vessel can last 30 to 40 years without being given the stroke of a pump'. What was requested were two builders and iron to be sent from France. Everything else was available in the forests and workshops of the Malabar coast. Rumours were ripe at the time that an agreement had been reached for the port of Chaul and the fort of Revdanda to be ceded to the French by the Pune Darbar. Saint-Lubin informed Versailles that he had a large scale map of the port of Chaul drawn by his own surveyor. In June 1778, when news of a conflict between England and France reached Pune, the Darbar informed Saint-Lubin that wood would be supplied free of cost to the French for building 'one vessel of 64 guns; four frigates of 32, sixteen corvettes of 16 with all the launches, cutters and other boats necessary for its service'.

I am not dealing here with the diplomatic imbroglio between the Marathas, Mysore, Hyderabad, the English, the Portuguese and the French, nor with the story of Saint-Lubin, when he returned to France. We are only interested here in the attestation given by French experts in the 1770's of Indian workmen's capacity in the Maratha ports in repairing or building man-of-war of a size and firing power similar to the ones in France in the third part of the 18th century.

The other Indian power which attempted to (and was prevented from) building a blue water navy was Mysore under Hyder Ali and Tipu Sultan. We know of the attempt on the main Maldive Island by Ali Raja, and the extensive trade Tipu developed with the coastal states of the Indian Ocean. But the tiny section *Mir Yam Cutchehri* or 'the Marine Department' in Mohibbul Hasan's Tipu Sultan is skeletal indeed. And in French archives, connection with Hyder Ali and Tipu in this field is not so well documented as the Maratha one. Still we have some pieces of evidence, which I would like to put before you.

The first navy built by Hyder Ali comprising about 30 man-of-war and many transports, was destroyed by the British in 1768. But it seems part of it had survived, or been reconstituted, in 1772 when Claude Hugau, a French Hussar officer sent with Colonel Hugel on a mission to reinforce the army of Mysore, resided for a week in Mangalore. He stayed at Desvaux' residence. Desvaux was a naval officer (Captain) of the French East India Company who had taken a brilliant part in naval operations off Britany during the precedent war. He belonged to Saint-Malo and had served in Surat. Hyder Ali appointed him Admiral of his fleet, giving him a 'patente' making him *General de la mer*. According to Hugau, who claims he saw the ships ashore, Desvaux commanded 14 warships, the flagship being *Le Tiger.* The manuscript is faulty, but it seems three of these ships (300 to 400 tons) were carrying 24 guns of different calibers. One ship had 16 guns, two ships 12 guns, two ships 11 guns. The crew on the main ships was 160 strong. According to Hugau, Hyder's standing instructions to Desvaux were not to risk the fleet because of Maratha high-handedness at sea, but to make specific expeditions mainly to arraign suspected vessels and bring them to Mangalore for verification.

We did not find who this Captain Desvaux was in the French archives, nor did Hugau know why such a good navy officer had opted for expatriation. We cannot follow the fluctuation of Hyder's naval power nor the shipbuilding activities Hyder managed to maintain in spite of so many odds. But when in December 1781 Suffren's powerful squadron sailed from Port-Louis (Mauritius) to India, he had with him 'Le Hyder, corvette de Hyder Ali Khan'.

Another information forwarded by Hugau to Sartine, the Minister of Navy and Colonies, was Hyder's offer in 1771 to give the harbour of Karwar to the French. Hugau gives a vivid description of Karwar, where he says 12 to 15 man-of-war could safely anchor during the monsoon. In fact, the last part of Hugau's 'Deuxieme et derniere partie' is entitled 'Notes particulieres sur Karwar... appartenant a Ayder-Aly-Cam...', a technical notice of the harbour and the surrounding heights were a modern fort could be built for protecting the fleet. Hugau's Memoire clearly says that Hyder was asking the French several things in exchange for Karwar. But the manuscript does not give any detail of this part of the bargain.

I do not think that we have the large map of Chaul done by order of Saint-Lubin in 1777 -1778 is preserved, but there are beautiful maps of Mangalore and Karwar among many maps of other Indian ports and important cities in French archives, both in Vincennes and Aix-en-Province. Saint-Lubin could not execute his orders of compiling a general survey of the Indian ports. But in 1777 orders were issued to Admiral Tronjoly, commanding the French fleet in the Indian Ocean, to embark on his flagship *Le Brilliant,* Lafitte de Brassier and to make ports-of-call to the potential allies to the French. Lafitte, an 'ingenieur geographe' was instructed to use these ports-of-call to draw fresh and accurate maps of the places where the fleet could land troops and harbour during the monsoon season. When he returned to France in 1785, Lafitte made two splendid albums of these maps for Marechal de Castries, the new Minister of Navy and Colonies. One year later, in 1786, Lamotte de Sominihac made a new album of maps for the same Minister, taking for several of them inspiration from Lafitte's work. These three albums have fascinating views of Indian ports and cities dating 1778-1781. In 2000-2001, some of Lafitte's maps (including one of Vijaydurg and one of Mangalore) were exhibited in the National Museum in Delhi. On the Mangalore map, the depth of the water and the two passes to the harbour are carefully indicated. So are the Fort and the fortifications of the city. The map also clearly shows the batteries protecting the shore, the docks for the ships, the building docks, the

Admiral's residence to the north of the fort and the powder magazine. This interesting map shows Mangalore as it was before the raid and destructions of Admiral Hughes in 1781. As for the map of Karwar, Lafitte reported in the captions the city, the fort, the island and fort of Champhy, the 'lie aux hultres' (Oyster Rocks) and another fort, with the inscription 'A Tipu (sic) Sultan' (in pencil) to which was added 'Aux Marathes' (in ink).

The first navy built by Hyder Ali, as we have seen, was destroyed by the British in 1768. After the destruction of Mangalore by Admiral Hughes in 1781, and with the 17" lone exception of *The Hyder* which sailed along with the French squadron the same year, Hyder and then Tipu did not have much of a navy capable of operating with the French against the English during the 1781-1784 war in India. As everybody knows here, the English squadron under Hughes affronted the French squadron under Suffren. Hyder Ali met Admiral Suffren on land, near Cuddalore, in July 1782. That was before the capture of Trincomale by Suffren. Hyder and Suffren discussed strategy, and in their last meeting the French Admiral disclosed to the Regent of Mysore the plan for his future operations as far as he could. Hyder in return asked him to sail on the Malabar coast to help him repel the English armies threatening from Bombay, the core of his country. But the season was too advanced, and Hyder acknowledged the fact that naval operations by the French on the Malabar coast had to be postponed till next year.

News of the signature of the preliminaries of the Treaty of Versailles reached Madras in early June 1783. In 1784 the English East India Company embarked upon a massive military build up in order to prevent India to follow the path of North America and escape colonial fate. When Admiral Souillac, the new Governor of Pondicherry in 1785, tried to explain to the Indian Darbar what the French had accomplished in North America, and why they did not succeed in doing the same in India, British reply to the same Darbar was that the French were once again bluffing since they did not achieve more in America than what they had done in India. No Indian power had the possibility to check these rather conflicting statements. Tipu alone knew what had happened at Yorktown and at the Chesapeake. It is probably the reason why he tried till the end to get from Versailles, and as late as 1788, a fleet and an expeditionary force which could do in India what had been achieved in America. To no avail, for reasons which cannot be discussed here. His three Ambassadors, who were received by Louis XVI at Versailles on 10th August 1788, visited the naval facilities in Toulon.

On their way back to India, they went to Brest, visited some ships of the line, assisted in the launch of a man-of-war and were shown some naval manoeuvres. Tipu's last instructions for the building of a 30 gun ship given by Piveron de Morlat: *'Le vingt-neuf [July] M. de Suffren revint au camp prendre conge du Nabab pour aller se remettre en mer. Illui communiqua a cet egard ses desseins autant Que'*. Powerful blue water navy must be considered in the framework of the present paper, and not as a kind of unrealistic dream of a lunatic or bombastic native ruler.

It is a paradox that in 1792 the most powerful blow to Tipu Sultan was delivered by Lord Cornwallis, who had surrendered with all his forces, flags and ships to Washington and Rochambeau at Yorktown in 1781.

This short survey raises more questions than it solves, and for obvious reasons. In early 19th century the future Admiral Paris could see, study and draw plenty of traditional types of ships all along the Indian coasts, but there was nothing left of any Indian blue water navy of the 18th century. How much of these traditional shipbuilding know-how so interestingly presented by Dr Rieth was incorporated into the modern man-of-war which the Marathas and Hyder/Tipu built and tried to develop, we know not. What did they borrow from French/European shipbuilding expertise and how did they adapt them to suit their own purposes, we know not. This is the missing link of Indian maritime history, which we might not be able to write about in the near future.

8

The British Strategic Vision of the Indian Ocean (1857-1947)

Andrew Lambert

The role of the Indian Ocean in British global strategy between 1857 and 1947 can best be understood as a second theatre of Imperial strength, and the pivot point for the entire Asian imperial/ commercial project. Secured by fortified naval arsenals, superior communications, the resources of the subcontinent and above all global naval dominance, the British saw the region as the base from which it could project amphibious and military power onto the shores, rather than a defensive naval concern. This situation changed very suddenly in 1941-42. The loss of naval dominance exposed India, Ceylon and shipping in the Indian Ocean to attack.

With the capture of Cape in 1806, Mauritius in 1810 and the removal of any lingering threat from the Dutch in Java after 1813 the Indian Ocean became a British mare clausum, on which the flags of rival powers could appear only fleetingly. Quite simply the British controlled the key access points, at the Cape and later Singapore, the northern and southern landmasses, India and Australia and they owned the only significant naval bases, Bombay, Mauritius, Simonstown, Singapore and later Sydney. Without access to a friendly base no hostile warship could operate in the area for long.

By the mid 19th century the occupation of Aden, domination of the Persian Gulf, and the development of steam communications with Europe had reinforced this dominance, a process enhanced by the submarine telegraph cable and the vital role of docking accommodation in the operational efficiency of iron screw steamships. These factors reduced the ability of rival powers to operate in the theatre.

Consequently, the strategic history of the region was, until near the end of the period, one of Indian based expeditions to secure, promote or extend British interests around the shores of the ocean and beyond. These ranged from Persia, Egypt and Abyssinia to Burma, China and Japan.

The low level of perceived threat, while the outer works of Suez, Cape Town, Singapore and Fremantle were in British hands simplified the task. However, the adoption of oil fuel for the Royal Navy, with the major supply coming from the region, and the brief foray of the *Emden* suggested that more attention was required. Between the wars, Admiral Richmond examined the strategic pattern, both as Commander-in-Chief, Eastern Fleet, and as a first rate historical analyst. The decision to pivot British strategy for the eastern empire on a new Singapore base reflected the importance of securing the ocean, while the Addu Atoll anchorage demonstrated a sound appreciation of possible problems. The loss of Singapore and a Japanese carrier incursion in 1942 marked the lowest ebb of British strategic power in the region since the 1780s. These disasters reflected the inability of Britain and the Empire to wage a truly global war against three major rivals in distinct areas of the world.

The Indian Ocean is a vast strategic theatre, some 27 million square miles of essentially landlocked ocean. Historically the access points were the Cape of Good Hope, the Malacca and Sunda Straits. The opening of Australia added a third point, through the southern Ocean. Steam ships and the Suez Canal added the Red Sea and the Torres Straits to the list. There are a limited number of major natural harbours, although usable anchorages abound. The greatest harbour, at Trincomalee, was ideally located to be the centre of a struggle for command of the Ocean, but possessed little commercial value, or strategic utility for operations outside the Indian Ocean. Instead the peripheral anchorages and man-made bases at the Cape, Aden, Bombay, Singapore, Batavia and Fremantle were the strategic centres while British command was not in dispute. Trincomalee and the man-made harbour at Colombo came into prominence only in 1942, when the British lost naval mastery and faced a battle to recover command of the Indian Ocean.

Like all ocean areas the commercial value of the theatre sea lay in the trade it carries, and any resources that can be extracted from it, mere possession provides no profit. This made the ocean important to the British, for whom maritime commerce and capital investment were the basic instruments of economic advance. They owned or controlled most of the trade that passed across the ocean, and in wartime a threat to that trade would have serious economic and political consequences. The ideal method securing control of the ocean was to deny any rival power the opportunity to base their warships on its littoral. This was adopted in the early 19th century, leading to the capture of the last French harbour, at Mauritius, in 1810. Where possible the British avoided unnecessary territorial expansion, preferring to control the ocean from small, insular locations that could be defended by naval force, rather than continental territory. They preferred to open trade with other states, not to occupy territory. However, there were always exceptions.

India was the greatest of these. However, British rule had to be profitable. The economic failure of the Company ended its role, while the growing need of India to find external sources of revenue to maintain its financial health sparked the second China War, which secured increased access to Chinese markets, while a less violent use of diplomacy and armed force secured improved access to the markets of Japan and Siam (Wong 1998: 386). After 1857 the external trade of the subcontinent, boosted by liberalised customs regimes, improved internal communications, and the Suez Canal began to grow rapidly. In addition to a burgeoning regional trade India became a key part of the British capitalist economy, and the most important imperial trading partner. However, the administration of India carried with it a serious problem, the security of the vast debt burden of the Indian Government. Much of this was held in Britain, to which large annual interest payments were made. Any large-scale default would have had catastrophic consequences for the international economy. To balance the budget, thereby safeguarding interest payments, required improved exports, reduced imports, and above all cuts in the military budget. It dictated that India must find the most economic form of defence (Cain and Hopkins 1993: 326-350). It was not the threat of invasion that really worried the Viceroy and his Council, but a loss of internal stability, which would cripple access) to Indian financial resources from which they fund government debt. When the military talked up the Russian threat their real fear was defence cuts, the feeble, far-distant Russians being the only threat that might impress politicians committed to the Gladstonian orthodoxy of free trade, sound money and balanced budgets.

In reality the security of British India was a product of her naval mastery in European waters (Lambert 1998: 102; Mahan 1900: 27). The defeat of her rivals there entailed the defeat of their outposts and detachments in the east, depriving local set-backs of long term significance. While Britannia ruled the waves, she had as much control over the Indian Ocean as she desired, or was prepared to pay for (Graham 1965: 31-62). Consequently the British only stationed large forces in the theatre when rival powers challenged their control. The French had posed a threat in the middle years of the eighteenth century, leading to a series of large-scale naval campaigns on the Indian coast. However, the defeat of the French, a combination of superior docking accommodation, resources application and commitment, reduced the threat to intermittent raiding by warships and corsairs during the Revolutionary and Napoleonic conflicts. When the only French base was taken in 1810 the danger evaporated.

The Treaty of Vienna enshrined British control of the Indian Ocean in 1815, by retaining the Cape, Mauritius and Ceylon. Indonesia was returned to the Dutch as part of European settlement. During the Napoleonic wars the British had developed a strategic vision for the Indian Ocean that reflected defensive concerns, they were anxious to prevent any hostile power assembling a large force to attack India from the sea, and at a lower level, to prevent stationing of effective cruisers in the region, to harass the trade that made India a key part of the British Empire. The occupation of key bases like the Cape, or the neutralisation of others like Mauritius, were not the strategic answer, they were merely part of the process. The key remained naval mastery, sea control, the ability to use the ocean for trade or war, and to deny such facilities to other powers. While the British were the undisputed masters of the seas their grip on the Indian Ocean would be secure. If Britain lost command of the sea, fortified posts would be of no avail, for they could not secure the easy passage of merchant shipping, or prevent sea-based attacks on India (Colomb 1896: 192). This concept remained central until the removal of British power from the region.

While the 1815 settlement had established British maritime hegemony, British policy-makers would have to apply a sustained effort to maintain their dominance of the region. By 1857 several new ports had been added to the list of British possessions, Singapore, Rangoon, Aden, Karachi, Hong Kong, Labuan, and Fremantle. Others had been abandoned, notably two naval bases on the inhospitable northern coast of Australia. To improve navigation the ocean route to India was surveyed by the Royal Navy in the 1820s (Ritchie 1967: 107-126). The penalty of total control was an ever open frontier, which drew the British into the Persian Gulf, largely to secure their trade and their client in Oman (Graham 1967). By the mid century oceanic trade passing through the region had grown to vast proportions, with important new Australian routes carrying gold and wool, while intra-regional trade was also expanding rapidly.

After 1815 the Bombay Marine and Imperial forces in the region were used extensively in support of military campaigns in Burma, the Persian Gulf, China and East Africa. These rarely required major units. The Indian Navy was abolished in 1862 (Lambert 1993: 9-18; Low 1877), shortly after the ending of Company rule on the subcontinent, being replaced by a small imperial squadron based at Bombay, limited in scale by the small, old docks of that port. Any major naval operations in the area would require enforcement from the other imperial squadrons.

With the removal of rival fleet bases from the region, and the change to steam powered warships the threat from hostile cruisers collapsed. This development was given legal form in the 1856 Declaration of Paris, which outlawed privateering, for so long the scourge of Indian Ocean shipping. Between the abolition of the Indian Navy in 1862, and the consequent separation of the China and East Indies stations in 1865, and Admiral Fisher's fleet redistribution of 1904 the Royal Navy's East Indies station was something of a backwater (Clowes 1903: 88). Although upgraded from a Commodore to Rear-Admiral in 1870, because it was a large station, the naval threat was minimal. Instead the squadron was occupied with second-line tasks, suppressing the East African slave trade, numerous punitive land campaigns and the support of commercial interests in the Persian Gulf and elsewhere (Fremantle 1904: 331-391; Kennedy 1910: 277-347). In an emergency forces would be shifted in from other theatres. Once the Suez Canal was available the Mediterranean became the central reserve (Fremantle 1904: 348-349). Diplomacy also served. The 1899 agreement recognising the independence of Kuwait blocked a potential threat to British trade and strategy posed by a German or Russian rail terminus on the Persian Gulf. Such a challenge could outflank the Himalayan defences of India, and open regional trade to rivals (Corbett 1920: 391; Mahan 1900: 56-60; Sluggett 2002: 789-795).

In the 1860s the British adopted a new global strategy. In view of the rapid improvement in the speed and reliability of communications they shifted the concept from significant local naval and military forces to little more than a local trip-wire and local support to the civil power. While these forces would be reinforced in time of crisis or war, the core of the policy was to threaten or strike the most sensitive point of an aggressor with forces directed from the centre. Thus, the defence of India against Russian encroachment was entrusted to the Royal Navy, which had the capability to conduct large-scale coastal offensive operations, and was familiar with the Cronstadt/St. Petersburg region. Russia was unlikely to enter Afghanistan if it feared a British attack on the capital. This far-sighted decision was driven by economic realities, and forced through by Gladstone when he became Prime Minister in late 1868. A year later the completion of the Suez Canal provided a far quicker route to India for British strategic forces, while the telegraph cable reached the subcontinent in 1870. With first class communications, control of all the dry-docks in the region, and almost all the coal, the British grip on the Indian Ocean was more secure than ever. The new policy enhanced the role of deterrence in British global strategy, and reduced the significance of any specific region. The Suez Canal shortened the sea-route to India, increased the use of steamships, and transformed Egypt into a key outwork of the Indian Ocean economic empire, a position it retained until after 1947.

Within this strategy India was the centre for eastern operations, providing military manpower, industrial facilities, revenues and bases. The 1867 Abyssinian Expedition, which combined the Bombay army with the Imperial naval squadron, directed from London and commanded by a number of Crimean war logisticians, demonstrated steam era strategy. No other power could have operated such a force in the region, or anywhere else so far from their metropolitan homeland.

British owned submarine telegraph cables encircled the globe by the late 1870s, and links between the main telegraph company, Eastern, and Government were close (Vincent 1978: 222; 1995: 117). In times of crisis, notably the occupation of Egypt in 1882, the company went beyond what might be expected of a commercial concern (Barty-King 1979: 73-74). In the 1878 Zulu War a new cable was laid from Aden to Durban to improve central control (Barty-King 1979: 69-70; Headrick 1991; Kennedy 1971: 728-752). Empire, however defined, was now defended as a single unit, rather than as a series of geographically and intellectually distinct areas. In 1899 it took only two months to lay 3,000 miles of cable from the Cape Verde Islands to Cape Town for the South African War (Barty-King 1979: 126). Little wonder the French considered the cable network more important to British power than the navy (Kennedy 1971: 748). The completion of the cable link to India in 1870 was only a stage in the Imperial system; within three years the wire had reached Australia. India was now a staging post for intelligence, her aid could be summoned by London or Sydney, in a matter of hours.

However, the maritime character of this strategy did not recommend itself to the soldiers who advised the Viceroy. They sought a 'military' threat that would justify their primacy in defence decision-making. Here the slow advance of Russia toward Afghanistan provided a cause for alarm which however unreal, swayed judgements in Delhi, in a way that would have seemed absurd while the subcontinent was still being governed from Calcutta. Soldiers, who had more time and the leisure to write on 'Imperial Defence' than sailors, were prone to view the problems in terms of land to be protected. Their empire, marked on a map, required a large army and many forts. Yet the real empire was the trade and capital that passed across the seas, they could not be protected by the largest army, or the stoutest fortresses. It is indicative of the negative mindset applied to the subject of 'Imperial Defence' that two powerful coast service ironclads, built in 1868-70 and stationed at Bombay, as a floating supplement to the shore defences, have always been seen as local fixtures, when they were demonstrably capable of steaming back through the Suez Canal to the Mediterranean, or anywhere else in the Indian Ocean region. Ideally suited to the attack of fortified harbours potential targets in the region were limited, but they were a valuable strategic reserve and were modernised in the 1890s (Parkes 1970: 166-170; Brown 1998: 57). As the Second Sea Lord minuted in 1888 'It might be of the greatest importance, for instance, to the C-in-C to be able to strengthen his squadron temporarily for some offensive operation by the addition of a part of the Indian coast and harbour defence vessels.'

While the Admiralty continued to argue that the sea was indivisible, and that overall responsibility for trade defence lay with the Navy, the Viceroy of India and his army dominated Council were trying to recreate their own Indian Navy. In the meantime the Admiralty provided key personnel for the Coast Defence ships. It also provided four vessels for Indian waters, at least one of which was stationed in the Persian Gulf in return for an annual subvention of close to £ 10,000. In 1888 the Viceroy was prompted to complain that the current vessels, an old iron frigate and a collection of obsolescent wooden and composite sloops and gun-vessels, were inadequate. Stung by this complaint, which had more than a little truth in it, the Admiralty was quick to hit back. They observed that a modern cruiser could not be sent until the docks at Bombay were enlarged, to provide the six monthly docking essential if it was to retain its speed. Furthermore the only rival warships in the area were French, two corvettes and two sloops. The exchange ended with better sloops being promised, and the Indian Government planning a conference on dry-docks.[1] The flagship was replaced by a new, sheathed second-class cruiser.[2] The Admiralty saw no reason to place a powerful warship in the Indian Ocean.

In 1882 the Carnarvon Commission on Imperial Defence provided a clear statement of priorities for the region. The Cape, with its important dockyard and coaling facilities was to be heavily defended, as might be expected when the passing trade, in wartime would be at least £150 million. Trincomalee, despite its strategic potential, was ignored, local effort focusing on the new commercial harbour at Colombo. Other coaling stations at Hong Kong, Labuan, Singapore, Mauritius, Calcutta and Bombay were given appropriate levels of defence. However, the principle of the report was that the navy would retain command of the sea, limiting shore defences to a level capable of driving off cruisers or small squadrons (Schurman 2000: 105-115). The suggestion that new coaling stations were required at King George, Sound and Thursday Island forced the Imperial Government and the Australasian colonies to consider the political and financial implications. This did not tend towards a generous or an early decision (Schurman 2000: 142-150). In truth the reluctance of the Imperial Government to spend more money so far from home was perfectly justified. Neither place was threatened unless hostile warships could operate in the Indian Ocean. The point of the defence system was not, as many land minded military and political leaders considered, to protect the ports, or even the coal, but the trade. This required, as the navy always stressed, sea control and secure basing for British warships.

The 1896 Colonial Defence Committee reasserted the principle that sea control was 'the basis of the system of Imperial defence', and the 'determining factor in shaping the whole defensive policy of the Empire' (Williams 1991: 7), This kept the army small, and the budget low, a situation aided by the use of Indian troops in all the campaigns east of Suez, at the expense of the government in Delhi (Williams 1991: 8). By 1900 increasing fear of a Russian invasion, or destabilising attack made the defence of India a core issue for strategic debate in London. However, the reality of Russian power, and the terrain, made this prospect unlikely, while the performance of the Russians at war with Japan, 1904-05 hardly warranted alarm. Their Baltic fleet had only reached the Indian Ocean on British sufferance, and was hastened out of Indo-China by British diplomacy.

In 1904 Admiral Sir John Fisher grasped the issue. Taking the logic of the 1868 policy to a new level he concentrated the China, Australia and the East Indies squadrons into a single Eastern fleet, centred on the modern docks at Singapore. Fisher recalled battleships and scrapped sloops to man a more mobile and modern force, based on powerful battle-cruisers. His 'Fleet Unit' concept was ideally suited to the vast expanse of the Indian Ocean theatre, where the scale of threat must be low, and the interests requiring protection essentially commercial. The strategic 'keys' of the region, as he termed them, meaning the communication, base and dock facilities, were all in British or Dominion hands (Lambert :97-108; Mackay 1973: 306-308). The creation of a Royal Australian Navy, with a battle cruiser flagship, together with imperial funds to purchase capital ships from New Zealand and Malay emphasised both the unity of the eastern arc of Empire and the modern warships required in such vast spaces.

Fisher's work was driven by the linked agendas of naval economy and the reduction of military influence on Imperial strategy. He called for a reduction in the army in India, while working with the Foreign Office to ensure British naval access to the Baltic in the event of war with Russia (Lambert 2003: 147-169). A master of deterrence Fisher was confident the Russians would not threaten India while the British had access to the Baltic. As this strategy was in place at the outbreak of the

World War I there were no major British warships in the Indian Ocean. The strategy for distant theatres was, as it had been in the Napoleonic era, to deprive the enemy of his bases, thereby ending any commerce raiding activity. The obvious Indian Ocean target was Dar-es-Salaam, which was attacked by Indian forces (Corbett 1920: 129). Cutting radio links via Kamina in the German Cameroons was a critical first step.

During the World War I, India continued to function as a second seat of Imperial power, providing troops and stores for operations in the Persian Gulf, Egypt and East Africa. The Indian Ocean acted as the strategic super-highway bringing Australasian troops rapidly into the European theatre of war, along with food and other vital raw materials (Offer 1989). The scale of the military effort rendered by India and Australasia proved vital to victory over Turkey, in Mesopotamia and Palestine, while food from the Dominions helped to feed the nation and the army on the western front.

The threat to these communications, already severely limited, was further reduced at the outbreak of war when two potential merchant raiders were captured in the Red Sea (Corbett 1920: 138). However, the priority for naval forces based at Singapore and the Cape was to escort troop convoys taking Indian and Australasian troops to the main theatre of war, or relieving Imperial units for service there (Corbett 1920: 128-130). The threat to unescorted merchant shipping posed by the German cruiser *Konigsberg* complicated theatre strategy, and the problem doubled when the cruiser *Emden,* detached from the China station, began to operate in the Indian Ocean. She bombarded Madras, sank a Russian cruiser at Penang and captured twenty-three merchant ships. On November 9th she was sunk off the Cocos Islands by the Australian cruiser *Sydney,* detached from a troop convoy following signals from the radio station (Strachan 2001: 479-480). *Emden* had been caught while attempting to cut British strategic communications, a fate that von Spee's squadron would follow off the Falkland Islands a month later. For a brief period the threat posed by the *Emden,* and the *Konigsberg* had paralysed regional shipping. The problem was resolved when a superior naval force bottled up and destroyed the *Konigsberg* in July 1915 (Strachan 2001: 587-589). She had distracted a sizeable force from the Cape Station, but with the end of the cruiser threat all the modern British warships went to more important stations, while the older vessels supported offensive operations in the region, from German East Africa to the Persian Gulf.

Once again these operations were based in India, and involved the Indian Marine vessels taken into the Royal Navy at the outbreak of war. By the end of 1914 the enemy had no bases or warships in the Indian Ocean.[3]

The only other German incursion, by the merchant raider *Wolf* in 1917 sank 120,000 tons of shipping (Marder 1969: 100). Her survival reflected the lack of radio-contact, she was not betrayed by her own signals. Without access to a base German forces could not operate effectively in the Indian Ocean region. With such a limited threat there was no need to interfere with the normal running of commercial shipping in the area. Only troopships were convoyed.

The post war settlement gave Britain and even stronger grip on the region, German East Africa was now a British mandate, along with Iraq, while Iran was effectively a client, the Russian challenge having collapsed. Yet this did not translate into improved security. In the short term the prospects were encouraging. Admiral Lord Jellicoe's reports while on his 1919 Dominion Mission had repeated the obvious point that in the event of war with Japan the imperial position around the Pacific was entirely dependent upon command of the sea. He also pointed out that this was a matter of practicalities. The latest capital ships could not be docked anywhere in the region, and therefore 'cannot be sent to the Far East except under the gravest disadvantages', disadvantages which made Britain practically speaking inferior to Japan. He called for a floating dock and improved docks in Australia capable to taking the latest capital ships (Patterson 1968: 339, 349). He had emphasised large docks just before he reached the Indian Ocean (Patterson 1968: 292), and again after reviewing the defective arrangement that placed the direction of the Indian Marine under the army (Patterson 1968: 296-297).

In his report on Indian naval situation Jellicoe pointed to the lack of anti-submarine defences, minesweeping organisation and inadequate gun defences, in addition to the damage caused by two German cruisers. He noted that while India paid for her own military defence, she did not make any contribution to the security of her sea-borne trade. He considered Singapore the 'naval key to the Far East', urging improvements to the docks and base facilities (Patterson 1968: 329). Jellicoe's imperial fleet in the Far East, with increased docking and base facilities was always a dream. While strategically correct, it was too costly and too 'imperial' for the nationalist mood of the age.

However, the limited threat of isolated German warships had been replaced by the growing presence of Imperial Japan, which was now the most obvious naval rival. Japanese bases were far closer to the Indian Ocean than those of Germany, and included a strong position athwart British trade routes to the China market. After Jellicoe's Mission the strategy for the eastern arc of Empire settled on the new base at Singapore, recommended in 1921. The Washington Treaty of 1922 averted a naval arms race involving Britain, Japan and the United States. It severely limited naval forces and limited the development of bases and fortified areas in the Pacific. However, the cost to British interests was high. The twenty-year old alliance with Japan was ended, and the fleet cut back to match the American Navy, which did not have anything like the global tasks that the British faced (Ferris 1989). Consequently British overstretch in 1941-42 was largely a product of inter-war limitation, not lack of underlying strength. Unlike the United States and Japan, both continental military powers with naval strength Britain was a global trading empire, utterly dependent upon sea control. Washington and the London Treaties marked the end of British maritime hegemony, and with that the Empire was doomed. They also created a temporary, but highly dangerous, reduction in the relative weight of naval power in global strategy, at a time when land and air forces were developing new strategic options, without limitation.

The new base at Singapore, chosen instead of Sydney, was intended to project British strength forward along the axis to Hong Kong, should offensive operations against Japan be required. It was the location for the only dock in the eastern arc of empire capable of handling a modern capital ship. Furthermore the Washington provisions also reduced American influence west of Hawaii, emphasising the need for Britain to create a dockyard, fuel reserve and operational base on the Straits. It was intended to serve as both an offensive opening, and a strategic barrier. However, the conflicting aims and policies of the three armed services delayed the development of the base. The leading inter war naval intellectual, Admiral Sir Herbert Richmond commanded the East Indies Station between 1923 and 1925, combining his contemporary strategic interests with serious historical research (Richmond 1931). He stressed the primacy importance of trade and communications, and the need for a unified Imperial strategy. He had inherited the one ocean one navy concept of his 19th century predecessors, within which bases and communications defined strategy.

However, such ideas did not sit well with the increasing independence of the Dominions. For Richmond the defence of the eastern empire was a unity, based on sea control, in which India, Singapore and China were mutually linked. He identified a Japanese land attack as the most likely threat, and stressed that holding the base until the main fleet could arrive from the Mediterranean was a job for all three services, making the escort of troop reinforcements from India the navy's first task. With Singapore secure the Japanese would be defeated. To ensure co-operation he proposed that the Commander in Chief, Eastern Fleet should sit on the Council of India. This long overdue measure was ignored. Consequently the 'defence' of India was invariably cast in terms of soldiers, and then aeroplanes, when the key remained secure sea-communication with the rest of the Empire. Richmond condemned the reliance on heavy gun defences, as the Japanese had little to gain from a bombardment of Singapore, while occupying it could cripple British power in the region (Hunt 1982: 132-148; Bell 1997: 123-141). The shift from coal to oil fuel was a mixed blessing. Local supplies, from the Persian Gulf and Burma were now available, and ships could refuel direct from tankers in sheltered anchorages, but storage would be costly, and needed to be secure from an *Emden* type bombardment. The increased availability of refuelling stages was a real benefit in this vast region, and the Admiralty took the opportunity to identify three isolated locations from the Arabian coast to the Sumatran archipelago where this operation might be conducted (Brown 2003). One of these, Addu Atoll, proved to be a great value in 1942. Another of Richmond's initiatives, to replace the Royal Indian Marine with an efficient Royal Indian Navy took some time to work through. The 1934 reforms were essentially the completion of his work (Hunt 1982: 144-145). This new force was equipped with sloops designed as convoy escorts (Hague 1996; Gibbs 1976: 834).

In 1939 the British, committed to French alliance and facing an European war, began to see that they could not sacrifice European and Mediterranean concerns for the security of the Far East. If, in the worst case, Britain and France faced Germany, Italy and Japan, without the United States joining the war, then the best that could be done was to rely on the deterrent effect of one or two capital ships at Singapore. The Chiefs of Staff concurred that it was 'essential for us to take what steps we could to protect our sea communications in the Indian Ocean against raids'. The French favoured a defensive posture, until Italy was knocked out of the war. Only in Europe would defeat be final, therefore the terms for the relief of Singapore were lengthened (Gibbs 1976: 422-429). The prospect of war had brought home to the policy-makers the impossibility of defending global empire with a battle fleet unnaturally restricted. In 1939 Britain had only ten battleships available for service.

After the outbreak of war in Europe the Indian Ocean remained relatively undisturbed, the odd raider cruise aside, an area where the threat to shipping was low. Indian and Australasian troop convoys were heavily escorted, but otherwise ships sailed independently. Fortunately Japan did not enter the war in 1939, allowing India to become, 'a base on major scale and a provider of troops and supplies'. With the fleet tied to the European war Singapore would have to hold out for at least 90 days (Butler 1971: 16, 30). In February 1941 Churchill, who had underrated, or ignored the threat from Japan since the 1920's (Bell 2003), confessed to Roosevelt that he was alarmed. This was not, he added by the threat of an attack on Singapore, but because he feared raids on the Indian Ocean, which would be more damaging than operations against the coast of the Dutch East Indies or the Australasian dominions. His reasons were primarily Mediterranean: 'Not only would this be a grievous additional restriction and derangement of our whole war economy, but it would altogether bring to an end all reinforcements of the armies we had planned to build up in the Middle East from Australasian and Indian sources' (1950: 157-158). His only hope was to revive the 1939 concept of a deterrent, using two fast capital ships to 'tie down,' the Japanese (1950: 768). This prompted the First Sea Lord to provide him with a brief lesson in global strategy: 'The Atlantic is the vital area, as it is in that ocean and that alone that we can lose the war at sea' (1950: 771).

As the prospect of war with Japan developed British planning, far from falling back on a defensive system, and trying to secure the Indian Ocean, was anxious to use Singapore, not as an outer bastion for the Indian Ocean, but as a forward base for an Anglo-American drive north, with hopes of rescuing or recapturing Hong Kong, and taking the war to Japan. These plans were the best option at a time when the British simply did not have the resources to defend their interests in the region, and needed to co-operate with the United States. Manila was considered the best location for the second phase of the plan, Admiral Tom Phillips and the American Admiral were discussing this option at Manila when the Pacific War broke out (Cowman 1996: 398-417).

That Churchill had not taken the possibility of war with Japan seriously was obvious, his reasons were equally clear. Britain would not lose the war in the Far East. At the last moment Churchill tried to fend off the impending disaster with a deterrent gesture. However, this was a case of far too little, and far too late. The loss of the two capital ships sent to deter war, *HMS Prince of Wales:* and *HMS Repulse,* was the most direct shock Churchill received in the entire war. He finally realised the meaning of overstretch, and that with the American fleet disabled 'we everywhere were weak and naked' (Cowman 1996: 551). Even so he did not see that this weakness would lead to the fall of Singapore and open the Indian Ocean to the enemy's naval forces. India remained the great base for the movement of men and equipment into fighting area. The neutralisation of Manila left Singapore seriously exposed.[4] Phillips knew it had not been designed or equipped to resist a major land attack, or to hold out without sea communications.

Once the Japanese attack on Malay began Phillips' only option was to assert sea control in the South China Sea with his two capital ships. In conformity with British strategic doctrine his target would be the Japanese invasion shipping. Unfortunately he was badly briefed on Japanese air capabilities, and believed they had no torpedo bombers capable of reaching his operational area. The loss of both ships on December 9th 1941 gave the Japanese undisputed command of the Southeast Asian theatre, without their main fleet being present. It was a striking endorsement of their pre-war thinking about land based naval air power. It also ensured the early fall of Singapore, the Philippines and the Netherlands East Indies, none of which was intended to hold out once sea control had been lost. This drove a wedge between the British and American forces facing Japan. It also left the British relying on five old battleships and a few modern, but unprepared carriers hastily assembling in the Indian Ocean. By contrast the Japanese, operating on internal lines, sent their carrier strike fleet from Pearl Harbour to Darwin. After the fall of Singapore in February a sortie into the Indian Ocean was highly likely. Vice-Admiral Sir James Somerville, the most experienced British carrier commander, was well aware that his scratch force could not compete with the Japanese, and only excellent intelligence work enabled him to avoid the blow (Hinsley and Strip 1993: 275).

For Richmond, the last great imperial strategist, the fall of Singapore reflected the failure to build an adequate fleet, prepare a combined arms strategy, and secure the key base. The British Empire remained a maritime creation, and once it lost that command of the sea it's mightiest outposts, be they really strong, or just giving the appearance of power, were doomed (Hunt 1982: 147). So too were its territorial possessions.

The fall of Singapore and Rangoon finally forced Churchill to recognise the danger that he was running to keep a grip on the Middle East, and Richmond prompted him to shift attention to Ceylon and Madagascar (Halpern 1981: 244-246). The enemy finally came knocking at the gates of India in April. Fortunately they withdrew after Admiral Nagumo' s carrier attack on Ceylon, 5th - 9th April and a sortie into the Bay of Bengal, sinking over 1,00,000 tons of shipping. Somerville's slow fleet also withdrew, to East Africa to cover Middle East shipping, while his best units were stationed at Bombay. Despite its limitations this fleet was so important that it was doubtful if it would have been used to defend Ceylon. Fortunately, the naval power that had driven the British back across the Indian Ocean was drawn into the central and South Pacific by American efforts, leaving the Japanese to campaign on land. The danger that they might return focused attention. Churchill stressed that the defence of Colombo and Calcutta, together with the overland link to China were the vital points. He was not overly concerned about Japanese landing on the Indian coast, but if Colombo were lost then the assembly of naval reinforcements would be futile. He considered Ceylon and China the only areas where Japan could secure any strategic success in 1942, appointing an overall island supreme Commander, Admiral Layton (1950: 163-164). Even so the whole theatre was exposed to a renewed Japanese attack for the next two months.

The British demonstrated their real priorities lay on other side of the Ocean. Fearing the Vichy regime in Madagascar would allow enemy submarines to operate against the Middle East convoys the British struck in early May 1942. They used troops en route to India and ships from the Eastern and Home Fleets to secure the key area in the north of the island, only mopping up the regime in September (Butler 1964: 483-503). The island gave the British control of the Indian Ocean convoy and supply routes for the Eighth Army in Egypt (Roskill 1956: 191). Only when their ability to use that route was under serious threat did the British reflect on the importance of the Indian Ocean to the cohesion of the empire and the Imperial war effort. Convoys carrying troops and equipment to the Middle East were Churchill's top priority, they were called WS for 'Winston's Specials' and occupied almost all the escort forces in theatre. Inter-imperial trade routes were equally sensitive, but had a lower absolute priority. The Empire needed oil from the Persian Gulf, which, along with coal from Bengal and South Africa were essential to the Egyptian war effort. Egyptian rice helped to feed India after loss of Burma, while Australian wheat fed the Middle East armies but relied, as did New Zealand, on Egyptian phosphates (Marder, Jacobsen and Horsefield 1990: 159; Willmott 1996; Behrens 1955). However, a total war requires resources to be allocated according to harsh priorities, and these needs could be overridden.

The Japanese rampage ended at Coral Sea, while Midway crippled her offensive power. Once the American offensive in the South Pacific began, and it was clear the Japanese would not come back in strength, the British ended the naval build up in the Indian Ocean. In early 1943 the Eastern Fleet lost almost all of its modern and heavy units to the offensive in Europe. However, they did develop base and harbour facilities at Colombo and Trincomalee, along with the vital airfields, so that when time came to switch to the offensive, they could deploy the air strength necessary to defeat the Japanese (Roskill 1956: 424-425). Although British command of the Ocean was never again threatened, German and Japanese submarines and surface raiders continued to operate in the theatre. This was the penalty for the loss of Singapore. While the enemy controlled access to the Indian Ocean, the British could only re-act to their moves. Defensive oceanic strategies possess no advantages, and numerous disadvantages.

The pivot of the war remained the North Atlantic, all other theatres sacrificed warships, merchant shipping and other resources to meet its needs. In early 1943 shipping for the Indian Ocean region was greatly reduced to support the opening offensive in Europe. This exacerbated a disastrous famine in Bengal (Behrens 1955: 345-353). It also precluded an amphibious offensive into Burma (Howard 1972: 398-399). Instead 48 units were detached for the invasion of Sicily (Roskill 1961: 219). Even when they returned at the end of the year the work of the Eastern Fleet was hampered, as it had been since 1941, by the scarcity of escort vessels and the priority given to convoy duties. The continued threat to shipping precluded any significant offensive operations by Somerville's fleet, even when the Americans requested a diversion. The Admiralty was anxious to end the convoy system, but although the Japanese fleet at Singapore did nothing, the Germans were more active, if only because the area was a relatively soft target (Roskill 1961: 222).

When the British secured absolute command of the Atlantic theatre, in mid-1944 and completed Operation Overlord they reinforced the Eastern Fleet. This, together with the end of significant enemy submarine activity the East Indies allowed the enlarged fleet to shift onto the offensive, supporting the re-conquest of Burma and attacking the Indonesian archipelago (Roskill 1961: 311). The main fleet was then shifted around the island barrier into the Pacific via Sydney, to take the war to Japan, while the second line forces advanced on Singapore.

Table 8.1

Allied Merchant Shipping losses in the Indian Ocean, 1939-1945

Year	*Tons sunk*	*Ships sunk*
1939	706	1
1940	173,416	24
1941	73,155	20
1942	724,485	205
1943	486,324	82
1944	322,802	50
1945	8,982	3
Total	1,789,982	385*

*8.3% of total allied losses (Roskill 1961: 479).

The end of Empire in the Indian Ocean region was inevitable. Britain had bankrupted herself to win the World War II, and could not afford Churchillian dreams of past glory. Instead a new fleet with superior afloat logistics would serve from more distant bases, the old Imperial bastions of Hong Kong, Singapore and Aden outlived the empire because they had been developed to serve oceanic trade, something that was largely unaffected by the passing of dominion in Asia.

NOTES

1. India Office Admiralty Correspondence 1888, see esp. Minutes on India Office to Admiralty 19.11.1887 and 9.5.1888: ADM 1/6936.
2. HMS *Bonaventure,* 4,400 tons, two 6 inch and eight 4.7 inch guns.
3. Corbett's book is a master-class in Imperial strategy.
4. This is the core thesis of Cowman's article.

BIBLIOGRAPHY

ANON (1950) *The Second World War*, Vol. III, London.

BARTY-KING, H. (1979) *Girdle Round the Earth: The Story of Cable and Wireless*, London.

BEHRENS, C.B.A. (1955) *Merchant Shipping and the Demands of War*, London.

BELL, C. (1997) How Are We Going to Make War? Admiral Sir Herbert Richmond and British Far Eastern War Plans, *Journal of Strategic Studies*, XX.

BROWN, D.K. (1998) *Warrior to Dreadnought*, London.

BROWN, W.M. (2003) *The Royal Navy and its Fuel Supplies, 1900-1939*, Unpublished Ph.D. Thesis, London University, London.

BUTLER, J.R.M. (1964) *Grand Strategy*, Vol. II, part II, London.

BUTLER, J.R.M. (1971) *Grand Strategy*, Vol. II, 2nd ed., London.

CAIN, P.J. and HOPKINS, A.G. (1993) *British Imperialism: Innovation and Expansion 1688-1914*, Longman House, London.

CLOWES, W.L. (1903) *The Royal Navy: A History from the Earliest Times*, Vol. VII, London.

COLOMB, P. (1896) *Essays on National Defence*, London.

CORBETT, Sir J.S. (1920) *The Official History of the War: Naval Operations*, Vol. I, London.

COWMAN, I. (1996) Defence of the Malay Barrier? *War in Modern History*, 3.

FERRIS, J.R. (1989) *Men, Money, and Diplomacy: The Evolution of British Strategic Foreign Policy, 1919-1926*, Cornell.

FREMANTLE, E.R. (1904) *The Navy as I Have Known It: 1849-1899*, London.

GIBBS, N. (1976) *Grand Strategy*, Vol. 1, London.

GRAHAM, G.S. (1965) *The Politics of Naval Supremacy*, Cambridge.

GRAHAM, G.S. (1967) *Great Britain and the Indian Ocean 1819-1850*, Oxford.

HAGUE, A. (1996) *Sloops, 1926-1946*, World Ship Society.

HALPERN, P. (ed.) (1981) *The Keyes Papers*, Vol. III, London.

HEADRICK, D.R. (1991) *The Invisible Weapon: Telecommunications and International Politics, 1851-1945*, New York.

HINSLEY, F.H. and STRIP, A. (ed.) (1993) *The Codebreakers*, Oxford.

HOWARD, M. (1972) *Grand Strategy*, Vol. IV, London.

HUNT, B. (1982) *Sailor-Scholar: Admiral Sir Herbert Richmond, 1871-1946*, Waterloo.

KENNEDY, P.M. (1971) Imperial Cable Communications and Strategy, 1870-1914, *English Historical Review*, LXXXVI.

KENNEDY, W. (1910) *Hurrah for the Life of a Sailor*, London.

LAMBERT, A.D. (1993) Dalhousie, The Last Flagship of the Indian Navy and the Question of Imperial Defence, In: R. Gardiner (ed.) *Warship*, London.

LAMBERT, A.D. (1998) *The Foundations of Naval History: Sir John Knox Laughton, The Royal Navy and the Historical Profession*, London.

LAMBERT, A.D. (2003) This is All We Want: Great Britain and the Baltic Approaches, 1815-1914, In: J. Sevaldsen (ed.) *Britain and Denmark: Political, Economic and Cultural Relations in the 19th and 20th Centuries*, Copenhagen.

LAMBERT, N., *Sir John Fisher's Naval Revolution*, Columbia, South Carolina.

LOW, C.R. (1877) *The History of the Indian Navy 1613-1863*, London.

MACKAY, R. (1973) *Fisher of Kilverstone*, Oxford.

MAHAN, A.T. (1900) *The Problem of Asia and its Effect on International Policies*, London.

MARDER, A.J. (1969) *From the Dreadnought to Scapa Flow, Vol. IV, 1917*, Oxford.

MARDER, A.J., JACOBSEN, M. and HORSEFIELD, J. (1990) *Old Friends, New Enemies: The Royal Navy and the Imperial Japanese Navy, The Pacific War, 1942-1945*, Oxford.

OFFER, A. (1989) *The First World War: An Agrarian Interpretation*, Oxford.

PARKES, O. (1970) *British Battleships: A History of Design, Construction and Armament, 1860-1950*, London.

PATTERSON, A. (ed.) (1968) *The Jellicoe Papers, Vol. II*, The Naval Situation in Far Eastern Waters, London.

RICHMOND, Sir H.W. (1931) *The Navy in India: 1763-1783*, London.

RITCHIE, G.S. (1967) *The Admiralty Chart*, London.

ROSKILL, S. (1956) *The War at Sea*, Vol. II, London.

ROSKILL, S. (1961) *The War at Sea*, Vol. III, London.

SCHURMAN, D.M. (2000) *Imperial Defence, 1868-1887*, London.

SLUGGETT, P. (2002) The Resilience of a Frontier: Ottoman and Iraqi Claims to Kuwait, 1871-1990, *International History Review*, XXIV.

STRACHAN, H. (2001) *The First World War*, Vol. 1, To Arms, Oxford.

VINCENT, J. (ed.) (1978) *Disraeli, Derby and the Conservative Party: The Political Journal of Lord Stanley, 1849-1869*, Hassocks.

VINCENT, J. (ed.) (1995) *The Derby Diaries, 1869-1878*, London.

WILLIAMS, R. (1991) *Defending the Empire: The Conservative Party and British Defence Policy, 1899-1915*, Yale.

WILLMOTT, H.P. (1996) *Grave of a Dozen Schemes*, London.

WONG, J.Y. (1998) *Deadly Dreams: Opium and the Arrow War (1856-1860) in China*, Cambridge.

Comment

Peter Hore

There are two themes, which I would like to explore; the first one I have called after a Sherlock Holmes story 'the dog did not bark'. Let me explain history to be relevant and must be interesting to us, must have some impact on our daily lives. If it is true, why is it as we have heard from several speakers yesterday in particular why is it that in thousand years of history in the Indian Ocean basin no navies were developed, few colonies existed. This is a situation, which is surely exceptional to have understanding of the concept of sea power. It counters surety to those understanding. It is a puzzle. As far as marine archaeology is concerned the naval officers faced enormous challenges in the explorations of marine archaeology in Indian Ocean in charting the seas in diving and lifting artifacts from the sea bed. But frankly this is not very relevant to a professional naval officer, whose professional job after all is to study, practice and prepare for war. With an exception, perhaps this unique question that why is it that the Indian Ocean basin did not develop navies and colonies whereas in a similar period of history and with similar pattern of trades in the north sea and the Mediterranean, organised state violence, did result in the creation of warship and of course navies. This is the dog that did not bark. Except as we understand in one period of history for that 200 years when the European nations fought each other in the Indian Ocean, a period which is taken by Cobert and Richmond as examples of their theories of Sea Power, the trade and flag go hand in hand.

Turning to the paper specially he says in some that during the 19th and 20th centuries which was the age of Pax Britannica as Admiral Tahiliani calls it, the British held key to all the entry points. The pax Britannica benefited all nations and even the colonies of the traditional enemies like the French were allowed to continue. They were'nt a significant threat to the British rule but nevertheless under none of this period did an indigenous navy arise.

My second theme in exploring Andrew's paper is this that if no indigenous navy did develop around Indian Ocean littorals and in particular no Indian navy developed except for those few years in the end of the 18th century, how is it by contrast since 1947, the Indian Navy has grown remarkably. Despite as we have heard the historical sea blindness in Delhi, the Indian Navy with the number of its ships, the quality of its people has far outrun the other navies in the region and what is future. There has been a remarkable continuant in Indian Ocean history. The sea lanes of communication ran the southern tip of the Indian continent have always been of immense importance to trade once, spices and porcelain and now, oil and global trade. Here is a small ridiculous example, that the British manufacturer of the retreaded tyres depends entirely upon the shipping via the Indian Ocean of used tyres from Japan, remarkable but true, and if that stops, the world system of global trade also judders to halt. But something in the years has happened with the rise of a strong regional navy, which must surely make a difference to the future history and the balance of power in the region. But the question, which I would like to leave you with, comes in two parts which continuum will now be followed. Will Indian Navy for example be strong enough to hold the key of the Indian Ocean and the corollary of this will Indian government use the resulting pax quite so well?

The second part is this, before that happens will the Indian Navy be close even to prevent the access or entry into one of the sea, which is adjacent to the Indian Ocean. Then perhaps repeat the influence which the Bombay Marine the Royal Indian Marine and for the short period the Indian Navy once exercised over the Arabian Sea.

9

The Stellar Compass and the Kamal an Interpretation of its Practical Use

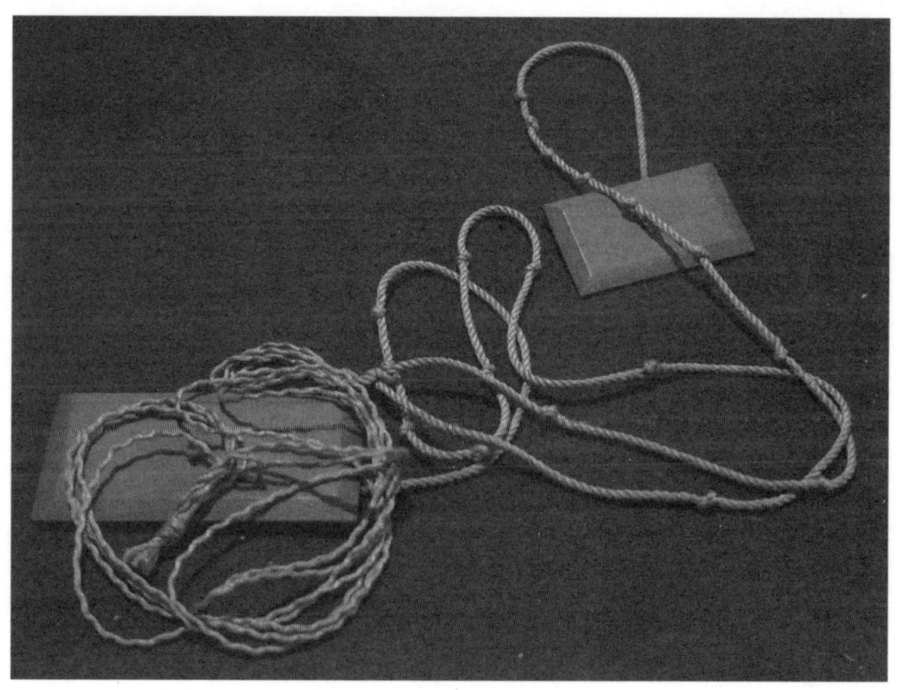

Jose Manuel Malhao Pereira

Introduction

This paper deals with the practical use of the stellar compass and the *Kamal* at sea. Although many scholars have extensively studied these two important instruments of Indian Ocean navigation, however their usage aboard ships has not been clearly explained by them. The author from his experience onboard sailing and motor ships has explained the use of these instruments which would be useful to historians to interpret the techniques of navigation in this area of the world.

Environmental Conditions of the Indo-Pacific Regions of Navigation and its Consequence to Navigation Techniques

Father Fernando Oliveira, a Portuguese sailor and also a religious man of the sixteenth century, sailed extensively in the Atlantic and wrote important manuscripts of navigation and shipbuilding (De Mendonca 1898). In *O Livro da Fabrica das Naus,* commenting on the Portuguese navigation of the period, and comparing them with those of the Greeks and Latins, who previously said that they were the inventors of the art of navigation Father Oliveira (1992: 139-140) wrote *"Ours deserve greater praise for this [the explorations of the high seas], than the Greeks or the Latins: because they have done more for navigation in 80 years than the others did in the 2000 during which they reigned. And more improvements have been brought to this art than they ever achieved"*. After these apparently Euro-centric words he added: *"However, not even this allows us to say that we are the first inventors of the art, as they have so arrogantly declared. And it is not just because they have claimed it that we must consent to the assertion: for, in many parts of the world where they have never gone and which were not reached by any doctrine of theirs, there are ships and the art of navigation: in some places better than in others, according to the maturity of backwardness of the people who live there. We find, in China and Japan, reasonably fine ships where the existence of the doctrines of the Greeks, or their Neptune, has never been heard of. In Guinea and Brazil, Greeks were never mentioned either and, without them, navigation exists, in its own style, anyhow as taught by nature".*

All people of any area of the world adapt their techniques of any kind, to their environment. The same happens with the techniques of navigation and it is explained briefly what are the physical conditions of the navigation in the Indian and Pacific Oceans,[3] compared to other areas (Tibbets 1971; 1979).

The first important fact is that the Arab, Persian, Indian, Malayan, Chinese and Polynesian navigation is performed mainly in inter-tropical areas.

Shaded areas of figure 9.1 shows where the inter-tropical areas of the seas of the world and schematically represented routes of navigation. It is immediately apparent that Asian routes are mainly inter-tropical, with the only exception of routes of northern parts of China. The Portuguese, Spanish, Dutch, French and English routes

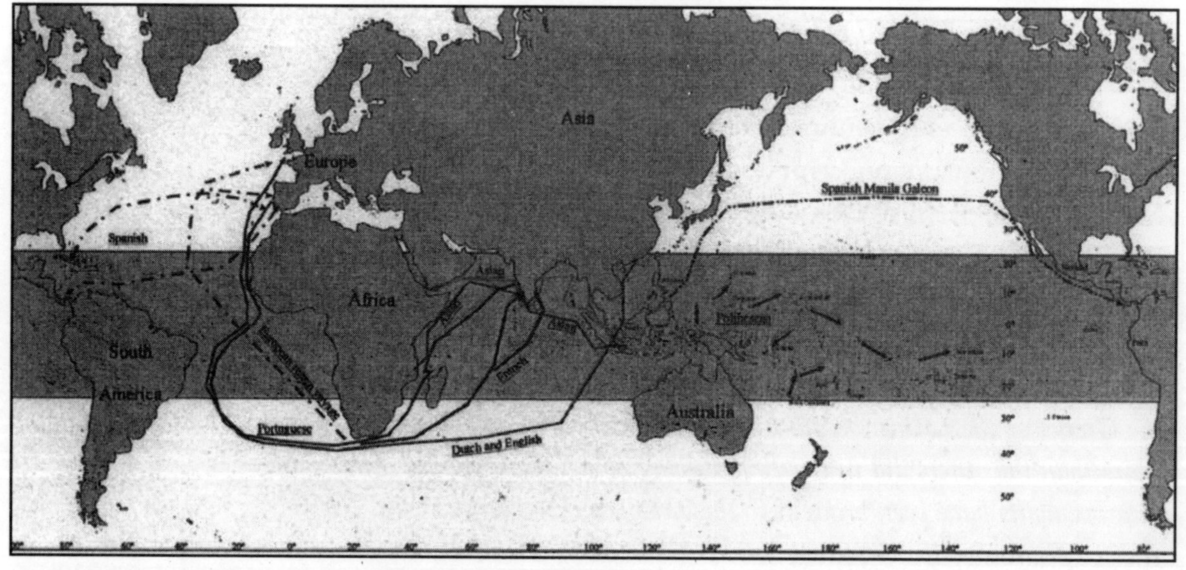

9.1. Intertropical areas at sea and European and Asian routes of navigation

have big proportions out of the tropics. It is also evident from the map, that—

—European routes are long distance ones with many months at sea and out of sight of land.

—Arabian, Persian and Indian routes have some medium distance ocean routes, but more coastal navigation.

—Malayan and Chinese routes are mainly coastal or short distance high seas navigation, with the exception of the probable but not very frequent direct routes to Africa.

—Western Pacific routes are mainly short inter-island routes, which never exceed the 350 miles distance, with the exception of the voyages to Hawaii and New Zealand.

Sailors find the necessary techniques and instrumentation for their areas of operation. They are very practical and only use or develop what is needed for the purpose of the navigation.

Taking into consideration what has been said above, it is necessary to imagine what does a sailor need to know to return to a previously discovered island or harbour in a continent, far away from his home place and out of sight of land which means that ocean navigation methods are needed, instead of coastal navigation methods.

The first thing is to find a way of recognising the direction he has to take to return there, admitting that the same system has been used to return to his point of departure. But for finding a direction it is necessary any reference to measure that direction.

Figure 9.2 shows two islands A and B sketched on the sand of a beach. The arrow is the path that any ship has to sail on, to go from A to B or vice-versa. For attaining this objective, it is necessary to materialise any means of finding the direction, for example from A to B.[2]

One of the ways is having a reference related to the direction of the swell, if it is constant during the period of navigation. Another can be the direction of the wind, again if it is constant in direction or the flight of birds, if they go always to the same place B.

These means have too many ifs, and people soon recognised that the heavens, with its stars and planets during the night, and sun and moon during the day, could be a good reference. From those, the best are the stars, because they maintain their relative positions for very long periods. Besides that they rise and set in well- defined positions

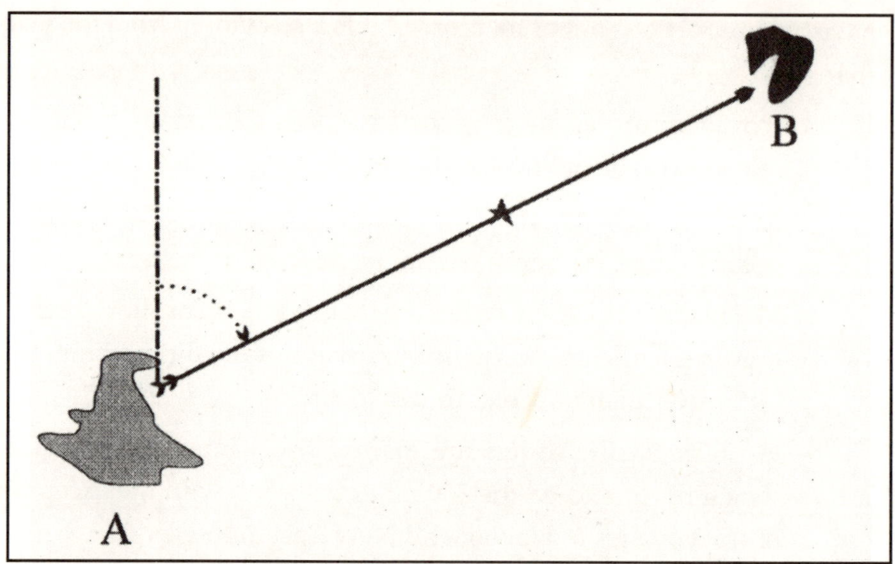

9.2. The need for direction

on the horizon. However, the effectiveness of stars to be direction references depends mainly on latitude.

On the North Pole the celestial sphere moves apparently in the opposite direction on account of the rotation of the earth around its axis. As an observer has the axis of the earth on his head, during 24 hours the celestial hemisphere, with all its visible celestial bodies, revolve around him and never rise or set. The astronomers and navigators call it parallel sphere, because the heavenly bodies have their movements along celestial parallels (fig. 9.3). As there is no rising and

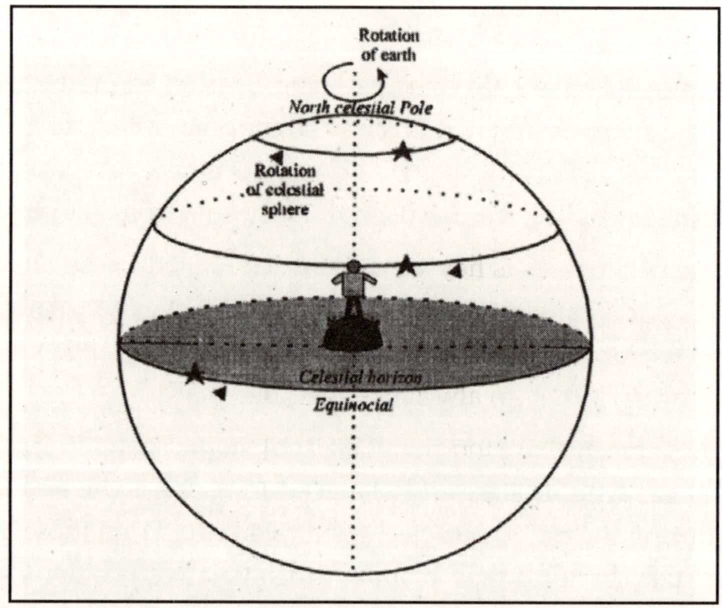

9.3. Parallel sphere - Observer at the Pole, stars never rise or set, only half of the celestial sphere visible throughout the year

setting, and the celestial bodies are always moving horizontally, they can never be good references for direction.

At the equator heavenly bodies trace perpendicular circles around the observer, and their rising and setting bearings are maintained during long periods, specially the equatorial ones (fig. 9.4). For example a star at the equator (or the sun during the equinox), rises in the east, maintains its east bearing till meridian passage and maintains its west bearing till it sets.

At high latitude (45^0 N.), heavenly bodies trace oblique circles relative to the horizon, and their rising and setting bearings vary rapidly (fig. 9.5).

At the same latitude an observer would have the same bearing of a given star during rising or setting throughout the year. But if he changes latitude, the bearing will change appreciably in high latitudes, but a small amount in places near the equator.

Table 9.1 clarifies the situation, where the bearings (Z) of the rising of four stars of northern declination and the settings of four stars of southern declination, are given for the latitudes 0^0 to 60^0 north, in intervals of five degrees of latitude. The 8 stars selected are part of the 16 that Tibbets identified in the Arab stellar compass, and cover almost the entire horizons (Tibbets 1971: 294-299; Ferrand 1928: 92-124).

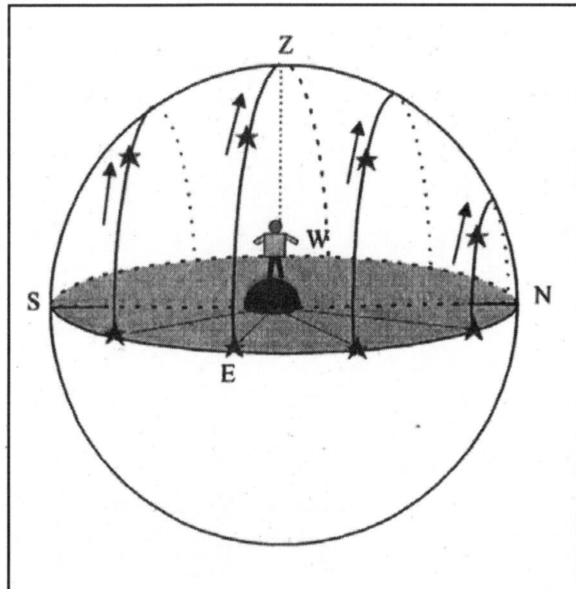

9.4. Right sphere - Observer at the Equator, stars trace circles perpendicular to the horizons, equatorial stars defining, during rising and setting, long periods of accurate bearing

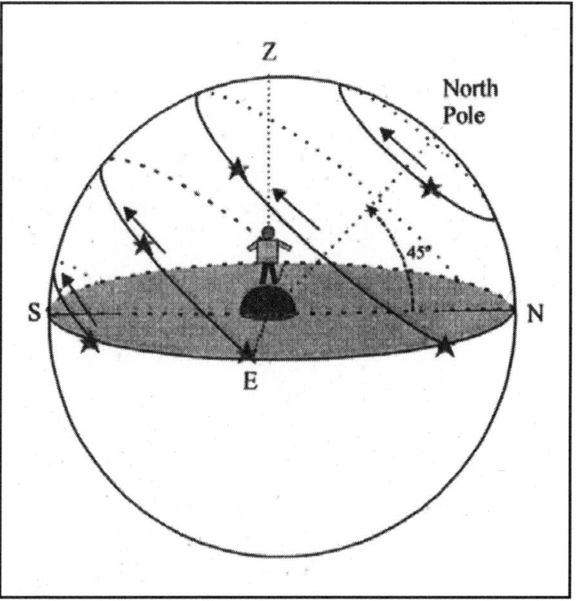

9.5. Oblique sphere - Observer at 45^0 of latitude, stars trace circles oblique to the horizons, defining, during rising and setting, very short periods of accurate bearings

Table 9.1

Observer in the northern hemisphere

Stars with northern declination									*Stars with southern declination*							
Bearings of rising									*Bearings of setting*							
Latitude	*Kochab*		*Schedar*		*Capella*		*Altair*		*Sirius*		*Antares*		*Canopus*		*Achernar*	
	δ	Z	δ	Z	δ	Z	δ	Z	δ	Z	δ	Z	δ	Z	δ	Z
0^0	74^{0}N.	016^0	56^{0}N.	034^0	46^{0}N.	044^0	9^{0}N.	081^0	17^{0}S.	253^0	26^{0}S.	243^0	53^{0}S.	217^0	57^{0}S.	213^0
5^{0}N.	"	015^0	"	033^0	"	044^0	"	081^0	"	253^0	"	243^0	"	217^0	"	212^0
10^{0}N.	"	012^0	"	032^0	"	043^0	"	081^0	"	253^0	"	243^0	"	216^0	"	211^0
15^{0}N.	"	004^0	"	030^0	"	042^0	"	081^0	"	253^0	"	242^0	"	215^0	"	210^0
20^{0}N.	"	—	"	027^0	"	040^0	"	081^0	"	252^0	"	242^0	"	213^0	"	207^0
25^{0}N.	"	—	"	023^0	"	037^0	"	080^0	"	252^0	"	241^0	"	209^0	"	203^0
30^{0}N.	"	—	"	015^0	"	034^0	"	079^0	"	251^0	"	239^0	"	203^0	"	194^0
35^{0}N.	"	—	"	—	"	028^0	"	079^0	"	250^0	"	237^0	"	194^0	"	—
40^{0}N.	"	—	"	—	"	019^0	"	078^0	"	248^0	"	234^0	"	—	"	—
45^{0}N.	"	—	"	—	"	—	"	177^0	"	246^0	"	231^0	"	—	"	—
50^{0}N.	"	—	"	—	"	—	"	076^0	"	244^0	"	226^0	"	—	"	—
55^{0}N.	"	—	"	—	"	—	"	074^0	"	240^0	"	220^0	"	—	"	—
60^{0}N.	"	—	"	—	"	—	"	072^0	"	235^0	"	209^0	"	—	"	—

Looking to the sky in January 2002, would be sufficient for the purpose of this study. The declinations and bearings are to the accuracy of the degree and the bearings are from 0^0 to 360^0, clockwise, which is the normal way used by navigators. It is clear that the bearings of the stars on rising (the setting is symmetrical), have a small variation with latitude, around latitudes 0^0 to 20^0 or 25^0.

By a similar operation (table 9.2), it is found that for the same stars but to South latitudes of the same amount, extended to 30^0 S., the bearings are the same.

This is a very interesting fact that sometimes is not recognised, even by experienced navigators (Lewis 1994; Nunes 1940: 165-174).[3] It gives an added advantage of using rising and setting stars for direction between the tropics because their bearings are equal if we are at the same latitude, even if we are in different hemispheres. This means that the entire area of the tropics is suitable for this purpose and not only half of it. It also shows clearly, that for observers between the tropics, the rising and setting of stars are an accurate way for direction, and that out of the tropics, they are not suitable for that purpose.

In consequence of what has been said above, it can be understood that inter-tropical navigators soon recognised that the bearings of rising and setting of stars were almost constant when they made their inter-island voyages, although they did not have the notion of latitude as we have today.

So, they began to recognise that the island B (fig. 9.2), was, as seen from island A or the continent, in the direction of the rising of some star. And this often repeated voyages. As other islands or harbours in continents were in other directions, they soon memorised other different stars that were suitable to maintain the direction required. As stars are only near the horizon for a short period, other stars were used on the same direction or bearing (this is nowadays the appropriate technical word).

As in the same bearing the amount of stars was not sufficient to cover all the horizon, soon practical navigators recognised that other stars would be associated with the others, if they made an image of them, altogether, on a piece of circular or square wood, where they marked all the stars useful for the purpose of finding different places.

From this stage, and taking into account that their voyages were in directions that covered the entire horizon, it is easy to understand

Table 9.2

Observer in the southern hemisphere

Stars with northern declination									*Stars with southern declination*							
Bearings of rising									*Bearings of setting*							
Latitude	*Kochab*		*Schedar*		*Capella*		*Altair*		*Sirius*		*Antares*		*Canopus*		*Achernar*	
	δ	Z	δ	Z	δ	Z	δ	Z	δ	Z	δ	Z	δ	Z	δ	Z
0^0	74^0N.	016^0	56^0N.	034^0	46^0N.	044^0	9^0N.	081^0	17^0S.	253^0	26^0S.	243^0	53^0S.	217^0	57^0S.	213^0
5^0S.	"	015^0	"	033^0	"	044^0	"	081^0	"	253^0	"	243^0	"	217^0	"	212^0
10^0S.	"	012^0	"	032^0	"	043^0	"	081^0	"	253^0	"	243^0	"	216^0	"	211^0
15^0S.	"	004^0	"	030^0	"	042^0	"	081^0	"	253^0	"	242^0	"	215^0	"	210^0
20^0S.	"	—	"	027^0	"	040^0	"	081^0	"	252^0	"	242^0	"	213^0	"	207^0
25^0S.	"	—	"	023^0	"	037^0	"	080^0	"	252^0	"	241^0	"	209^0	"	203^0
30^0S.	"	—	"	016^0	"	034^0	"	079^0	"	251^0	"	239^0	"	203^0	"	194^0

the development of a compass rose that showed to the navigator the different directions which in the future could be the reference for future voyages. In consequence of this the stellar compass was introduced.

Stellar Compass of the Indian Ocean

The navigators of the northern Indian Ocean (Persians, Indians, Arabs, Malays, Chinese), who sailed in northern latitudes between the 7th and 23rd parallel mainly in an east-west direction soon recognised (like the Portuguese after recognised in the Atlantic), that the pole-star which was always above their horizon and near the North Celestial Pole, was a good way to find direction and very appropriate to maintain an east-west course.

But as the pole-star or the other stars of the Little Bear could be covered by clouds, other stars were chosen and drawn on a piece of wood, separated each other by the angles that separated them on the horizon. So, a star compass of the Indian Ocean[7] was idealised being its details explained later (fig. 9.6) (Ferrand 1928: 91).

All of this has been deeply studied by many scholars like Gabriel Ferrand (1928: 91), G.R. Tibbetts (1971; 1979), Laguardia Trias (1963),

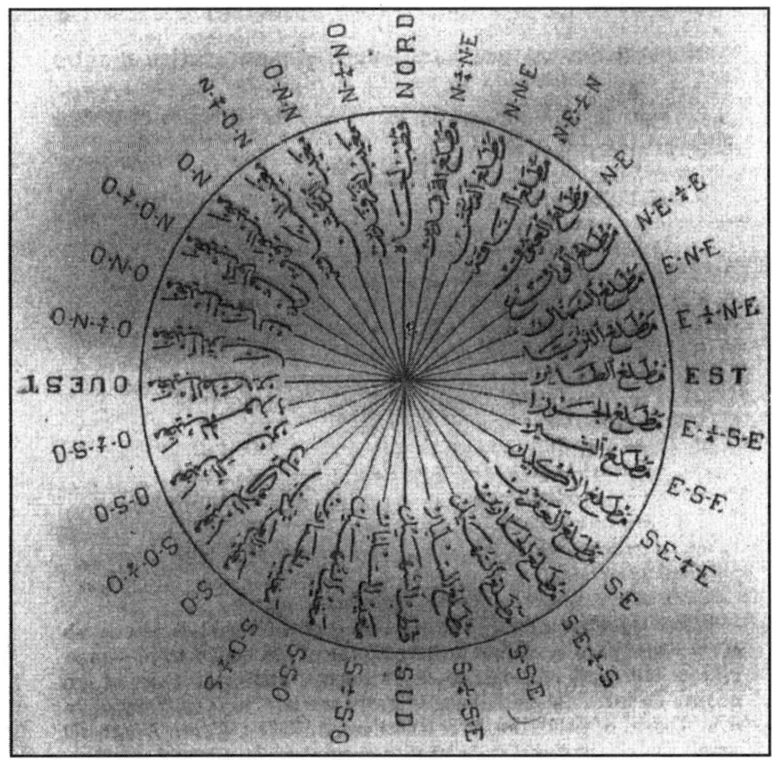

9.6. Stellar compass of the Indian Ocean

Custodio de Morais (1960). Vitorino Magalhaes Godinho, Moura Braz (1962), David Lewis (Braz 1962), William Kelsalka (1987), Teixeira da Mota,[16] among many others. Lewis and Kelsalka made interesting sailing experiences, with extremely important results.

Leopold Sassure has deeply studied this subject and his considerations were extensively explained and justified in an article published in 1923 in Geneve that Gabriel Ferrand included in his *Instructions Nautiques et Routiers Arabes et Portugais.*[17] Tibbetts refers also to this subject, mainly following de Sassure's ideas (Tibbets 1971: 295-312).

The stellar compass rose, as previously explained came into use on account of the specific conditions of the seas navigated. In consequence of the need of a multiple of 4 directions (the result of the previous definition of the 4 cardinal points), 32 directions were found which were related to the directions of the rising and setting of 15 conspicuous asterisms.[4] The remaining directions needed were the North Pole (defined by Polaris), and the South Pole.

The stars were primitively chosen mainly according to their bearing of rising and setting, the times of this phenomenon, and magnitude. It was necessary that after a star sets, becoming invisible, another be already above the horizon but sufficiently low to be a good direction indicator. This last star will be useful during a short period, after which another star will rise or is setting in a convenient position for direction reference.

So the idea was to have at any time of the night at least one star sufficiently low. All the others should follow in the same condition, in different positions of the horizon and obviously in different times of the night. With this array of asterisms, there will be always the possibility of orientation of the compass in relation to the north-south line, because the different stars make known angles with the north-south line that is mainly represented by Polaris and the stars around the North Pole.

Scholars studied deeply this subject, trying to interpret its origin and date of implementation, the different stars or groups of stars used, etc. Based on de Saussure's work[5] the names of the stars and the compass itself are shown (fig. 9.6-9.7).

The table 9.3 shows, for 15th of January 1300 CE, in a place of latitude 10^0 N. and longitude 80^0 E. (just north of Sri Lanka), the stars or constellations indicated by de Saussure[6] with their magnitudes, declinations, bearings, altitudes and times of rising and setting and also times of meridian passage.

On the columns of rising and setting times I highlighted the times that are in good condition for usefulness of direction indication and numbered them in sequence of observation during all the night. The ones not highlighted are not useful during that time of the year, because they rise or set during the day. But we have to understand that the stars not useful in January will be useful in other times of the year on account of the annual movement of the earth around the sun.[7] So, taking into consideration that a star near the horizon can be utilised during a long period for direction indication,[8] the intervals of time shown are an indication that at the same time there was always one star available.

The compass included in the work of Leopold de Saussure with its 32 divisions, is divided in two halves by the north-south line, materialised by the pole-star, and has inscribed the names of the stars

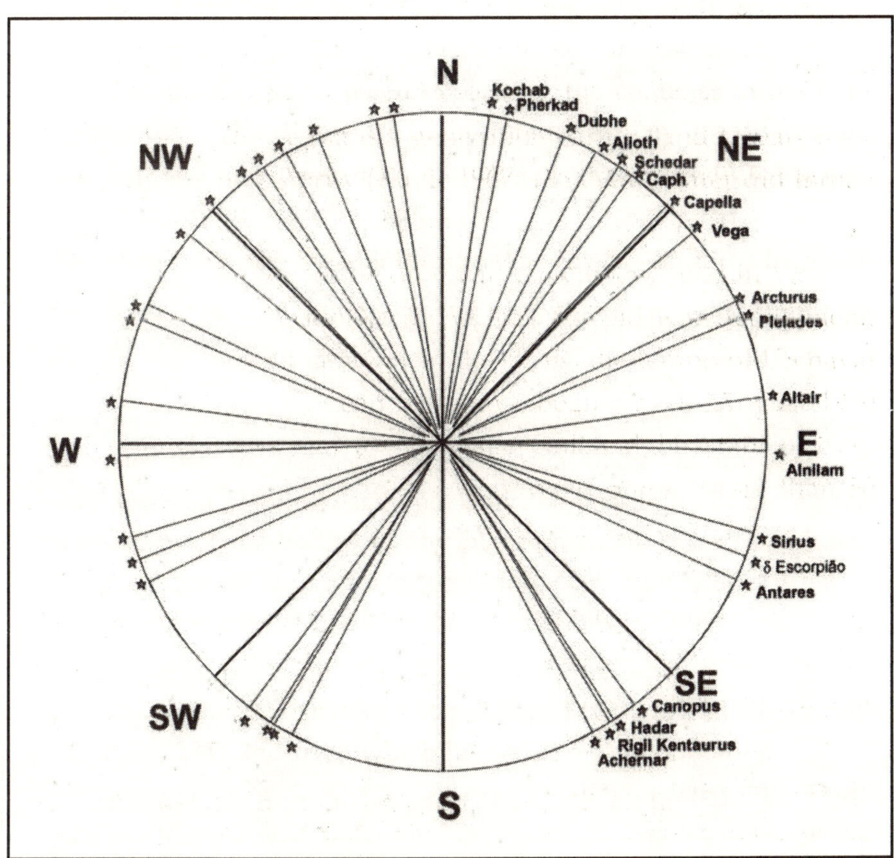

9.7. A graphical representation of the bearings of the stars included in the table

Table 9.3

The bearings of the stars of the stellar compass at 10^0 N., 70^0 E. on 15 January 1300

Star/Const.	*Magnit.*	*Declinat.*	*Rising*	*Bear*	*Setting*	*Bear*	*Time m.p.*
Polaris	2.1	85^0 28'.1 N.	—	—	—	—	1536
β Ursa menor / *Kochab*	2.2	76^0 59'.9 N.	2053 (3)	008^0.5	1605	351^0.4	0637
γ Ursa Menor / *Pherkad*	3.1	74^0 18'.9 N.	2210 (6)	011^0.6	1544	348^0.3	0659
α Ursa Maior / *Dubhe*	2.0	65^0 25'.1 N.	1813 (7)	022^0.3	0928	337^0.6	0153
χ Ursa Maior / *Alioth*	1.7	59^0 48'.3 N.	2039 (2)	028^0.4	1112	331^0.6	0357
α Cassiopeia / *Schedar*	2.5	52^0 39'.3 N.	0840	036^0.0	2232 (7)	324^0.0	1536
β Cassiopeia / *Caph*	2.4	55^0 15'.2 N.	0805	033^0.2	2209 (5)	326^0.7	1507
Capella	0.2	44^0 50'.6 N.	1316	044^0.1	0245 (15)	315^0.9	1958
Vega	0.1	38^0 21'.3 N.	0314 (17)	050^0.8	1622	309^0.3	0948
Arcturus	0.2	22^0 33'.9 N.	2255 (8)	066^0.9	1136	293^0.1	0518
Pleiades	—	24^0 07'.0 N.	1300	065^0.0	0141 (13)	295^0.0	—
Altair	0.9	7^0 18'.9 N.	0444 (19)	082^0.5	1657	277^0.5	1051
Cint. Orion-*Alnilan*	1.8	1^0 54'.3 S.	1433	091^0.8	0237 (14)	268^0.2	2033
Sirius	- 1.6	16^0 12'.9 S.	1556	106^0.4	0340 (18)	253^0.6	2146
β, δ, π, Escorpiao- δ	2.5	20^0 22'.9 S.	0108 (11)	110^0.6	1240	249^0.4	0654
Antares	1.1	24^0 36'.2 S.	0138 (12)	114^0.9	1305	245^0.1	0721
α Centaurus	0.1	57^0 35'.2 S.	0025 (10)	148^0.8	1023	211^0.2	0524
β Centaurus/*Hadar*	0.9	56^0 50'.5 S.	2347 (9)	148^0.0	0953	212^0.0	0452
Canopus	- 0.9	52^0 25'.2 S.	1631	143^0.4	0255 (16)	216^0.6	2141
Achernar	0.6	60^0 52'.5 S.	1154	152^0.3	2135 (4)	207^0.7	1644

or star groups in one side, which are repeated on the other side, all the designations being preceded by the Arabic word meaning setting in the western part and rising in the eastern part.

We can understand that the separation of the useful stars by a regular angular interval will be impossible to achieve in nature. A representation of the "real" compass for the year 1300 CE, (fig. 9.7) shows that the bearings of the stars that correspond to the table 9.3 are not distributed regularly.

This apparent problem has intrigued scholars who tried to find the epoch when the compass was introduced through the calculation of different arrangements of the same stars that, on account of precession, change slightly their positions in the sky.[9]

But even de Saussure arrived to the conclusion that" ... l' usage de la boussole a rendu les noms de rumbs purement conventionnels (comme, par exemple, le nom de notre mois de septembre qui n'indique plus le septieme mois".

Tibbetts (1971: 297-298) also found that Ibn-e-Majid himself commented the inaccuracy of the bearings for direction finding. It is very useful to quote the two passages of the *Kitab-al-Fawaid* referred by Tibbetts and by him translated in his work. The first one is as follows:

These stars or rhumbs [of the compass] are all of them approximate as we have shown in the Hawiya:[10]

"These stars and rhumbs with the Arabs are only approximate, Oh my captain. If you set course exactly on them in a narrow place, then you will have difficulty".

For they are only used for their names and not for their actual position in the heavens. Their positions are used for the compass rose which is divided into divisions and we have spoken and commented on this. Then we said on the *Hawiya*: The rest of the rhumbs of the compass are alike: It is a numerical division which can not be misleaded.

The other quotation referred to another chapter of the *Kitab al-Fawaid* that dealt with compass rhumbs (Tibbetts 1971: 139). It is as follows: "Aiyuk [Capella], is used to divide the heavens exactly in quarters for it 15.45^0 from the east point and 45^0 from the North Pole. It is the only star used as rhumbs which is accurately in position".

It is very interesting to note in figure 9.7 that Capella in fact, has the bearing of 45^0 NE, which fits exactly with the words of Ibn-e-

Majid. If there were any doubts about this subject, this only statement from this most famous Arab pilot would be sufficient.

Ibrahim Khoury calls also the attention to this problem in his work *As-Sufaliyya. 'The Poem of Sofala'* by Ahmad Ibn-e-Majid. When commenting *Al-Hawiya*, the poem of Ibn-e-Majid where this pilot referred 'The Circle of Ship' (which is the stellar compass), says: 'The basis of this choice [the division of the compass in equal parts] would be, theoretically, the apparent rising and setting of these stars in the sectors. In fact, this is not always true, because the distance between two consecutive stars or groups of stars, is never 7 fingers as it should be ($11^0 15'$), and that Cen (*al-himaran*) sector comes before Canopus and Achernar sectors, although its declination is superior to theirs, and this fact is well known among the Arab seamen. In this conjecture, it seems that the Point-Polaris is of the utmost importance for drawing the sectors, which lead to the guiding stars. Thus the sectors are believed to be a kind of frame reference to show the beginners how to find and learn the positions of the stars of navigation that give sailing directions' (Khoury 1983).[11]

The way to use the compass was not so complicated as sometimes is suggested.

Navigation in the Indian Ocean and the Adequate Techniques

Before analysing the kind of navigation practised by the sailors of the Indian Ocean let us complement the environmental conditions that were explained in the beginning of this work.

The north Indian Ocean, except during the height of the south-west monsoon which corresponds roughly to the months of June, July and August,[12] has light to moderate winds and waves. There are generally clear skies and especially clear nights, very appropriate to the observation of heavenly bodies.

The voyages from west to east were performed from the beginning or the middle of April till the middle of May. This was the first period, taking in advantage the south-west wind of the beginning of the south-west monsoon. The second period was the crossing of the sea at the end of the south-west monsoon, to arrive at the coasts of India in mid-September.

The return voyage, with the north-east wind was performed after the middle of October till the beginning of April.

It is clear that the voyage in the north Indian Ocean was mainly along a parallel of latitude and sometimes it was necessary to cross narrow passages like the Lakshadweep Islands that were also spread in a north-south direction (fig. 9.8). The coasts of the Indian subcontinent have also more or less the same orientation. It is also interesting to note that the navigation area was mainly between the equator and around the 23rd parallel of north latitude.

So, the safer and more adequate technique was to find the parallel of the destination, which was previously known, and stay on it until arriving near the coast. Figure 9.8 shows an example of voyage from Cochin to Socotra. This was the kind of technique used by all the modern navigators till the better knowledge of longitude in the end of the 18th century.

To find the parallel of destination a course had to be maintained for which the stellar compass was being used. To find ashore the latitudes of the harbours and to maintain the ship on the same parallel for long periods the stars were also utilised.

For this the pole-star was the most useful and accurate star to be used, because it was recognised by the experience of observation that

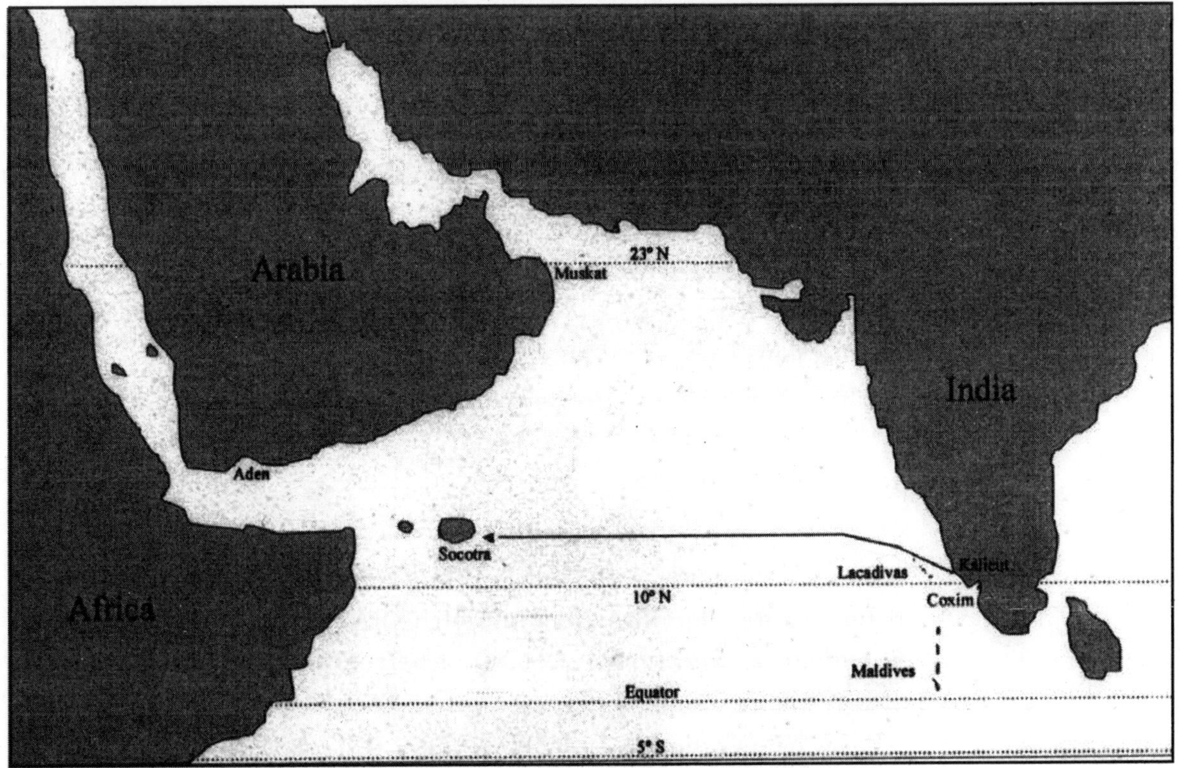

9.8. The north Indian Ocean area of navigation

all the heavens circled around a point very near this heavenly body. The height of this star over the horizon was the main reference in the area, as for example is clearly explained by Ibn-e-Majid in the *Faawid.* It was visible from around 6^0 N., covering all the area navigated from this latitude to the north.[13] Sometimes the heavens were obscured by clouds, the positions of other stars relative to the pole-star were used for parallel navigation being the height of the pole-star on its inferior meridian passage.[14]

This last situation is represented in figure 9.9 where the Little Bear is represented with the pole-star in its tail, the movement of the heavens counter clockwise for an observer looking north, the celestial pole and the polar distance of pole-star during the beginning of the 16th century.

So, having already seen the way to find direction, the remaining operation would be to measure with an acceptable accuracy the height of a star above the horizon.

The Kamal

It seems that the first approach to this problem has been the measurement of this angle with something that could be practical and relatively uniform. The more intuitive instrument has been the human body. So soon sailors found that a stretched arm with the four fingers

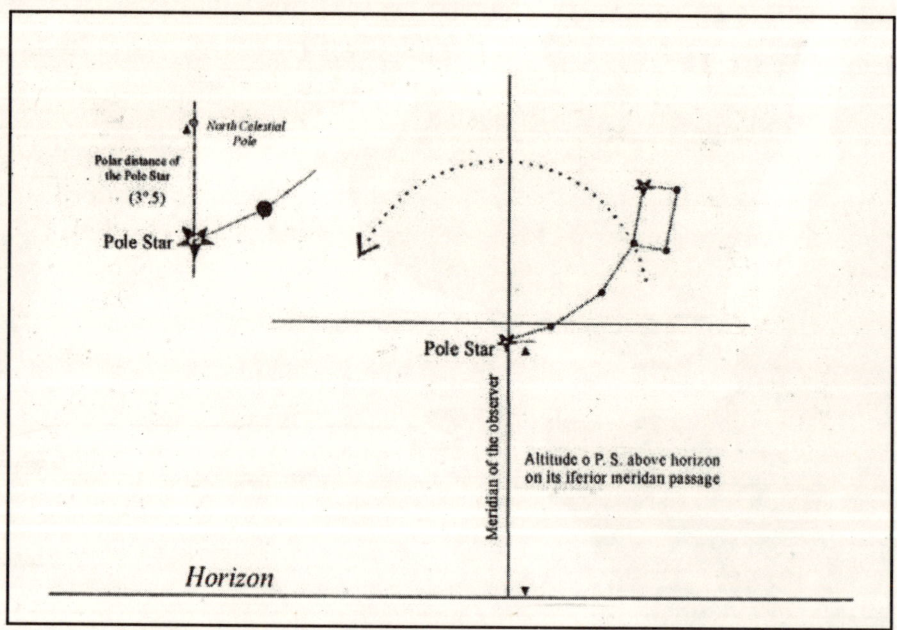

9.9. The Little Bear, the pole-star and the main operation needed for navigating along a parallel

of the hand perpendicular to the line of sight could be a reference for measurement of an angle.

With the little finger in coincidence with the horizon and one, two or three fingers more superimposed to it, as the altitudes to be measured were very small, an altitude of a star could be measured. If the altitude exceeded four fingers it was possible to estimate how many fingers more were necessary to measure the altitude of the star.

As the angle subtended by the fingers of a human body are more or less constant because there is a proportion between the fingers and the length of the stretched arm, this simple instrument was probably the first instrument in the history of navigation, after the lead line.

Measurements taken by the stretched arms, height of the forefinger and height of four fingers with the expectation of the thumb of nine persons and angles calculated by simple trignometry are shown in table 9.4.

Table 9.4

Length of arm (mm.)	*Forefinger height (mm.)*	*Forefinger's angle (⁰)*	*Four finger's height (mm.)*	*Average of four finger's angle (⁰)*
730	21	1⁰37′	78	1⁰31′
650	23	2⁰00′	82	1⁰48′
645	19	1⁰41′	72	1⁰36′
630	18	1⁰37′	72	1⁰37′
660	21	1⁰48′	73	1⁰36′
600	21	2⁰00′	79	1⁰52′
660	23	2⁰00′	76	1⁰38′
700	22	1⁰48′	80	1⁰38′
640	20	1⁰47′	74	1⁰39′
	Total Average	1⁰48′	Total Average	1⁰39′

It is very interesting to verify that the results have not a great discrepancy in the measurements. The last column is the angle subtended by 4 fingers divided by 4, which corresponds to the angle subtended by a finger with the average height of the 4 fingers.[15] The total average is $1^0 39'$, value that will be very interesting to analyse later.

This system originated the Arab or Indian Ocean unit of measurement of angles, which is the finger or *isba* in Arabic. In the beginning sailors probably were not preoccupied in having a uniform unit, or at least did not converted it to any constant unit. As soon as navigational activities increased, it was found useful to standardise the finger measurement and the value agreed by scholars now gave to the *isba* the value of $1^0 37'$.

Ibn-e-Majid, who says in his First Faida that"... the distance between two rhumbs is 7 *isba* and between each two mansions 8 *isba* making a total of 224 *isba*.' (Tibbetts 1971: 76, 314-315), admits a constant value for the *isba*, which will be $360^0/224 = 1^0 36'$.[16]

James Prinsep gave very interesting information about the evolution of the simple instrument that followed this anatomic way of measuring angular altitudes. His article published in 1836 in the *Journal of the Asiatic Society* is very important to clarify the situation (Ferrrand 1928: 1-24).

According to Prinsep this anatomic way was first followed by the use of nine tablets of wood, the first and smaller one with four horizontal equidistant divisions, each one of one *isba*. The other 8 tablets, of increasing size, represented 5 to 12 *isba*.[17]

These 12 tablets were all connected by one string, which had the main purpose of maintaining all the tablets together. In fact, the principal of the measurement was that the tablets, which had a uniform height, could substitute the fingers for measuring an angle with the arm completely stretched, without the need to superimpose fingers. So the navigator had always with him the possibility of measuring an angle from 1 to 12 fingers more accurately than with his fingers.[18]

Prinsep described the *modern* instrument, brought to him, by a sailor of the Maldive Islands as: 'a small parallelogram of horn (about two inches by one) with a string (or a couple of strings, in some instances), inserted in the centre. On the string are nine knots. To use the instrument for taking the height of Polaris, the string is held between the teeth, with horn at such distance from the eye, that while the lower edge seems to touch the oceanic horizon, the upper edge meets the star: the division or knot is then read off as the required latitude' (Ferrand 1928: 2).

Luciano Pereira da Silva published an excellent work in 1924 about this subject. The objective of the work was mainly to comment the meeting of Vasco da Gama with the Muslim pilot in Melinde and the instrument that he showed to him consisting of three tablets (Da Silva 1924: 31-41).

Trying to understand the instrument he finally found in the Ethnographic Museum of Hamburg of two instruments called *Kamal* (fig. 9.10), from where he got a drawing. The instrument on the left was offered to Pereira da Silva by an Hindu pilot Captain Doher of the Steamship line of Calcutta in 1892.

The instrument is in wood of 4 mm. thickness and has 6.65 x 4.8 cm. size. The rope has 16 knots, which correspond to the height of the

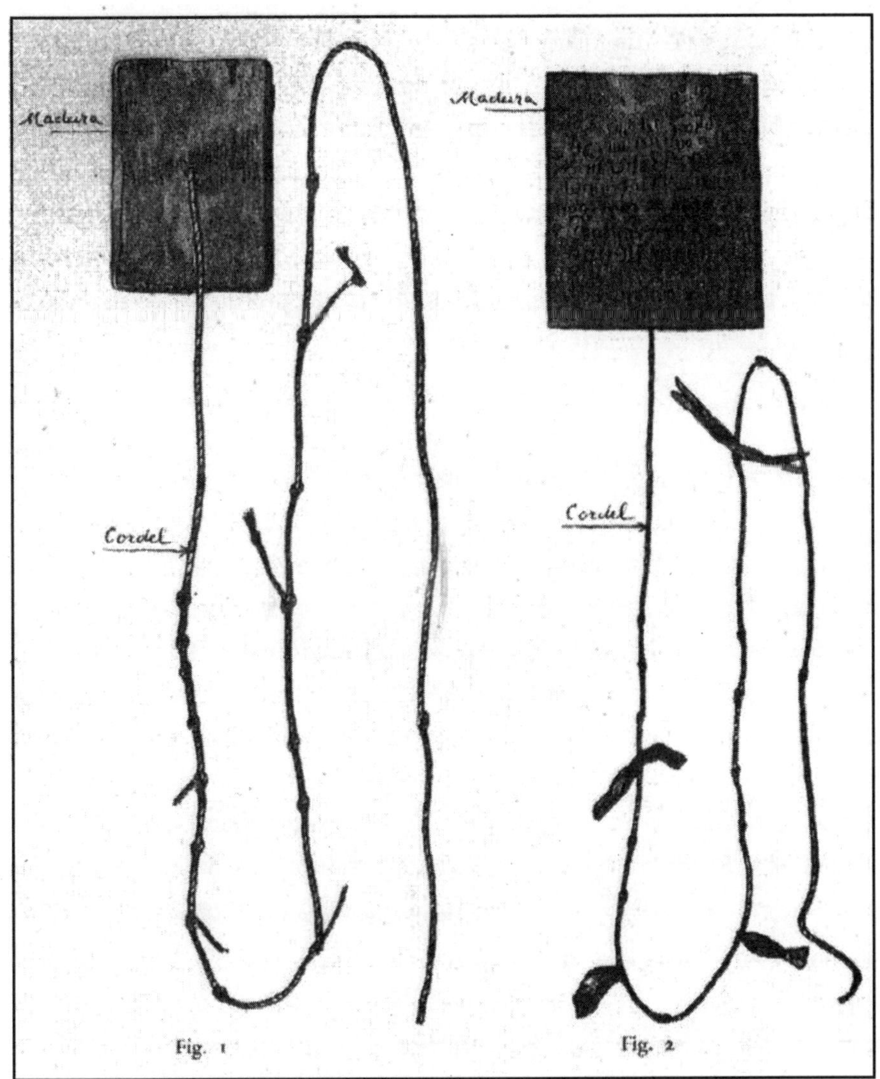

9.10. Two examples of *Kamal*

pole-star in 16 harbours of the Bay of Bengal (Da Silva 1924: 32). The instrument in figure 9.10 is similar in nature, have in one of its faces a list of harbours in Tamil. The designation *kamal* to this simple instrument is probably modern. Tibbetts comments that 'the measurer's instrument is never called this by the navigators' (Da Silva 1924: 315-316).

Experiences with replicas of the Kamal

Experimenting techniques of the ancients is very important to understand history. Replicas of the *kamal* were made on the basis of the description by Luciano Pereira da Silva. It was made by pieces of hardwood of 4 millimeters of thickness and of 70 x 50 mm. being its width and height very near the 6.65 x 4.8 cm. of the original.

The principle of graduating the instrument with the appropriate knots, which now will represent *isbas,* and is equivalent to the principle of graduation of the cross staff, is explained in figure 9.11.

Taking into consideration the dimensions adopted for the instrument and the value of the *isba* of 1⁰37′ adopted by de Saussure,[19] the table 9.5 has been produced.

Graduations for the *Kamal* used in its vertical position (values for the 35 mm.) and horizontal position (values the 25 mm. column) were obtained.

Instead of one instrument with two strings attached with the two different graduations the author used the other string with the different graduation in another piece of wood.

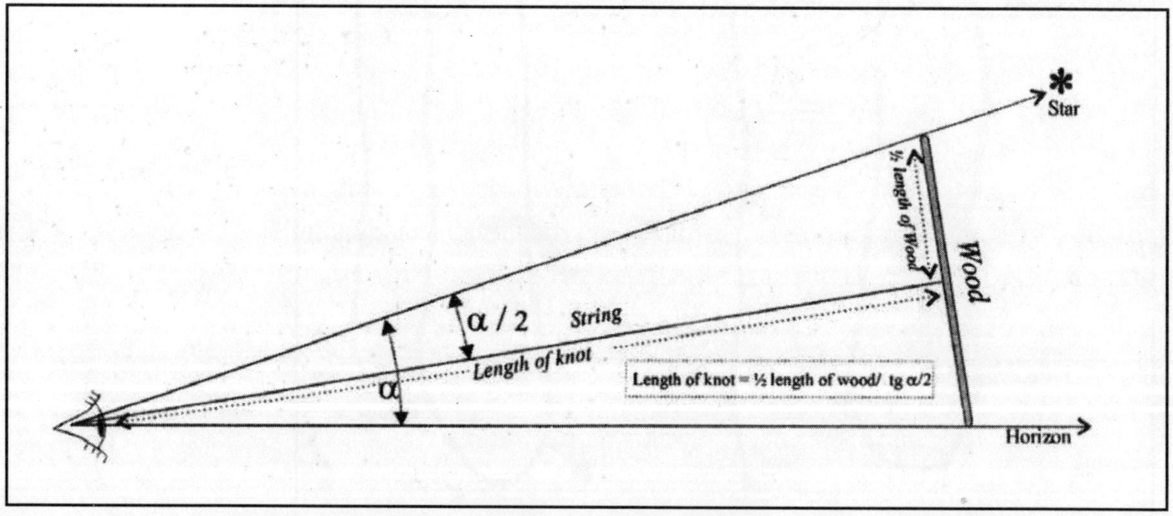

9.11. Geometrical principal of the *Kamal*

It is clearly seen that the interval among graduations has not a linear variation but in accordance with the variation of the co-tangent of the angle.

Table 9.5

Isba (97′)	*(1/2 Isba) (°, ′)*	*1/2 width (mm.)*	*Length of the knot (mm.)*
1	0°43′	25	1998.6
2	1°37′	25	885.8
3	2°25.5′	25	590.3
4	3°14′	25	442.5
5	4°02.5′	25	353.8
6	4°51′	25	294.6
7	5°39.5′	25	252.3
8	6°28′	25	220.6
9	7°16.5′	25	195.8
10	8°05′	25	176.0
11	8°53.5′	25	160.0
12	9°42′	25	146.3
		1/2 height (mm.)	
1	0°43′	35	2532.92
2	1°37′	35	1240.09
3	2°25.5′	35	826.46
4	3°14′	35	619.55
5	4°02.5′	35	495.35
6	4°51′	35	412.49
7	5°39.5′	35	353.25
8	6°28′	35	308.79
9	7°16.5′	35	274.17
10	8°05′	35	246.44
11	8°53.5′	35	223.72
12	9°42′	35	204.76

The author made the experiences aboard sailing boat at sea last month. Two stars of low altitude and with a magnitude similar to Polaris were chosen for observation and for all the needed information computer program Starry Night was used. The aspect of the night sky during the two nights of the observation was schematically the one shown in figure 9.12. It was observed that the stars *Adhara* and *Diphda*, fulfill the desired conditions.

We sailed near the coast, being the wind light to moderate and some light swell from the south-west. There was moonlight, which was not a good thing for observation because the somewhat faint stars observed were not very clear. Ibn-e-Majid gave a big amount of suggestions for good conditions and procedures for observation and moonlight was not welcomed (Tibbetts 1971: 319-324).

The author made all the observations with the help of his wife and a friend. The instrument used is the one graduated for the upright position, because the altitudes were sufficiently high for it. The altitudes and the observations made on two consecutive nights are given in table 9.6. The heights of the stars are in *isba* and tenths, estimated by eye. Each reading is converted in degrees and minutes making each *isba* to 97' ($1^0 37'$). The correct value for the altitude of the star was obtained from Starry Night, introducing on the computer the co-

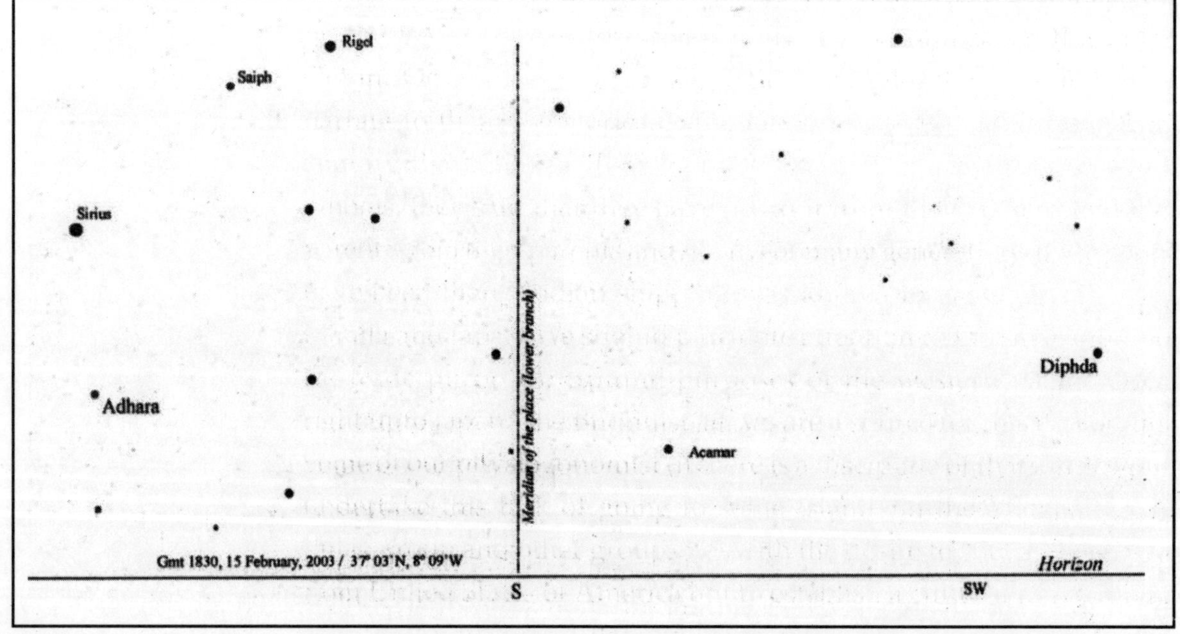

9.12. Some of the stars of the southern sky during the observations

ordinates of the place, height of the observer and time of the observation.

While studying all the collected data it was found that almost all the observations were uniformly too high, which corresponded apparently to any systematic error. After some thinking about this problem it was recognised that the string was held by mouth to make the observation and the graduation was calculated for the eye of the observer so the readings should only be correct if these two distances were equal (fig. 9.13).

With our observations there was a difference of around 15 millimeters between the two.[20] We were observing the star nearer than the reality and so we had to apply a correction to all the observations. The result of that is shown on the appropriate column and all the corrections were made graphically.

The errors found are very scattered in absolute value, but except for two values of more than one degree ($1^0 17'$ and $1^0 06'$) the other ones are acceptable and within the range of the errors. It is noteworthy that it was the first time that the instrument was used and with some training the errors will be lesser.

Besides that, there is a handicap for old people, which is long sight. The author had difficulty in focusing at the *Kamal*, which was some centimeters away and at the star and the horizon, at the same time. So, a young person will have better results and that would be checked in the future.

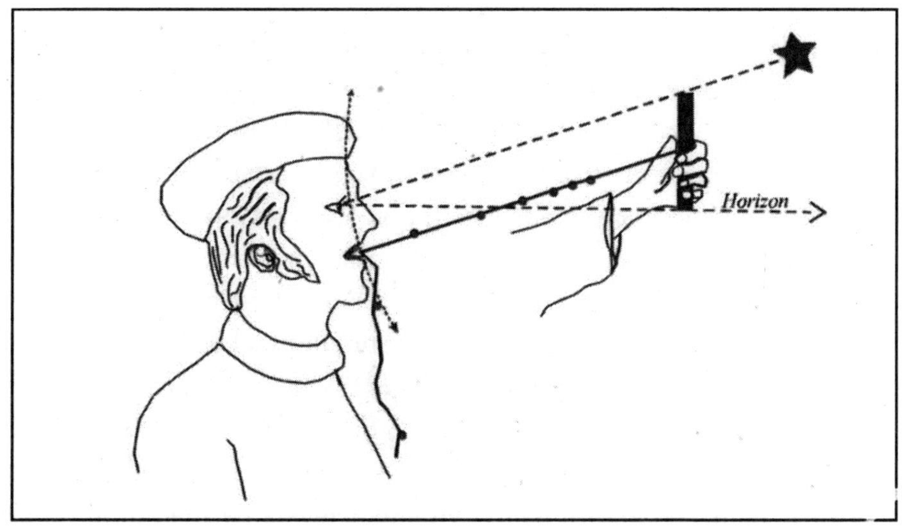

9.13. The *Kamal* and its use at sea

Table 9.6

GMT	*Star*	*Altitude (isba)*	*Altitude corr. (isba)*	*Altitude corr. (^{0})*	*Altitude Calc.(comp.)*	*Error*
			15th February 2003 (37^0 03′ N., 8^0 09′ E.)			
18 50 24	Adhara	7.8	7.5	12^{0}07′	11^{0}46′	- 21′
18 55 44	Adhara	7.7	7.4	11^{0}58′	12^{0}26′	+ 28′
19 01 24	Adhara	8.5	8.1	13^{0}06′	13^{0}08′	+ 2′
19 03 29	Adhara	8.4	8.0	12^{0}56′	13^{0}22′	+ 26′
19 05 43	Diphda	10.2	9.75	15^{0}46′	15^{0}57′	+ 11′
19 07 00	Diphda	10.1	9.85	15^{0}55′	15^{0}45′	- 10′
19 08 35	Adhara	8.9	8.65	13^{0}59′	13^{0}58′	- 1′
19 10 24	Diphda	9.8	9.3	15^{0}02′	15^{0}13′	+ 11′
19 12 00	Adhara	9.1	8.9	14^{0}23′	14^{0}22′	- 1′
19 16 32	Diphda	9.6	9.1	14^{0}43′	14^{0}14′	- 29′
19 17 33	Adhara	9.4	9.0	14^{0}33′	14^{0}59′	+ 26′
19 18 52	Adhara	9.4	9.0	14^{0}33′	15^{0}08′	+ 35′
19 20 07	Diphda	9.4	9.05	14^{0}38′	13^{0}39′	- 59′
19 21 00	Diphda	9.2	8.9	14^{0}23′	13^{0}30′	- 53′
19 21 55	Adhara	9.8	9.3	15^{0}02′	15^{0}28′	+ 26′
19 24 46	Adhara	10.0	9.75	15^{0}46′	15^{0}47′	+ 1′
19 27 26	Diphda	8.2	7.95	12^{0}51′	12^{0}27′	- 24′
19 31 14	Adhara	10.4	9.9	16^{0}00′	16^{0}28′	+ 28′
19 32 12	Diphda	8.0	7.8	12^{0}37′	11^{0}40′	- 57′
19 33 23	Adhara	10.6	9.95	16^{0}05′	16^{0}41′	+ 36′

contd.

GMT	*Star*	*Altitude (isba)*	*Altitude corr. (isba)*	*Altitude corr. (^{0})*	*Altitude Calc.(comp.)*	*Error*
19 34 52	Diphda	7.7	7.5	12^{0}07′	11^{0}13′	- 54′
19 36 13	Adhara	10.9	10.25	16^{0}34′	16^{0}58′	+ 24′
19 37 16	Diphda	7.1	6.95	11^{0}11′	10^{0}49′	- 22′
19 40 57	Diphda	6.7	6.45	10^{0}26′	10^{0}12′	- 14′
19 41 56	Adhara	11.0	10.6	17^{0}08′	17^{0}32′	+ 24′
19 44 06	Diphda	6.3	6.1	9^{0}52′	9^{0}40′	- 12′
19 47 07	Diphda	6.2	6.0	9^{0}42′	9^{0}10′	- 32′
		16th February 2003 (37^0 03′ N., 8^0 09′ E.)				
18 55 29	Diphda	11.1	10.65	17^{0}13′	16^{0}56′	- 17′
18 56 53	Diphda	11.0	10.55	17^{0}17′	16^{0}47′	- 30′
19 03 19	Adhara	9.4	9.0	14^{0}33′	13^{0}49′	- 44′
19 04 29	Diphda	10.0	9.6	15^{0}30′	15^{0}31′	+ 1′
19 06 06	Adhara	9.6	9.05	14^{0}38′	14^{0}08′	- 30′
19 08 36	Diphda	10.2	9.8	15^{0}50′	14^{0}52′	- 58′
19 09 37	Adhara	10.2	9.8	15^{0}50′	14^{0}33′	- 1^{0}17′
19 10 46	Diphda	10.0	9.6	15^{0}30′	14^{0}31′	- 59′
19 12 42	Diphda	9.5	8.95	14^{0}28′	14^{0}13′	- 15′
19 13 31	Adhara	10.0	9.6	15^{0}30′	14^{0}59′	- 31′
19 16 29	Diphda	8.7	8.2	13^{0}15′	13^{0}36′	+ 21′
19 17 25	Diphda	8.7	8.2	13^{0}15′	13^{0}27′	+ 12′
19 23 41	Diphda	8.4	8.0	13^{0}06′	12^{0}26′	- 40′
19 24 41	Adhara	11.0	10.55	17^{0}17′	16^{0}11′	-1^{0}06′

The Practical Use of the Stellar Compass

The interpretation of the stellar compass's practical use, based on the reading of some of the known books of the ancient Indian Ocean navigators, also the experimental voyages of David Lewis, the idea briefly stated by de Saussure and experience of the author, is materialised with this simple instrument.

It is a disc of wood with one handle, the stellar compass being bonded to its surface. Two small pinulas will allow sighting the pole-star.

After that the course previously wanted relative to the north will be found. Any star in the bow or the stern will be used to maintain the desired course during long periods.

The instrument can be mounted on the rail and used when necessary. If clouds cover the pole-star other stars will be available for finding direction, although they were not the ones used in the compass. The stars of the compass were mainly used as the winds of the magnetic compass; that is they will only have the purpose of defining a course with a known angle relative to the real direction finder, the pole-star.

The correct course by day will be maintained having as a reference the sun, the swell and the wind. Everyone knows, for example, that nowadays cruising sailors use the wind vane that maintains a course relative to the wind. In fact, as the wind in the high seas maintains its direction for long periods even electronic auto pilots are utilised maintaining the course relative to the magnetic compass. So, simple methods were chosen according to the needs of the sailors.

NOTES

1. There is similarity of conditions between these two oceans; it is why I connect one to each other. Besides that the sailors of both areas should have been in contact, and developed similar techniques, although with slight differences, in consequence of some differences in their respective environments. It is this also for example the opinion of Tibbetts, expressed in a paper in 1979 he says that: "My own opinion is that in both areas [Indian and Pacific Oceans] there must have been a general mutual awareness of the way that navigational problems were solved. In the days when the Malaysian peoples travelled to Madagascar, the Pacific peoples may have been in contact with Southeast Asia and similar navigational systems may have operated on both Malayan archipelago." (Tibbetts 1971: 13).

2. We can also consider A as a continent. We can for example imagine a route from the African eastern coast to the island of Socotra.

3. The formula for finding the bearing Z of any celestial body at rising or setting is: $\cos Z = \mathit{sen}\ \delta \sec \varphi$. As the latitude is always less than 90^0 its secant is always positive, so cos Z will always be of the same sign of sen δ. So the bearing Z will always be of the same name as the declination. This is also very clearly shown on the excellent work of David Lewis (1994). In its Appendix I is a clear study about this subject. To the same conclusion arrived Pedro Nunes (1940), the well-known Portuguese mathematician of the sixteenth century and cosmographer major of the Portuguese crown, when answering to the many questions of Martim Afonso de Sousa, a captain of an Armada recently returned from Brazil. To the question of finding in 35^0 S. during sunrise and sunset the bearing of the sun SEl/4E and SWl/4W, respectively, when the sun had its most southern declination (in the Tropic of Capricorn), the same bearing that he had already experienced in the same latitude in 35^0 N., Nunes (1940) explained it clearly, not only with text and images but also mathematically.

4. Ibn-e-Majid also says, when explaining the compass in his First *Fahida*, that "The rhums are 32 in number as they are divided on the ship and they make even numbers of them without using odd numbers." (Tibbetts 1971: 76).

5. Saussure has made the identification of the stars with the help of a mobile globe of the Observatory of Geneve, which had sufficient accuracy for the purpose. This was made around 1920, but nowadays we have the enormous help of astronomic computer programs, which have great accuracy and flexibility. They allow us to materialise the heavens and its celestial bodies in any year and date we want, having easily not only the possibility of seeing the movement of the bodies in any place on earth, but also to have, among many other useful information, their coordinates. One of the programs I have used for the calculations is Cyber Sky.

6. De Saussure bases his identification on the work of Antoine d' Abadie, a French scientist that published his work in the *Journal Asiatique,* 1841 and from which he makes a long quotation.

7. On account of the translation of the earth around the sun, this celestial body moves apparently around the earth as seen from the earth. So, the stars seem to move relatively to the sun (the projection of the apparent movement of the sun around the earth is the called the ecliptic). One complete revolution of the earth around its axis

and having as reference a star (sidereal day), is completed before the earth completes its rotation around the sun (solar day), because it is moving in space. As 360^0 translation takes a little more than 365 solar days the retardation of the solar day in relation of the duration of the sidereal day is a little less than one degree of arc ($360^0/365$) or about 4 minutes of time. So, we can say that the stars rise every day 4 minutes earlier. This will also means that in a certain place on earth and at the same day of the year we never have the same star in the same position on the visible celestial sphere.

8. David Lewis gives an indication of optimum altitudes of steering stars, as per his observations at sea and changing information with the local seamen. One of the indications says that a star can be used from near the horizon till around 15^0 altitudes. A sailor from the Pacific says that in Tonga they suggest to "steer by a star until it has reached a height the same as the sun has at 10 a.m.". This means in tropical areas an altitude of around 50^0, which is too high (Lewis 1994: 97-98). My personnel experience, navigating in the Atlantic in inter-tropical areas in the square-rigger *Sagres* of the Portuguese Navy, shows that a star of the magnitude used in the stellar compass can be seen above the horizon with an altitude above around 6^0. Its usefulness for direction can go to 20^0 of altitude. It is useful to recognise that my information and the one of David Lewis are related to the Pacific and the Atlantic, where atmospheric conditions are not as good as the ones in the Indian Ocean.
9. Not only the above-mentioned scholars who studied Arab navigation tried to find the date when the bearings agreed with the equidistant rhumbs. An excellent work of Marina Tolmacheva tried to interpret this problem with an extensive use of different calculations for different epochs. Tolmacheva gave finally her stellar compass composition for the year 300-310 BCE, "... resolved in retrospect by a selective process bringing together a star and a rhumb showing best agreement of the present and projected azimuth" (Tolmacheva 1977).
10. Ibn-e-Majid was quoting the passage in one of his previous works also referred by Tibbetts (1971: 74-75) who mentioned and listed 40 of those works.
11. In this work, the author tries to demonstrate that Ibn-e-Majid was not the pilot of Vasco da Gama. One of the reasons is the inclusion of false verses in this known work of Ibn-e-Majid.
12. Ibn-e-Majid explains all of the strategies employed by the pilots to choose the best season for crossing the Indian Ocean from all the

harbours and directions (Tibbetts 1971: 225-242), which includes Tibbet's translation of the Eleventh Faida, *Monsoons and Connected Matters*. Commenting the anchorage in harbours after the crossing Ibn-e-Majid says: "Because of this [the strength of the Dabur wind], intelligent men never make this journey during the 3 months or 90 days for then it is a gamble" (Tibbetts 1971: 227). The 90 days corresponded to the months of June, July and August.

13. At least, because the Indian Ocean atmosphere is much more pure than the Atlantic one. See note 7 above.

14. In all the texts of the pilots (Ibn-e-Majid or Suleiman al-Mahri, for example), it is clearly referred that this position of the pole-star was the main reference. All the other star altitudes were reduced to this same reference. All the scholars referred recognise also this principle.

15. This anatomic measure was called by the pilots a *dhubban* and Ibn-e-Majid explains this saying that "Four *isba* is one *dhubban*, the *dhubban* being the place occupied by the sword handle in the palm of the left handle as far half the nail of the little finger of the left palm" (Tibbetts 1971: 76). This explanation is not very clear to me. De Saussure also gives an interpretation for this that is also not very clear to me. I think that Ibn-e-Majid tried to explain what was the horizontal reference to measure vertically an angle with the 4 fingers, because their height decreases to the extremity of the fingers. I think that the reference was the vertical aperture of the 4 fingers passing through the middle of the little finger (Ferrand 1928: 162).

16. The value of the *isba* is also clearly explained in the *Livro de Marinahria de Andre Pires*, where this Portuguese Pilot refers in his manuscript (written between 1517-1540), that "Se caso for que achares alguma carta de mouros e a quiseres graduar a nossa usanca, tomaras 5 polegadas e reparti-las-as em 8 partes, que sao 8 graus….." This means that one *isba* is $8/5^0$ or 1^0 36′ (De Albuquerque 1963: 135). He studies very deeply the precious information about the techniques of navigation in the Indian Ocean included in the manuscript.

17. It was the *Muhit* of Admiral Sidi Ali Celebi that described this instrument. This nautical guide was produced by this Turkish Admiral who was in charge of conducting the Turkish fleet in its return to Egypt after being attacked by the Portuguese in the mouth of the Persian Gulf being after scattered by a storm. While anchored in Gujarat he produced this important nautical guide based mainly

in the Ibn-e-Majid and Sulaiman al-Mahri works, although many comments of his own were introduced.

18. The explanation given by Prinsep or his interpretation of the *Muhit* is very confusing. The following text, which I am not sure if it is a quotation of the translation of the *Muhit*, says: "The method of taking the measure is as follows: You take the table with the left hand and the thread that passes through the middle in the right; you stretch your left hand firm and take the elevation which gives..." I do not understand the need to take the thread with the right hand unless for separating the undesired tablets from the one that was being used.

19. We could adopt the value of 1^0 36′ as was the one used by the navigators, but this is irrelevant to the experiences, because I will find the errors in degrees and the graduations will be in any way proportional, although each knot will have a slight difference in position relative to the piece of wood (Ferrand 1828: 162).

20. It is necessary to note that the image in the eye is formed on the retina, which is the anterior part of the visual organ, so the correction can probably be higher than 15 millimetres. About this problem and when dealing with observations at sea with the cross staff, I studied deeply this problem (Pereira 1994: 165-192; 2000: 11-18; 2001: 201-220).

BIBLIOGRAPHY

Braz, C.A. Moura (1962) *O Encontro das Marinhas Oriental e Ocidental na Era dos Descobrimentos*, Sociedade de Geografia, Lisboa.

D'Abadie, Antoine (1841) *Journal Asiatique*.

Da Silva, Luciano Pereira (1924) Kamal, Tabuas da India e Tavoletas Nauticas, In: *Obras Completras de Luciano Pereira da Silva*, Vol. III, Agencia Geral do Ultramar, Lisboa.

De Albuquerque, Luis Mendonca (1963) *O Livro de Marinharia de Andre Pires, Junta de Investigacoes do Ultramar*, Lisboa.

De Mendonca, Henrique Lopes (1898) *O Padre Fernado Oliveira e a Sua Obra Nautica, Memoria, comprehendendo um estudo bigraphico sobre o afamado gramtico e nautographo, e a primeira reprodurcao typographica do seu tratado inedito Livro da Fabrica das Naus*, Academia Real das Sciencias, Lisboa.

De Morais, J. Custodio (1960) *Determinafiio das coordenadas geograficas no Oceano indico pelo pilotos portugueses e pilotos arabes no princfpio do seculo XVI*, Universidade de Coimbra, Coimbra.

FERRAND, Gabriel (1928) *Instruction Nautiques et Routiers Arabes et Portugais des XV erne et XVI erne Siecles*, Tome III, Librairie Orientaliste Paul Geuthner, Paris.

GODINHO, Vitorino Magalhaes, Navegacao oceanica e origens da nautica astronomica, *Ensaios*, Sa da Costa, Lisboa.

KELSALKA, William (1987) *An Ocean in Mind*, University of Hawaii Press, Honolulu.

KHOURY, Ibrahim (1983) *As-Sufaliyya - The Poem of Sofala by Ahmad Ibn Majid*, Junta de Investigacoes Cientificas do Ultramar, Coimbra.

LEWIS, David (1994) *We the Navigators — The Ancient Art of Land Finding in the Pacific*, 2nd ed., University of Hawaii Press, Honolulu.

NUNES, Pedro (1940) *Obms*, 4 Vol., Academia das Ciencias, Imprensa Nacional, Lisboa.

OLIVEIRA, Fernando (1992) *Livro da Fabrica das Naus*, Academia de Marinha, Lisboa.

PEREIRA, Jose Manuel Malhao (1994) Experiencias com Instrumentos da Epoca dos Descobrimentos, In: Mare Liberum no. 7, Comissao Nacional para as Comemoracoes dos Descobrimentos Portugueses, Lisboa.

PEREIRA, Jose Manuel Malhao (2000) *Experiencias com Instrumentos e Metodos Antigos de Navegacao*, Academia de Marinha, Lisboa.

PEREIRA, Jose Manuel Malhao (2001) Norte dos Pilotos, Guia dos Curiosos, de Manuel dos Santos Rapos, *Um Livro de Marinharia do Seculo XVIII, Estudo Critico*, Tese de Mestrado, Lisboa.

ROLANDO, A. (1963) Laguardia Trias, *Las Mas Antiguas Determinaciones de Latitude en el Atlantico y el Indico*, Instituto Historico de Marina, Madrid.

TIBBETTS, G.R. (1971) *Arab Navigation in the Indian Ocean before the Coming of the Portuguese*, The Royal Asiatic Society of Great Britain and Ireland, London.

TIBBETTS, G.R. (1979) *A Comparison of Medieval Arab Methods of Navigation with those of the Pacific Islands*, Centro de Estudos de Cartografia Antiga, Junta de Investigacoes do Ultramar, Lisboa.

TOLMACHEVA, Marina (1977) An Analysis of the Arab Wind Rose, University of Toledo, paper presented to the 1977 Convention of the Middle East Studies Association.

Comment

 *Lotika Varadarajan*

This paper marks a welcome departure from most scholarly presentations on the subject. Not only has the author provided an astronomical background to the usage of the *Kamal* but has also fashioned models and experimented with these to measure stellar altitude at sea. This approach would be particularly useful to mariners of Lakshadweep who are rapidly forgetting their own expertise in such areas. Such methodologies also need to be replicated in this country to derive maximum benefit from ethnological source material, which in the Indian context can provide valuable insights into customary procedures of the past.

The background to this paper is the East - West encounter at the plane of knowledge systems, a necessary concomitant to the Portuguese discoveries. The Mediterranean, which since remote times, had served as the bridge linking ancient Greece and Rome with Asiatic centres of civilization, was to be the base from which the Portuguese were to expand their maritime ventures into the Atlantic. As such there were entries to Oriental usages with which the more northerly Atlantic seafaring groups were much less attuned.

From the intellectual and cognitive points of view Portugal had been greatly enriched by the Arab-Moor presence in *Al Andalus.* In the area of maritime endeavour it was through Arab intervention that knowledge of the magnetic needle was added to the Mediterranean Sea compendium based on tidal movement. Circa 1200 CE under the influence of the sailors of Amalfi, another innovation is recorded - the magnetised needle was placed on a wind rose of thirty-two rhumbs. Circa tenth century CE the Arabs introduced the planispheric astrolabe into Europe. The Portuguese were to transform this into the sea astrolabe circa 1480 CE.

The Jews, active in the textile trade, also made distinct contributions particularly in map making and the earliest reliably documented reference to a Portolan chart is dated by Tony Campbell to 1270 CE (Harley and Woodword 1987). Portolan charts did not have parallels and meridians. After the fifteenth century the coastline was indicated by thirty-two magnetic rhumb lines. It was to this base that further refinements were added as the Portuguese inched their way across the Atlantic hump of Africa towards its southern point of termination at the Cape. It was at this point that the waters of the Indian Ocean mingled with those of the Atlantic.

The Portuguese were aware that once the Cape had been rounded they would enter into the well -traversed seaways of the Indian Ocean. The navigational experience which the Portuguese had garnered and recorded in the course of their attempts to find a passage to India by coursing down the African coast had taught them a great deal about stellar navigation and simple methods of direction finding at sea.

Unlike the temperate zone, which had shaped European seafaring traditions, commercial traffic in the Indian Ocean in which the Persians, Arabs, Indians, Malays and the Chinese had participated, had a more facile contour. The geographical setting within the tropical zones brought into play a different set of circumstances - predictable seasonal changes presaging reversal in wind and current. The visibility of different sets of rising and setting stars within a single night facilitated latitudinal night time sailing. Malhao Pereira has illustrated these facts very effectively by means of charts. However, he has not mentioned that longitudinal deviation could also be accommodated within this nautical tradition, which had obviated the use of compass and maritime chart.

Unlike the uninterrupted expanse of water characteristic of Atlantic crossings, sailings in the Indian Ocean were segmentalised. The Persians and Arabs had a particular comprehension of the Red Sea and Persian Gulf areas. The Indonesians, who had peopled Madagascar at the turn of the Common era, were acknowledged for their skill in sailing and are said to have developed a complex cartographic tradition. Chinese expertise tended to be restricted to the area east of Malacca. Gujarati presence was linked to Java - Sumatra while the Cholas (who could be broadly identified with the group known as "Kling"), were more visible in Malaysia. As the Arabs ventured further eastwards they synthesised and recorded the existing nautical expertise of the Indian Ocean. Arabic treatises became important sources of reference for European scholars. The latter tended to overlook the synthetic nature of these works and attribute the totality of information to the Arabs.

The Javanese and the Chinese had developed an azimuthal compass rose in which the horizon was divided into twenty-four directional units. The thirty-two rhumb Arab compass rose, to which Malhao Pereira makes a reference, was of composite origin and differed in orientation from area to area being modified in conformity with local conditions. In Lakshadweep the thirty-two points of compass directions are associated both with constellations bearing Arabic nomenclature as well as with *yogataras* associated with Hindu astronomy. Available evidence both in Gujarat as well as in Lakshadweep point to the existence of this sidereal azimuthal compass rose only in chart form to be memorised and used as a mnemonic device rather than as technological tool as is suggested by Malhao Pereira.

A brief review of South Asian developments becomes necessary at this point. Indian achievements in mathematics and astronomy in antiquity are well attested. It was this tradition, which absorbed Greek inputs, consequent to the invasion of Alexander of Macedon. Circa 5th century CE. Aryabhata I, born 476 CE, determined the fixation of the pole-star and other prominent stars in terms of *desantara* (longitude) and *akshamsa* (latitude). The term for distance covered in a given run in sailors' vocabulary is *chamam* or *yam*. This necessarily encapsulates functional aspects of the meridian. Further, Varaharamihira, born 505 CE, provided the method of calculation of longitude on the basis of observing the timings of eclipses at two different places.

Between 753-774 CE Sanskrit and Pali material was translated into Arabic under the aegis of Caliph Mansur. Haroun Al Rashid continued this enterprise between 786-808 CE. It was through this route that Ptolemy entered the Arab world of learning. Since the method of transmission in South Asia was very different from that which was customary in Europe, the Greek input remained an identifiable component within the Indian astronomical tradition.

As has been pointed out by Malhao Pereira, the genesis of Indian Ocean instrumentation is to be sought in the basic unit of finger measurement. However, the progression did not stop at the *Kamal* but was carried further to the *balisti* (the balestilha in Portuguese and cross-staff in English terminology) and *kolepalagai* (the term used in Lakshadweep), or the so-called Davis Back Staff/ "Davis Quadrant" (English) or Double Arc Quadrant (Portuguese) (Varadarajan 1998: 691-693). The term "Davis" quadrant was used by the English as it was attributed to the Englishman, Captain John Davis, known to have been active in the Indian Ocean circa 1597 CE. If our concept of progression is accepted, and this in no way negates the improvements and refinements introduced by the Europeans, the logical conclusion would be to place the origin of these instruments in the Indian Ocean. It is the *kolepalagai,* which furnishes the most interesting data. Indian terrestrial methods of time keeping have been based on the shadow instrument, the *sanku* or the vertical gnomon, in which the shadow falls along a horizontal axis. In the *kolepalagai,* however, the shadow falls along a vertical axis. The 6th to 7th century Sanskrit text, *Vrddhavasistha Siddhanta* (III.61-62), has described this horizontal gnomon and has provided formulae for calculating lengths for shadow and hypotenuse. This information relating to the *kolepalagai* would certainly carry important implications for the other associated tools and methods.

The Asian spread of progression from finger width to instrumentation is also to be noted in the information provided in the Chinese Wu Pet Chih, Records of Military Preparation, dated 1628 CE. Among other material these also included a reproduction of the charts used in the Cheng Ho expeditions, 1405-1433 CE. Apart from compass and chart, the Chinese utilised instruments similar to the *Kamal* and *balisti* for measuring stellar altitude in units of chili, finger widths (Jun 1988: 83-96; Needbam 1978: 560-584).

Placed within this context the communication of Malhao Pereira is of great interest as the experience of a sailor is brought to bear on the material presented. The tables and computer projected medieval star positions in relation to the stellar compass also provide fascinating data.

Having stated this, a few additional comments need to be made. The list of scholars could be expanded by including the articles by David King, Regis Morelon and Henri Grosset Grange (Roshdi 1996). While the work of Tibbets remains the most accessible one available to the English speaking world, the translation is known to be faulty and should be used along with the publications of the late Ibrahim Khoury (1971: 249-386; 1985-86: 163-276; 1987-88: 191) of Institut Francais d' Etudes Arabes de Damas.

BIBLIOGRAPHY

HARLEY, B. and WOODWORD, D. (ed.) (1987) *The History of Cartography*, Vol. 1, Chicago-London.

KHOURY, I. (1971) La Hawiya de Ahmad b. Majid, *Bulletin d'Etudes Orientales*, Journal of IFEAD, XXIV.

KHOURY, I. (1985-1986) Les poemes nautiques d' Ahmad Ibn Majid, 2e partie: les poemes a rime unique, *Bulletin d'Etudes Orientales*, Journal of IFEAD, XXXVII-XXXVIII.

KHOURY, I. (1987-1988) Les poemes nautiques d' Ahmad Ibn Majid, 3e partie, Les poemes a rime variable, *Bulletin d'Etudes Orientales*, Journal of IFEAD, XXXIX-XL.

NEEDBAM, J. (1978) *Science and Civilization in China*, 4, Part III, Cambridge.

ROSHDI, Rashed (ed.) (1996) *Encyclopaedia of the History of Arabic Science*, Vol. I, Routledge, London and New York.

VARADARAJAN, Lotika (1998) The Question of the Anti-Meridian-Economic and Technical Aspects, In: A.T. de Matos and L.F.R.R. Thomaz (ed.) *A Carreira da India e as Rotas dos Estreitos*, Proceedings of VIII International Seminar on Indo-Portuguese, History Angra do Heroismo.

YI JUN, Zheng (1988) As tecnicas de navegacao nas armadas de Zheng He e a sua contribuicao para a ciencia nautica, In: *Actas do Seminario Ciencia Nautica e tecnicas de Navegacao nos Seculos XV e XVI*, Instituto Cultural de Macau, Centro de Estudos Maritimos de Macau, Macau.

Comment

B. Arunachalam

The star topics and the *Kamal* have been fascinating to me. For the last two decades I have been chasing them and there are a number of points I would like to stress upon as Dr Lotika particularly pointed out and let us remember the sailors never worry about the source from which they were getting the methods. What is important to them is the availability of the method and availability of a tool. This brought in two important results one is the diverse guarding of the wisdom that the individual community has. At the same time there is also a very keen desire to know how the others were doing it, for example when Vasco da Gama came into the Indian Ocean he would not have taken the Malviya Kanika whoever he may be a Gujarati, Omani or a Nigerian. It doesn't matter much. But what he was keen to know is what was he trying to do or how he was going to do this. Only when he gets convinced he takes him as his pilot. So far as the *Kamal* is concerned it is a highly versatile hand tool, and over a period of centuries it is evolved through successive stages. It is not that there was one single *Kamal* coming up at a particular point of time with a particular utility. It is not so. It was an instrument, which got progressively evolved, improved, perfected and made finer and finer. So that if you look into a *Kamal* of 17th century and if you look into the what you call a *Kavali palagahi* of the 11th or the 12th centuries you will find that there is a marked difference. I have come across not less than eight proto types of this equipment whatever name by which it may be called. This is because they belong to different period of time and it has been progressively evolving. About four years back I saw at Chetlat a very interesting *Kamal* which was not only using the string as has been rightly pointed out on the nasal bridge was fitted, what was more significant was it was using the blank stem cells for subdivision of a unit. The unit that is being used all over the Indian Ocean, it may be the *Isba* of the Arabs, it may be the Kerala of the Tamilians it may be even in China the Chi you must have heard about the famous 'Upaipechart' is all based upon the finger measurement. Now, the finger measurement essentially started as direct observations using the God given instrument the human hand. I have seen this even today being used on the Coromandel coast between Pondicherry and Cuddalore.

The second point which must be kept in memory is that a hand tool is a very small hand tool. That is the very value of it so that it can be very easily carried, it can be easily used even on a rocking boat. The result of it is its range is very limited. It can't measure from zero to ninety at a time. It can measure only a limited area, it can measure only 15 fingers. That's quite substantial for them and the other important point to note with the *Kamal* is the use of the *Kamal* for the purpose of latitude measurement is perhaps a very late stage. If you look to for example at Kamrives paper or the Maldivian sailor's paper it is not the *Kamal* that they describe is one, which is not in uniform intervals distance. It is fixed for specific ports. It was more for port identification rather than for latitude measurement. Latitude measurement is a corollary which comes out of it. So, that is one important usage which you must keep in mind. Similarly, it could be used in other great circle facility also.

The third point which I want to make is about the star compass. The Star compass has become versatile to such an extent in the Indian Ocean traditions that they would not trust the magnetic compass when it was introduced. They would not trust it because they felt that star compass is much more accurate although there are differences in the methodologies and in the stars used say between the Arab shores, the Mediterranean down to the Indian shore and even the east coast even the stars might be slightly different, many of the stars are common but there are few which are unique for specific traditions.

Another point which you must remember about the star compass is that the stars which are chosen the 16 of them which are used both for the rising side and for the setting side, these are stars which are not exactly on the correct azimuth as you would consider it for the present measurement. They are only indicative and secondly they knew that no star remains in the horizon all over, it shifts in position. And as it shifts in position azimuth changes, therefore you need other stars in the same azimuthal sectors. So, these are aspects which are to be kept in mind though a lot of interest has now come up from various scholars. Still we have to go a very long way to understand fully the mechanism and there is nothing wrong with the traditional methods although they may not have the same instrumental precision.

10

The Indian and Pacific Oceans in World Maritime History

Joseph J. Krol

The paper delves into the history and current activities of US Navy in the Indian Ocean.

The Growing Economic Importance of the Indian Ocean

The Indian Ocean today is of enormous and growing importance, not only to India, but also to the many countries that depend on it as a lifeline of trade and communications. With 69 million square kilometers, it is the third largest ocean in the world, 23 times the size of India itself. Twenty-eight states posses coast or port facilities on the main body of Indian Ocean. Together, they have a population in excess of two billion people, and a gross national product well in excess of a trillion dollars (CIA 2002; Bouchard 2002).

Historically, the Indian Ocean has provided access to merchant ships from distant lands. For centuries ships have plied the waters connecting the subcontinent with Asia, the Middle East and Europe. They included Arab traders, Portuguese explorers, French entrepreneurs and the Dutch and English East India Companies — all seeking a share in the prosperity their trade might bring. The opening of Suez Canal in 1869 greatly accelerated that trade.

Yet despite this long tradition of maritime commerce, it was not until after World War II that the international trade really mushroomed and became essential to the livelihood of the peoples and economies of the nations of South Asia. Three separate factors drove this phenomenon: the dramatic growth of the economies of Asia, the emergence of the Middle East as the world premier source of energy, and modernisation of the shipping industry to include more specialised ships.

Let us consider each of these as they pertain to India and the United States.

First, the economic growth in South and East Asia has had a huge impact on the Indian Ocean and India Itself. In 2001 East Asian trade through the sea lanes of the Indian Ocean exceeded $700 billion, of which $25 billion was with India (Anon 2002).[1] A study of trade conducted in 1990s revealed that 1.5 billion deadweight tons of shipping pass through the Strait of Malacca each year, with 80 million tons destined for India (Noer 1996: 10). Part of that trade is due to India's 'Look East' policy, as India's trade with Southeast Asia more than tripled from 1993 till 1999. But Indian trade to the west is even more important, totaling some 72% of all Indian trade. Indian trade now exceeds $ 100 billion annually and the United States has become India's largest trading partner (Anon 2002; CIA 2002).[2]

Second, the increasing global demand for energy has made Middle Eastern sources, all of which transit the Indian Ocean. China also increasingly depends on Middle Eastern oil. In 2002 its imports of crude oil rose to 15 percent to reach 69 million tons, the figure is expected to rise significantly in the coming years.[3] Given the growing dependence of world economies on Middle Eastern and Central Asian petroleum resources, the already high volumes of trades is likely to rise substantially, with bulk of those imports traversing the Indian Ocean. A large part of that trade will also be with India, whose net oil imports are expected to double within ten years.

Third, the quantity and size of commercial ships today dwarf those of even few years ago. Each day an average of 134 ships of 1,000 deadweight tons or more past eastward through the Indian Ocean. These includes six Very Large Crude Carriers (VLCC) of over 160,000 deadweight tons transporting oil to Asia's hungry markets. They also include small tankers, large dry bulk ships, general cargo ships and cellular ships loaded with easily loaded twenty foot equivalent units (TEUs), that transport vast quantity of merchandise trade (Noer 1996: 9, 27, 56 57). India partakes heavily in this trade, about Merchant Marine boasts over 300 ships of 1,000 tons or over.[4]

Of course the Indian Ocean is of growing economic importance for its resources as well. It is estimated that 40 percent of the world's offshore oil production comes from the Indian Ocean (CIA 2002).[5] Fish resources are abundant, and are being exploited by many nations (Principal among them are Japan, China and South Korea). Beach sands rich in heavy minerals and offshore deposits are also of considerable value.

Given the growing importance indeed, the essentiality of the Indian Ocean to the increasingly interdependent nations of the world, not to mention India itself, it is only natural that security issues need to be addressed. Along with those of other nations, US naval forces have an important role to play in this regard.

The History of the US Navy in the Indian Ocean

For many years the US Navy paid little attention to the Indian Ocean. During the 19th and early 20th centuries, US Navy presence in the Indian Ocean was small and in India was even smaller. The famous Great White Fleet of President Theodore Roosevelt never visited India on its famous world cruise (However, during its one transit of the Indian Ocean, the Fleet stopped at Colombo). During World War I the US Navy was preoccupied with Europe and North Atlantic and until World War II had little time or interest in deploying to South Asia. Besides, Washington considered the region a British sphere of influence and felt that British naval power could take care of any contingency that could come up in South Asia.

In fact, however, London felt that the tyranny of distance would provide maritime protection to India. The major British efforts, as manifested in plans for the Defence of India, (the Blue Plan of 1920 and the Pink Plan of 1931) were directed against land attack from the northwest, with the little consideration for the security of India's coastline. Britain did take some steps to develop an indigenous navy. However in 1887 it amalgamated the non combatant Bombay Marine with other naval forces to establish Her Majesty's Indian Marine, subsequently renamed the Royal Indian marine, and in 1934 the Royal Indian Navy. This force provided the nucleus for the modern Indian and Pakistani navies, but at the outset of World War II it consisted of only eight vessels (Kavic 1967).

With the advent of World War II, and in particular the Japanese attack on Pearl Harbor the perception that South Asia was secure because of tyranny of distance began to change. The war exposed the vulnerability of the region to Japan in particular, it showed that a strong maritime power could dwarf the problem of distance that for so long provided protection. In 1942 Japanese naval units and aircraft appeared in the Bay of Bengal. It was then, and only then, that the United States Navy began to participate in regional defence as part of the China Burma India theater of operations. The history of that conflict is well known. During the war over 50,000 servicemen from modern India and Pakistan lost their lives. The indigenous Navy increased its ship level to 12 and played an important role in protecting India's coastline. Moreover, India provided the strategic ground for the allied counteroffensive into Southeast Asia, and led, ultimately, all the way to Tokyo.

The end of the war bought a sense of relief and a focus on internal matters. The World War was over and India reduced the size of its military forces. So, too, did the United States, which largely demobilised after the war. However, these were critical years for both the countries. For the US, it heralded the beginning of Cold War, with Winston Churchill in 1946 speaking of an "iron curtain" having fallen across Europe, and George Kennan in 1947 articulating a policy of containment of an expansionist Soviet Union. In 1947, South Asia, saw the independence of India and the establishment of Pakistan and in 1948 the independence of Sri Lanka (then Ceylon) and Myanmar (then Burma). Just prior to partition of the British Indian Empire, however, Jawahar Lal Nehru reminded its people of the strategic position of India with the following words: 'The Indian Ocean region depends for its defence greatly on India, which is strategically situated in the center'. Thus both Southeast Asia and Middle East defence arrangements will partly depend on India (Kavic 1967: 27).

Likewise, the United States Navy began to refocus on these two regions. In 1948 Admiral Richard Conolly, famous for his amphibious operations during World War II, cruised to India and Pakistan, opening the modern era of US Navy interest in the region. Subsequent US Navy deployments to the Indian Ocean were frequently linked to support of either the Central Treaty Organization (CENTO), and the Southeast Asia Treaty Organization (SEATO). During the 1950s and 1960s Washington considered these organizations to be key pillars of multilateral security in the cold war.[6] The US Pacific Fleet conducted bilateral and multilateral exercises with the SEATO nations, including exercises in the Straits of Malacca.

Between NATO and SEATO lay CENTO. Although not a member, the United States provided considerable support for the organization as well. Since Pakistan was both a member of CENTO and SEATO, American military assistance naturally flowed to Pakistan. There were limits on the types of assistance, however, such as US opposition to Pakistani navy acquisition of submarines.[7] There were also temporal limits, as SEATO and CENTO were dissolved in 1977 and 1979 respectively.

India, meanwhile, had developed its own naval forces. New Delhi desired an independent type naval force, not an appendage to the British Navy. In this context India began naval modernisation with British cooperation. The US Navy was not heavily involved in these matters. In 1957 India purchased the British carrier, *HMS Hercules,* and after modernisation, commissioned it as the *INS Vikrant* in 1961. During the 1950s British officers were gradually phased out, both in the Indian and the Pakistani navies, but a British Admiral continued to serve in the Indian Navy until 1961. On the eve of the conflict with China (1962), virtually all Indian Naval ships were of British origin, but with an entirely Indian crew and officer corps. Indian naval modernisation, however, suffered in the wake of China conflict. As with the British Defence plans a generation earlier, land threats from the north had higher priority (Kavic 1967: 125).[8] Only later did Indian naval modernisation seriously begin.

During this time the US Navy was heavily engaged in Vietnam and across the Atlantic and Pacific Oceans in cold war deployment. With the end of the Vietnam war in the 1970s, the tempo of US naval operation in the Indian Ocean increased. Beginning in August 1990 Fleet assets began deploying to the Persian Gulf and the Arabian Sea in support of operations *Desert Shield* and *Desert Storm.* More than 50 Atlantic and Pacific Fleet ships, including the carrier battle groups *Independence, Midway* and *Ranger,* along with the battle ship *Missouri* and 20 amphibious ships were deployed. After their initial assignment in support of economic sanctions, they participated in the liberation of Kuwait. US Naval Forces also participated in humanitarian operations for example in 1991, the *USS Tarawa* Amphibious Ready Group provided considerable assistance to cyclone devastated Bangladesh. Two years later naval forces assisted in the effort to end starvation in Somalia. In subsequent years, the US Navy has been permanently present in the Indian Ocean, often in connection with contingencies in the Persian Gulf region.

The US Navy and the Indian Ocean Today

The US Navy today maintains a strong interest in the Indian Ocean and its littoral. Ships from both Atlantic and Pacific Fleets deploy there, although it is the Pacific Fleet that most often deploys near India. The Pacific Fleet now encompasses some 192 ships, 14,000 aircraft and 2,22,000 sailors and marines spread between the west coast of the United States to the Indian Ocean (Fargo 2001). Together they keep the sea lanes open, deter aggression, provide regional stability, and support humanitarian relief operations. The Pacific Fleet area of responsibility reaches from the west coast of the United States to and including the Indian Ocean up to water south of Pakistan. At that point the United States Central Command and its US navy component, the US Navy Central Command, assumes responsibility. In fact, ships from one command are frequently transferred to the other. For example, on September 11, 2001 the *USS Carl Vinson* carrier battle group out of San Diego reported into the US Central Command.

One of the noteworthy features of recent years is the renewal of cooperation between the Indian and the US Navies. The process of defence had gained momentum in 2001, even before the events of September 11 that year, stimulated by the visit to Washington by the External Affairs and Defence Minister Jaswant Singh in April, and subsequent visit to India by the Chairman of the Joint Chiefs of Staff, General Huge Shelton and the Deputy Secretary of State, Richard Armitage. Of course the events of September 11 and afterward accelerated that cooperation. US PACOM and US PACFLT officers met with their Indian counterparts and agree on areas of cooperation focusing on peacekeeping, counter terrorism, special operations training and naval activities.

The Indian Navy plays an important role in these areas. For example, it recently contributed to the global war on terrorism by escorting high value ships through the Strait of Malacca, one of the world's busiest sea lanes. The escort operation were divided between US Navy and Indian Navy. The *INS Sharda*, an offshore patrol vessel of the Southern Naval Command, deployed in April 2002, and was relived three months later by the *INS Sukanya* of the Eastern Naval Command (PACOM and PACFLT sources). By escorting high value ships through Strait of Malacca, one of the world's busiest sea lanes, the Indian Navy provided protection against piracy and terrorism at a considerable distance from base.

The resumption of the annual Malabar exercise in the Indian Ocean also highlights the cooperation between the US Navy and the Indian Navy. In October last year the United States and India renewed this exercise, which included surface, submarine, and air intero perability training. Three previous exercises were conducted in 1992, 1995 and 1996. In 2002 the Indian Delhi class destroyer, Godavari class frigate and Shisumar class submarine, Sea King and Alouette aircraft joined US Naval ships and aircraft in the exercise (PACOM and PACFLT sources).

The future presents many challenges to India and the Unites States in the Indian Ocean area. US and India share common goals in upholding and promoting democracy, resisting terrorism and maintaining and enhancing regional security. Also shared is recognition of the importance of maintaining sea lanes through vital choke points such as the Straits of Mallacca, the Straits of Hormuz, the Babel Mandeb (Dijbuti Yemen), and across the Indian Ocean. Keeping these sea lanes open and free from terrorism, piracy, or other threats is vital not only to global commerce, but also to India, with its 14,000 nautical miles of coastline, numerous major ports and 163 minor ports.

Fortunately, leaders in both the countries are promoting cooperation to achieve these goals. This includes, for the US Navy, the Commander of the Pacific Fleet, Admiral Walter Doran. A graduate of the Indian Defence Services Staff College in Wellington, Admiral Doran earned a Masters Degree in Defence Studies. Later, as Commander of the Seventh Fleet, and presently as Commander of the Pacific Fleet, he has done much to emphasise the importance of the Indian Ocean and India for the United States. This points to the importance of the Expanded International Military Education and Training program whereby Indian and other officers attend schools like the National War College and the US officers attend the Indian Defence Schools.

India also has its share of leaders. One of them, Admiral Madhvendra Singh, the Chief of the Naval Staff, well appreciates the importance of Indian Ocean in all our lives. As India and the United States move forward in pursuit of common goals of democracy, stability, and prosperity. I recall the great Indian born woman Kalpana Chawala who was an inspiration to us all. Just before the Columbia disaster she sent the following message from space: 'The path from dreams to reality does exist. May you have the vision to find it, the courage to get on to it, and perseverance to follow it' .

NOTES

1. It is estimated that 90% of total trade is by sea.
2. The direction of Indian trade is 12% to the Persian Gulf, 32% through the Red Sea, 28% to and around Africa, 26% via Malacca, and 2% to Australia/New Zealand.
3. China crude oil imports 15% in 2002, *People's Daily*, February 11, 2003.
4. The CIA lists 319 such ships.
5. Hydrocarbons are being tapped in the offshore areas of Saudi Arabia, Iran, India and Western Australia.
6. SEATO was organised under the Manila Treaty of 1954. Its members were Great Britain, France, Philippines, Thailand, Pakistan, Australia, New Zealand and the United States. France left in 1968 and Pakistan in 1975, the millitary organisation was dissolved in 1977.
7. Pakistani Chief of the Naval Staff as cited in *Sentinels of the Sea: The Pakistani Navy*, Islamabad, 1997.
8. Some Indians quoted an earlier Indian editorial that stated — The most comforting thing about our navy is that the British and American fleets guard the western and eastern entrances to the Indian Ocean (Kavic 1967: 125).

BIBLIOGRAPHY

ANON (2002) Direction of Trade, *Statistics Year Book*, The International Monetary Fund.

BOUCHARD (2002) Emergence of a New Geopolitical Era in the Indian Ocean: Characters, Issues and Limitations of the Indian Oceanic Order, paper presented at the Inaugural Conference of the Indian Ocean Research Group in November 2002 at Chandigarh, India.

CIA (2002) *Fact Book on the Indian Ocean*.

FARGO, Thomas (2001) The US Pacific Fleet in the Century of the Pacific, *Asia Pacific Defence Review*, Summer.

KAVIC, Lorne J. (1967) *India's Quest for Security*, University of California Press, Berkeley.

NOER, John (1996) *Chokepoints: Maritime Concerns in Southeast Asia*, National Defence University Press.

Comment

M.K. Rasgotra

This last day and a half, we had been discussing marine archaeology. Admiral Krol, today brought us on the surface, brought us face to face with the current problems and his paper gives a futuristic perception and most parts of it, I am in full agreement with him. One or two minor diversions which I want to get out of the way. He says in his discussion of British strategy in the Indian Ocean, London felt that the tyranny of distance would provide maritime protection to India. The tyranny of distance did not protect India against British depredation. It is utter hypocrisy. The British strategy, not in operation but in its manifestation for the world is littered with these poetic phrases and this is one of those. What the British actually did was to take care of their early competitors, the French, the Dutch, the Portuguese in the Indian Ocean. The only remaining potential threats in those early years of British occupation of the Indian Ocean and India was from China. And China they took care to emasculate with massive supplies of opium gathered from Malwa in western India. So, there was no threat left. When the Japanese threat loomed on the horizon, they knew the American navy was already there, to take care of them in many future conflicts. So I wouldn't take these assertions of British strategy too seriously and I plead with you to be on caution, to be on guard.

Earlier today, we heard or we were feasted on most eloquent charming disarmingly amnesiac sweeping sanctification of British occupation of India. Prof Lambert, said that the British Empire was the empire of trade. The trade existed in these regions, before British came on the scene. And in the context of time, the trade was massive and what actually happened was that when the British came here, the Indian trade was very substantial. When they left, the total trade both exports and imports were worth one billion dollars in 1950, we have a figure compiled for that year. In 1947, it was probably somewhat less and this quantum was increased thankful to the war, supply of British textiles, supply of Indian medicines and textiles, etc. So to say that they suffered the heat and dust of this country in those trample uniforms, totally unnecessary in India's weather for 200 years, just to carry on trade or to promote trade in the region. The trade did increase, did expand for other reasons. They used trade to occupy to build up a territorial empire. Now their trade has increased as Prof Lambert said trade has expanded globally. India's trade has also expanded from one billion dollars worth in 1950 to 100 billion dollars today and India is way behind the other countries of the world even in Asia. There is no need for pretence or sanctification of the past. Indians are very tolerant in their history and have accepted Britain as part of Indian history. Indians are now colonising Britain in return giving the taste of Chicken tikka and things of that kind. This is the civilisation process in the reverse. They had a mission there, Prof Lambert said he didn't mention anything about the British civilising mission of India. The pretext for the empire was trade was one, trade they could have carried on without occupying India, just as the Arabs, Portuguese and other nations were trading and they didn't have to occupy, fight those many battles. To what limits they went to suppress India, Jalianwala Bagh comes to my mind you know the other movements which were suppressed brutally, ruthlessly that was not necessarily in the interest of the establishing a trading empire. Now this was one point, I thought, I would take issue with you, in a friendly spirit.

He has said many things in which I am in full agreement. For example that the stakes in the Indian Ocean today are much larger than they were in earlier times and this has many reasons. Trade was expanded as he has pointed out by giving evidence of facts and figures, statistics.

Another reason I don't know why we feel shy of stating this, oil. Modern civilisation more specially in the West, but also in India, China and Japan is dependant on oil and likely to remain dependant on oil for decades to come. Now therefore a security aspect comes here, the security of the Indian Ocean today is a very different proposition from the security of the Indian Ocean in earlier times.

Modernisation of shipping and industry, economic growth especially in Asia has added to the importance of the Indian Ocean as a pathway for the movement of goods and I mentioned oil. The other security issues, the need because of these economic and other factors of stability in Indian Ocean region, in the littoral countries is of vital importance not only to India today but to US, to Britain, to Europe to the whole world in fact and to these regions themselves. And these are aspects to which no one can take exception or disagree with. Some Indians are hesitant in this regard and we should have no hesitation whatsoever, in acknowledging that US has very valid genuine security and economic interest in the Indian Ocean and in the Indian Ocean littoral, in India and in other countries of this area. In the last 50 years both the US and India, committed serious blunders which could have been easily avoided if the two countries were in constant contact with each other, if they had exchanged their thinking, their viewpoints. From the US side the blunder was the forging of these alliances in our region SEATO and CENTO. In these organisations pow wow goes on all the time and a lot of wrong advice get funneled into thinking heads. If this South East Asian Treaty Organisation had not been there, United States involvement in the tragic Vietnam war, could probably have been avoided. There is no time to elaborate on this but this is one area in which some research is needed. Closer to us Pakistan's involvement in this organisation and in CENTO, did great harm to India, did great harm to Pakistan because it was that involvement which encouraged Pakistan to embark on a career of invasion of India. India made a mistake that revered figure Pandit Jawahar Lal Nehru, President Eisenhower was very regretful about the US having entered into arms supply agreement with Pakistan. He is on record about this, in his letters to Prime Minister Nehru and other writings on this subject. He said to Nehru when he came here, while we got involved in this and I can't get out of it now and many of my people especially in armed forces feel that this involvement with Pakistan in military matters is vital for US security etc. I don't know what to do about it, but Prime Minister why can't I give you the arms that you need. And our Prime Minister declined, refused that offer I personally think in retrospect that if we had accepted Eisenhower's offer, the war of 1965 would have been avoided and if that war had not been there subsequent conflicts with

Pakistan might not also have taken place and our relationship with US would have been of a very different character and that would have been good for Asia, good for India, good for US and the kind of bonhomie that has come about in our relationship today might have been achieved earlier. For the rest I am in general agreement with this paper and I commend this paper to not only men of our navy but also to the scholars as it is a futuristic paper which give hints of possibility of cooperation between India and US specially in maritime matters, in naval strategies in cooperation in this area.

A word about the Indian involvement in the Pacific Ocean although it was a part of our subject but it has not been dealt with any degree of thoroughness in this session. In fact Pacific Ocean has not been covered at all. Our involvement in the Pacific Ocean was of a cultural kind, which extended India's cultural frontiers there. In 1960, as India's representative on the Trusteeship council in the UN, I was sent out on a mission to the United Nations Pacific Islands Trust territory. The group visited Pulao and at one particular island I was struck, this was something instinctive that held my attention, the physiognomy, the dances, music they were reminiscent of faces I saw in certain parts of India, the music I hear and the dances I see. And I had a strong feeling that probably these people had something to do with India. They are far away, Java was one of our cultural frontiers, Java is on the Pacific, it is also on the Indian Ocean. The Indonesian archipelago is on the Pacific Ocean to the north and the west also on the Indian Ocean. On approaching some of these people, they wanted me to narrate to them the stories of the *Ramayana* and *Mahabharat*. On my enquiry that whether they learnt about the *Ramayana/ Mahabharat* in schools, they said that they have heard it from their parents and their parents from their parents and so on. For many generations these people have been there. Indian shipping was known to have gone as far as Manila and Japan. We should pay more attention to Indian exploration for trade purposes, cultural purposes of the western Pacific Ocean right upto Japan. The Buddhist *Jatakas* are a source for this exploration. Some of our physiogonomist (if there is a discipline of that kind) would undertake this task of going to some Islands in the Marshal group, Pulao group and other groups not with the desire to snatch them away from United States of America but to establish a cultural link with that country. And I hope a day will come when we do that.

And one last word about the Indian Navy, I have been an advocate of a strong Indian Navy throughout my diplomatic career. Kautilya's *Arthashastra* tells you something about the necessity of a Navy and a merchant marine. Panicker's book *India and the Indian Ocean* it has come into some controversy quite unjustifiably, but Panicker what he basically says is let's not look at the past for the future if India is to sustain itself remain an independent country become strong it will have to have to be a naval power which enables it to guard the entrances to the Indian Oceans.

Our navy is not in a position to do that. For that reason and other reasons, the US presence in the Indian Ocean should be welcomed by India, so long as, we are on cooperative, terms and in the next 25 - 50 years nothing should come between India and United States. Therefore, my hope is that the Indian Navy will continue to grow, and it will, in the real sense, become a blue water navy which has a reach to the West and the East because the future threat to India in the ocean is not from Pakistan, not from Britain, not from US but from China, and Chinese have already embarked on a course of making inroads into the Indian Ocean.

11

Maritime Heritage, Mutual Heritage: Research Beats Collecting

Thijs J. Maarleveld

Introduction

What matters in underwater archaeology is the thinking not the sinking.

In this contribution, an occasional speech, two, quite separate themes are explored and underscored. These two themes are thought to be of interest to the participants of the international seminar on a marine archaeological perspective of the Indian Ocean. In the contributions to the seminar proper both themes figure implicitly at the most (Tripathi 2004: 93-104), whereas in a perspective of development of the discipline it may be useful to stress them. Now dealing with two themes that are not directly related is always slightly awkward. When the author was invited to submit the ideas that were put forward in that particular speech for inclusion in the present volume, he decided to stick to the original duality. It is more practical to combine the two, as both aspects are equally relevant for the situation in Asia. Both can do with a bit of stressing, both for the region of the Indian Ocean and for South East Asia. However, the reasons for doing so are not specific for these regions. They apply elsewhere as well and relate to the general development of the discipline. Neither aspect has been neglected in every respect or in every instance, neither in India nor in other parts of Asia or elsewhere in the world. To highlight them once more in Asia, may nevertheless be useful.

The first theme has to with the public nature of the underwater cultural heritage and therefore with the public responsibility for its management or curation. As it touches upon the role of government, it is evidently politically laden, although the reason to touch the issue derives from scientific interests. The nature of discovery, the vulnerability of archaeological contexts and the way in which the first contact with a site determines whatever any scientist can derive from them in the way of unique historical data is the true reason to touch upon this sensitive issue.

The second theme is the exceptional importance of estuarine and tidal flat environments as the potential producers of completely un-anticipated archaeological information. Admittedly it is a hobby-horse of mine. My roots are in the Netherlands. Having worked for twenty-five years in the murky (fig. 11.1), shallow, dirty, sediment-laden waters of the subsiding North Sea basin in which such rivers as the Rhine discharge, I have really come to like such areas. Not that the work conditions are attractive. They are appalling and therefore easily dismissed; too easily in fact, because such areas have the most wonderful archaeology in store. They can live up to their promises, but only if one is prepared to adjust to their dynamics.

If one wants to see a relationship between the two themes, then it lies in concern for the exceptional archaeological information value of well-preserved organic materials. That is where careful management and informed observation can really make a difference. Integrity of context, including the organic component is after all what makes the

11.1. Underwater archaeology in North Sea basin

underwater heritage so exceptionally valuable. That, perhaps, is the overarching theme.

Additional Considerations

Before embarking on either of the two subjects it is adequate to acknowledge the fact that India, in organising a series of international seminars that result in published proceedings has reached a very important landmark in the development of its underwater archaeology. Having started at a relatively early date and having produced relevant results (Rao 1991; Kuppuram and Kumudamani 1996), underwater archaeology presently witnesses an important change. It is only in the last decade that underwater archaeology is truly being integrated into the Indian system of heritage management. The management of underwater cultural heritage is the logical step that follows after initial exploration and assessment. Its implementation promises to strengthen the existing tradition of maritime studies as part of history and ethnology in India (Ray and Salles 1996; Swamy 2000; Patniak 2003). The close co-operation of Archaeological Survey of India and the Indian Navy, both in the excavation of shipwreck off Bangaram Island and in the international seminars is an icon of the fact that the subject itself and its implications are taken seriously. The Indian Navy, Archaeological Survey of India and India itself can only be congratulated.

A PUBLIC RESPONSIBILITY

Development of the Discipline

Archaeology under water is a relatively young discipline. In 1972 the United Nations Educational, Scientific and Cultural Organization (UNESCO), published a volume called *Underwater Archaeology : A Nascent Discipline*. Although quite a bit had happened by then, in 1972 the discipline was deemed nascent rather than mature, and quite rightly so. In the same year the Convention concerning the Protection of the World Cultural and Natural Heritage (usually called the World Heritage Convention) was concluded. It implied a major structuring of the heritage philosophy world-wide. Joint responsibility for heritage of outstanding universal value and mutual collaboration in protection schemes are the mainstays of the Convention. Acceptance of mutual interest in heritage, acceptance of the mutual origin of "mutual heritage" and acceptance of the fact that heritage protection is not for the benefit of one or two groups but for the benefit of humankind are the deeper philosophies behind it.

The nascent discipline of "underwater archaeology" was definitely to be integrated in that philosophy. It was not, however, very well defined. The book (fig. 11.2) that UNESCO aided by Honor Frost and Angela Croome edited, contains no less than 27 articles. In the vision of the editors, the subjects and the 25 authors that discuss them belong together. On hindsight, all of them have indeed some relevance for the discipline as we see it today. Nevertheless, there is extremely little coherence between the subjects, and one would wonder whether the one author knew the other. There is relatively much technical substance. For instance, there is a history of underwater photography; there is an impromptu on fossilised sessile organisms as a geological feature dating sea-level changes in the long (very long) past. What there is on archaeological research, addresses very diverse aspects of the past in very diverse ways. In fact, some aspects dealt with are truly experimental, both technically and in substance, whereas others are clearly part of a long-standing and mature tradition of management and research (UNESCO 1972; Rao 1991).

On hindsight it is relatively easy to recognise three or four, basically separate traditions within the development of the maritime archaeological discipline. Even within these three there are substantial differences. Such differences are the logical outcome of the fact that archaeological deposits are different in themselves, are related to different periods or different aspects of society, with the consequence that research questions related to quite different scientific discourses are being prioritised. However, in between the three or four traditions the approaches, the developments and the way of thinking are absolutely distinct. This even results in different approaches in legal protection. For discussion's sake I have once labelled three of these traditions (Maarleveld 1996: 251-265). The fact that I was looking at the European situation is reflected in the labels I used. However, these labels may not have been well-chosen, as I presently think the

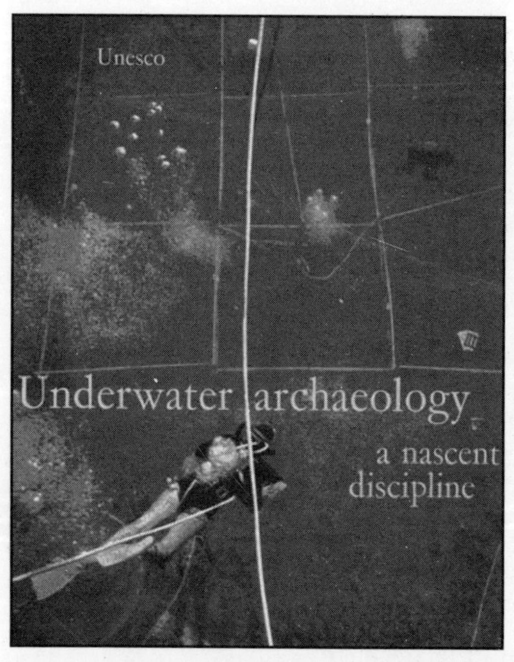

11.2. UNESCO publication

three traditions that I distinguished then, are as relevant to the Indian Ocean region as they are to Northern Europe.

The first tradition, labelled Mediterranean, although "classical" would be as adequate. The tradition combines with classical and Middle Eastern archaeology. It has an emphasis on the "classical" world and the "classical" period. Also, it developed in an area where the overall archaeological tradition saw wholesale looting by antiquarians and art-dealers feeding preponderantly northern European markets and museums ever since the Renaissance. Consequently, a set of restrictive regulations have been put in place after the emergence of the modern nation-state at the end of the nineteenth century. For an understanding of the shifts in ethics, the position of museums and the roles of archaeologists in this tradition, a recent essay by Lord Renfrew (2000) is very revealing and valuable. Firmly embedded in this "classical" world, a specific tradition of underwater maritime archaeology developed, even when diving was still restricted to hard-hat diving. It boomed as soon as Self-Contained Underwater Breathing Apparatus (SCUBA) was developed and Jacques-Yves Cousteau started his promotional work for the silent world world-wide (Cousteau 1953). Much of the archaeological work in India, Sri Lanka and Male that John Carswell quotes in his article on the Arab seagoing in the Indian Ocean falls firmly within this same tradition (Carswell 2004: 57-66). On hindsight, in maritime archaeology this tradition features

- comprehensive survey;
- repressive protective legislation;
- a sharp division between archaeologists and 'clandestini';
- forty years of significant archaeological excavations.

Let us highlight a few significant results by just mentioning site and excavator: George Bass, Çemal Pulak, Cape Gelidonya, Ulu Burun; the sheer quality of the deposits in combination with the sheer quality of their meticulous research added significantly to our understanding of the Bronze Age economy (c. 1400 / 1200 BCE) (Bass 1967; Bass et al. 1989: 1-29). In that economy, as Bisht has so aptly explained, the maritime dimension is one of the most important dimensions, but a dimension as good as absent in previous historical research, a dimension as good as absent in the presently available source material (Bisht 2004: 19-34). Maritime archaeology reveals a source of knowledge of incomparable nature, offering vista's on the past which are more detailed or fine-grained than those produced by many other sources. In fact, quite a bit of the early work in India falls within this classical tradition (De Romanis and Tchernia 1997), labelled as Mediterranean for reasons of convenience, not of principle.

The second tradition evident in maritime archaeology in its present configuration is perhaps the tradition that is most familiar to the maritime professions, including the Navy. It is the tradition exemplified by the recovery of the *Vasa* and the *Mary Rose*. For convenience sake, labelled as northern European, 'maritime historical' would have been as adequate. It finds its basis in historical maritime research rather than in an understanding of archaeological deposits, of archaeological taphonomy or an understanding of archaeological theory and archaeological source-criticism. It is characterised by

- a historic approach, where the documentary sources dominate what to look for in the archaeological record;
- an approach in which to search for a particular wreck is seen as more meaningful than the survey of other, as yet unidentified sites;
- a legislative approach based on rights as opposed to responsibility;
- a sliding scale of commercially and/or academically interested parties;
- emphasis on display rather than on research.

The upside of this tradition is that in the good examples it produces wonderful, empathic displays. The downside is that it has quite a few faults to it. Some of the problems are to do with the theory of scientific discovery or epistemology, the way in which new knowledge is being generated. The approach means to look for what one already knows, in order to illustrate that fact. That is exactly what Anders Franzén (1960) did. That is exactly what Alexander McKee (1982) did. Unfortunately, I know too little about archaeology in India, but I can imagine that besides many examples to the contrary, some archaeological activities have been undertaken in order to show a story already constructed, rather than to be open to whatever evidence archaeological investigation may produce. Anyway, in international maritime archaeology there are many examples of archaeology as a producer of illustrative material rather than archaeology as a discipline with its own specific limits and assets. There can be no doubt that it is my conviction that these assets should be put to right use: archaeology as a discipline is specifically strong in questions of *longue durée*, whereas in this tradition the emphasis is on the *évènement*, the event (Braudel 1969). Improperly the epistemological quality of anonymous - or still anonymous - deposits of wreck is assessed relatively negatively in this tradition. That is another side of the same coin (conclusions in the Netherlands experience).

Within this tradition one often sees that people frantically try to identify a wreck-site; put a name to it; and wreck everything before even assessing what is there; before being able to conclude that it concerns another (and perhaps for other reasons more interesting) archaeological deposit. We see awful examples of that. Often the examples occur in very good faith. Perhaps the so-called Tudor warship off Alderney in the Channel is a good example (Monaghan and Bound 2001). The present awkward dealings regarding the presumed wreck of *HMS Sussex*, that sank in deep water close to Gibraltar in 1694 are inconclusive and unacceptable on the same basis.[1] For an archaeologist the disquieting aspect of this approach is that it is so 'easily graspable', that no one needs to think (see motto), that one hardly needs to be informed about the first thing of archaeological data production and archaeological source-criticism to go for it. This has as its consequence that most - if not all - of the mistakes made in two (or more) hundred years of development of archaeology are liable to reoccur. In fact they constantly do. These include interpretative aspects and aspects of methodology, but also those self-defeating antiquarian approaches to collect goodies for goodies sake, to sell off bits and keep others instead of viewing an archaeological assemblage as an 'organic' whole. Despite, the downsides of this tradition, it would be unfair not to mention the fact that it has produced interesting scientific discussions on the integration of archaeological and historical data and that some of its results can be reprocessed to produce significant archaeological data (Hocker and Cederlund 2006).

The third and again different tradition is labelled "Prehistoric" in quotation marks. Once more, the label is not really apt, as it is not meant to have chronological meaning. It was chosen for the simple reason that at least until recently there tended to be a discrepancy in the archaeologies that were being taught at northern European universities between 'classical archaeology' and 'prehistoric or Northwest-European archaeology'. A similar division can be seen in the American traditions. It seems that the division does not reflect the situation in India, although most of the archaeology related to grand monumentality definitely seems to relate to the 'classical' tradition. Medieval and Post-Medieval archaeology have absorbed aspects of both traditions. A lot of its source-related theory, however, is derived from 'prehistoric' archaeology. That is the reason for this label. Maritime archaeology as developed from inductive reasoning and trying to cope with deposits that need to be managed generally falls within this third tradition, regardless of what period it is specifically concerned with. This is so at present, but it was no less true in 1972 when UNESCO published its book. The tradition is characterised by

- a geographical, ecological and stratigraphical approach in field studies;
- a comprehensive geographical approach in protective legislation;
- a significant contribution by non-vocational archaeologists.

The last characteristic holds true for the other, northern European tradition as well.

Another tradition of the trade found its origin in the management of parks and therefore limited areas. The author did not include this tradition in his original analysis, as he did not think it was very relevant for northern Europe with which he was most concerned. For a wider survey it is important to include this tradition, which in the meantime also has gained a lot of importance in northern Europe. In line with northern American terminology, this tradition is called the Cultural Resource Management or CRM tradition. Far less, than the other three traditions is it geared to research and excavation. Just getting to know what we have got, in order to develop a management policy for it, is the basis. As such the tradition is characterised by the following terms

- thorough, meticulous
- descriptive
- desk-based research
- non-destructive
- focussing on a limited area
- geared to multiple use

With the latter characteristic, it does not necessarily give central place to the archaeologist's research interest. Instead it sees many interests as going hand in hand or competing.

1972 and Coherence

Looking back at 1972 we may conclude that some of the work presented then stood in firmly mature relationship with established traditions of research. As a whole, however, maritime archaeology as a whole was nascent and not mature at all. The 1972 book shows little coherence. This results for instance from the lack of coherence between the three earlier mentioned approaches. The fourth one is absent in its pages. What then was the unifying factor? What was and is the reason to bring such disparate information together? What is the reason that it is useful that the three (or four) traditions interact? The answer is simple: it is not research, but it is the challenge to formulate a comprehensive approach to the protection and management of a newly available heritage resource. This is the challenge that Archaeological Survey of India and the Navy start to meet. The source quality of the underwater cultural heritage is incomparable. It is, however, being opened up with such uncontrolled force and such uncontrolled speed, with all actors homing in on what is in it for them, scavenging its until then pristine nature, that one should speak of a very finite resource indeed!

A.J. Parker of Bristol University compiled a histogram (fig. 11.3) concerning the maritime archaeological discoveries made in the Mediterranean during the first 35 years of SCUBA exploration. Every diver, professional, amateur, wholesale looter or 'clandestino' homed in on the until then un-revealed amphora-mounds ensuing from wreck of classical cargoes. The amphora abundance quickly became legendary. Of all the more than twelve hundred discoveries on which Parker could trace some information, more than 90% was raised to legend, without proper on-site consideration (Parker 1992). Only a few scores of such magnificent sources of information are properly researched, with questions asked, hypotheses formulated, etcetera. The use and intricate conversion of raw materials in the relevant shipbuilding tradition - timber of strangely rich varieties - is one aspect on which research (especially French research) has brought wholly new data (Rival 1991). Retracing the spoils of amphora-heaps, Parker is able to link them with longue-durée developments. He even stands up for

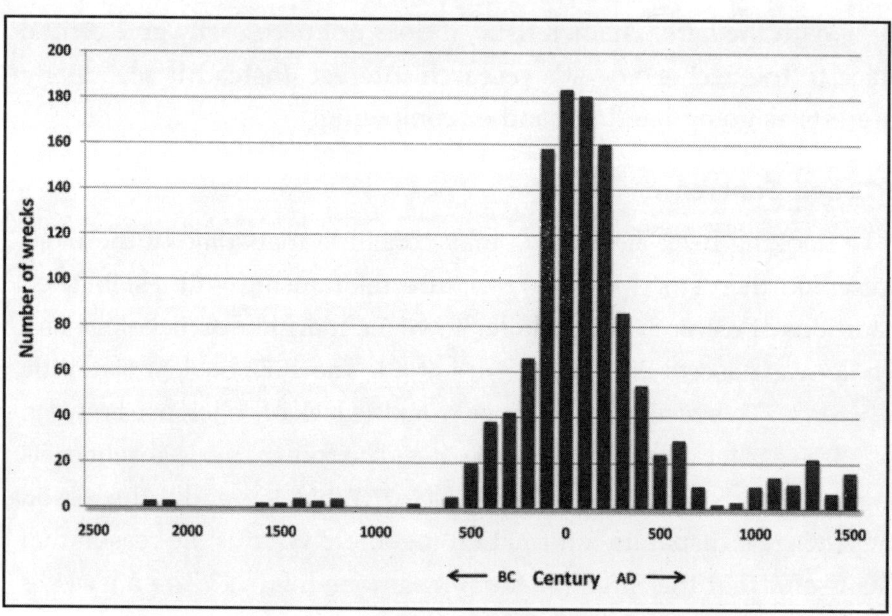

11.3. Histogram showing maritime archaeological discoveries in the Mediterranean (after Parker 1992)

detailed research-oriented, archaeological documentation of dispersed and looted sites. Now that we are better equipped to formulate such research we should stand up and protect the next lot from undergoing the same inconsiderate destruction for goodies, in which each artefact is just an artefact, but loses its context and most of its informative nature.

It is one of the major challenges to which that other event of 1972, the conclusion of the World Heritage Convention committed so many states, India among them. It recognises a common responsibility for the common heritage of humankind. This is an important principle, not only for known heritage, but especially when dealing with as yet unknown archaeological values. This joint responsibility puts the management of the archaeological resource in the public domain: the philosophy extends to where no one previously went: to the underwater world and the maritime spaces. Maritime heritage is mutual heritage and research beats collecting.

Evaluation and Evolution

If we review developments in maritime archaeology over the past thirty years, the friction between the various traditions and the efforts to bring them in line with the World Heritage philosophy are perhaps the most significant. One sees that more and more States take responsibility for the underwater archaeological heritage. From this, significant new research ensues. India has done so. The joint project of Archaeological Survey of India and the Indian Navy off Bangaram in the Lakshadweep Islands is just one example. Heritage management and the necessary research are being combined on the basis of it being a public responsibility and one sees that in the maritime context as on dry land that philosophy works. In other countries in the region governments have been lured into seemingly attractive proposals of private commercial companies offering the search and exploitation of sites on the basis of split profits. The sale of artefacts is what attracts these companies. They hope for a percentage of the proceeds; a percentage, however, of what? Of scientific information? Of organic material? Of the mud displaced in an archaeological project? The approach is self-defeating. "Why sell of your heritage if you can have it all" is the motto with which to advise such governments. India has likewise been approached, but has chosen to take its public responsibility for the benefit of mankind instead.

Why, if nascent in 1972, the discipline as a whole can still not be called mature 35 years later? This is because, despite all the examples above and despite the World Heritage Convention, despite longstanding traditions of mutual co-operation in archaeology, despite guidance from such international guidelines as:

- UNESCO Recommendation on International Principles Applicable to Archaeological Excavations (New Delhi, 5 December 1956);

- ICOMOS International Charter on the Protection and Management of Underwater Cultural Heritage (Sofia, October 1996); and

- 2001 UNESCO Convention on the Protection of the Underwater Cultural Heritage, for which India has been a strong proponent;

despite all this, the conviction that heritage management is a public responsibility is not yet fully engrained everywhere.

It is especially in the context of particularistic maritime exploration that much of the hard-learned experience of over two hundred years of archaeology, of coping with diverse interests, of coping with the interest of objects and the interest of sites, in coping with co-operation between the management of immovable monuments and the management of collections, that much of that hard-learned experience which resulted in guidelines and ethical codes, that much of that hard-learned experience is forgotten for convenience sake or just out of ignorance.[2]

In view of this, it is essential to be aware of an important political fact. The 2001 Convention on the Protection of the Underwater Cultural Heritage may not have entered into force, but it nevertheless has some clout. The Convention contains operative clauses and an Annex in which good practices for the management of the underwater cultural heritage are codified. At the vote on the Convention, on the 2nd of November 2001, it was not only those states fervently in favour of the Convention, such as India, that subscribed to "the Rules of the Annex". Each and every State that supported the Convention, voted against it, abstained or had observer status committed itself to the application of the Annex, unilaterally if need be. By this important political fact such States as the USA, Great Britain, the Netherlands and France have all agreed to regard archaeological heritage as a public responsibility, to be managed according to a coherent philosophy and according to specific rules.

ESTUARIES AND TIDAL FLAT ENVIRONMENTS

What eminent founding father of geology, Sir Charles Lyell (1830-1833, Vol. II: 16), expressed "*..for it is probable that a greater number of monuments of the skill and industry of man will, in the course of ages, be collected together in the bed of the ocean than will be seen at one time on the surface of the continents.*" might well be the earliest predictive statement on the archaeological potential of the underwater world. Much of Lyell's theories have long been challenged. The current model including continental drift has replaced it. Nevertheless Lyell's statement is worth considering at the eve of taking responsibility for heritage in the seas washing one's country. It means, for instance that Archaeological Survey of India and the Indian Navy face a great challenge at sea. The richness of the area may even be greater than on land, which is quite something to say considering this grand and rich subcontinent.

Is there any way, however, in which this enormous task can be limited? Is there any way, in other words, that priorities can be set? Of course there are. One way is to just react to what happens and to act accordingly. This, however, is hardly a strategic approach. Another way is systematic survey with acoustic and other geophysical methods.

Such surveys are highly promising. Although they help underwater cultural heritage management, they always serve other purposes as well. Synergy can therefore be accomplished. Here again the partnership between Archaeological Survey of India and the Indian Navy is to be applauded. In complement specific environments can strategically be taken into account, as they promise to reveal the most astonishingly new information on the past. A third way is to think first.

To stress the latter point it is useful to build on the expectations brought forward in for instance Bisht's paper on seaborne transport in the Harappan culture (Bisht 2004: 19-34; Swamy 2000). Finding a wrecked ship from such a period will really add to our present understanding, especially if it is well preserved and if it contains inventory and cargo. The time-capsule aspect of such a ship-find, the "Pompeï premise", is one reason. The fact that a ship is a mirror of its parent culture, as well as of its parent economic system, and the fact that maritime technology represents the cutting edge technology of any seafaring nation is another. All such aspects contribute to make such discoveries extremely valuable. Where, however, to expect them? Chakravarti (2004: 35–56) addresses a later period, but for instance points to the silting of harbours and river mouths.

Lyell's statement, even if made more than 170 years ago, is also the answer to Carswell's preoccupation with the fact that maritime trade needs presently to be inferred from ceramic and other non-organic evidence. Important sites such as the "emporium" Mantai at the north-west tip of Sri Lanka, that functioned for more than a thousand years, do not feature any organic material (Carswell, 2004: 57-66). The sediments of the seas supplement this complementary evidence. Thereof one can be sure.

Even though the experience in the Netherlands (Maarleveld 1997: 35-57) cannot boast to discoveries relating to such grand ancient cultures as can be expected in India, it may perhaps nevertheless be relevant to a strategic approach in India. For instance the oldest boat-find (fig. 11.4) dates from around 5000 BCE (Louwe Kooijmans 2001). The discovery was made in a silted over, estuarine environment, at around 11 m. below the present datum. It is such environments that determine underwater archaeology in the Netherlands. They may prove important for India. If submerged, they do not feature nice and attractive

11.4. Prehistoric boat found in silted estuarine environment, 11 m. below sea-level

diving conditions: mud and currents mostly set the scene. The preservation conditions, however, are better than anywhere else. It is that aspect that warrants great archaeology.

Strategic attention for estuarine (and tidal inlet) environments and their well-preserved content has been a mainstay in the development of archaeological research and archaeological heritage management in the Netherlands coastal waters during the past 25 years (Maarleveld 1998). Recently we have started to model the processes involved. Understanding what happens when a ship comes to grief, understanding what happens when it becomes embedded in sediment and understanding what happens upon discovery informs choices regarding protection and mitigation of impacts. In fact we have started to use modelling for predictive purposes. There is good reason. Very often in archaeological heritage management far-reaching decisions have to be taken on the basis of meagre evidence. Under those circumstances, predictive models can be a useful tool (Van Leusen and Kamermans 2005).

Among others, such approaches were tried by the National Service for Archaeological Heritage, ROB (now Netherlands Department for the Preservation of Archaeological Heritage, Cultural Landscapes and Historic Buildings). In the development of a predictive management tool, different models had to be used for different situations. A quantitative approach was feasible for that part of the Netherlands where the research tradition and cumulative knowledge extend over two centuries. For the submerged, underwater areas we can only hope that such a longstanding tradition will develop in the future. At present we have to model on the basis of limited archaeological information. Geological and oceanographic information, however, is available. It helps to distinguish between different environmental situations (Maarleveld 1995: 313-328). These varied environments correlate with varied circumstances for archaeological site-formation, varied circumstances for long-term preservation and varied circumstances for discovery. Without going into too much detail here, it may be understood that analysis of such factors may lead to well-founded considerations on the relative archaeological importance of different environments. In fact the Netherlands Department for the Preservation of Archaeological Heritage, Cultural Landscapes and Historic Buildings, has developed a predictive archaeological map to the background of such considerations for archaeological heritage management in the Netherlands. It is called the indicative map of archaeological values or IKAW and a detailed account on the reasoning behind it has been published elsewhere (Deeben, Hallewas and Maarleveld 2002: 9-56). The present map is a tool for present emergencies, based on present perceptions. It will have to be emulated. Hopefully it will. Hopefully it will stir up such discussion that our present perceptions will be improved and that better tools, with greater resolution can be prepared in the future.

In fact, the approach was found so promising, although amenable to improvement, that since author moved from management to university and from the Netherlands to Denmark, he has, together with his colleague Bo Ejstrud, started similar work in relation to Danish waters, starting to model parameters, hoping to integrate them later (Ejstrud 2007: 23-25).

Some Theoretical Considerations

The quantitative drift of Lyell's statement, quoted above, largely finds its basis in the specific qualities of both the cultural and the geological systems. On the one hand one could say that humankind's dependence on water and the fact that "anything that can submerge will submerge"[2] is the guarantee for a rich original distribution of remains. On the other hand, the fact that water basins large and small are accumulative rather than erosive environments will warrant preservation. However, the existence of 'monuments of the skill and industry of man' is not the same thing as the distribution of finds. For the latter, it is as much essential that things be found as that they exist. If they are missed when the original sediment is disturbed, they will never figure in the records. And that is just one of the biases.

Any distribution of finds and observations is subject to a whole series of processes (Schiffer 1987). Some of these are simple and easily understandable. Others are more complex. We are only starting to objectify those factors in theory. It is, for instance, easily understood that nothing is found where nobody ever looks. This may be a truism, but it is not a trivial factor in underwater archaeology! Nevertheless, it is just a simple observation. A whole range of factors contributes to the formation processes of the 'maritime' archaeological record. Their is an intricate interplay. Understanding it, is a major challenge of present-day underwater archaeology.

In order to assess the distribution of 'archaeological values', the present distribution of finds is indicative only in a very limited way. In fact, it can only be accepted as predictive if two serious conditions are met. For one, the original distribution of phenomena over a large area should be a random one. At least, one should be able to surmise that it approaches randomness. For two, a substantial part of that same area should be opened up as completely as to dependably reveal the distribution. Under those two conditions, a simple extrapolation can be used to make quantitative predictions for other parts of that area. Neither condition, however, is unproblematic. Neither is met in the Netherlands. Neither is met in Denmark. Neither is met in India.

The example above well illustrates the fact that the distribution of finds can only be interpreted in relationship with an interpretation of paleo-geography. Moreover, one could say that reconstruction of successive geographical developments is the inevitable prerequisite for any predictive statement regarding the potential archaeological wealth of a particular landscape, or seascape for that matter. One can address this in various ways. One can, for instance, combine environmental reconstruction with the modelling of behaviour and cultural preferences of past societies and give the latter a lot of weight.

The cultural dimension of the location of sites resulting from accidents (such as shipwrecks) may not immediately be evident, but can neither be repudiated. In fact it is rather influential, both in preferred trade routes and in so-called 'curative behaviour': in order to avoid shipwreck, to reduce risk and to reduce loss in cases where disaster is unavoidable a sailor will take deliberate action. As a consequence he chooses or influences an eventual place of loss. In dealing with those factors in predictive assessment one could advocate a 'nautical uniformitarianism'. Nautical uniformitarianism is the proposition that a mariner's interaction with environmental factors like tides and winds, may be variable according to his experience and according to the size and propulsion of his craft, but will essentially be the same, regardless of his position in time or culture.

Adopting this uniformitarian approach has the consequence (and the practical advantage) of giving even more weight to (palaeo) geography. On rocky coasts a concentration of wreckage and dumped material may be surmised at the cliff-foots of shipping hazards such as reefs, isolated islets, projecting headlands and the like. Such environments, however, are not necessarily the best environments for long-term preservation. The shipwreck off Bangaram in the Lakshadweep Islands (Tripathi 1996: 111-117; Tripathi 2004: 93-104; Patniak 2004: 85-92) proves the point.

11.5. Roman shipwreck find in Netherland, 6 m. below sea-level

The Netherlands Experience

The Holocene geology of the Netherlands and the exceptional density of shipwreck-finds (an 'embarrassment of riches'! (Maarleveld 1993: 31-36; Schama 1988) that reclamation, dredging and erosion brought about (fig. 11.5), have urged archaeological heritage management in the Netherlands to look for other criteria than reports of findings alone. It is not quantity of remains that is taken as indicative, it is integrity and the preservation of fragile organic materials. It is the greatest density of high quality that is to be given the highest rating on the indicative map.

To the background of experience in the collection of material and in archaeological excavation, the most preserving environments have been identified and the processes that led to optimal preservation have provisionally been modelled. The eustatic sea-level rise is an important formative factor for Netherlands geology (Nigam et al. 1991: 116-121). It is combined with tectonic subsidence. The North Sea on which the Netherlands borders only emerged through these processes. They are the result of global warming that started some 18,000 years ago. The Netherlands is located where a range of rivers, the Ems, the Rhine, the Meuse, the Scheldt discharge into the North Sea. Consequently the subsidence and drowning have been offset by rapid sedimentation.

Several coastal environments have emerged. Although their specific locations have shifted, and although specific conditions varied through the Holocene, general characteristics have been rather constant (Maarleveld and Peeters 2004: 102-112). Tidal inlets and tidal basins have succeeded each other (Van der Spek 1994). Deltas have formed. Long-shore bars, sand barriers, beach ridges and dunes have formed. Peat has grown in areas protected from direct marine influence. Riverine and marine sediments have added to the system.

Within this general environment specific maritime archaeological niches can be identified. Especially important prove to be the tidal basins. The theorem of nautical uniformitarianism is supported here as they show to have served as the best alternative to save, naturally sheltered harbours, which are absent in the Netherlands. Important also are the fan-shaped ebb-tidal delta's that have a generally convex cross-section perpendicular to the shore (Sha 1990). It is the channel-deposits that have been preserved underneath that feature the most important archaeological surprises. Such channel deposits are all recent on a geological time-scale. They all have different ages in historical / archaeological time. They all share the characteristic of eminent preservation. Of course there are differences between regions within the Netherlands, just like we may assume differences between the Netherlands and Denmark, or between the Netherlands and India. Tectonic subsidence, for instance, has been far more pronounced in south-western Netherlands than in the North. As a consequence deeper sedimentation has occurred. In addition, the tidal range is much greater. This is a consequence of the general layout of the north Sea and the British Channel. The channel-deposits can be surmised to contain wreck-sites of their respective post-Mesolithic ages. Organic materials are excellently preserved.

RELEVANCE FOR INDIA

India is fortunate in that heritage is seen as a public responsibility. In quite a few other Asian countries the underwater cultural heritage is at constant peril as a result of wheeling and dealing between government officials and private individuals.

The reason for mentioning predictive modelling and the endeavour to prepare a predictive map, IKAW, in the present context is that the Dutch experience has shown that it is especially a sedimentary environment that gives the best scores. India, with its large river estuaries, with its deltas with convex cross-sections, boasts environments that conform exactly to the criteria that have been allotted the highest predictive score. Silted tidal basins, filled in channels in areas of subsidence add to these.

Generally, these environments are not associated with diving and diving exploration. They are not the environments that the adventurers, collectors and treasure-hunters alluded to in the first part of this contribution, are looking for. Nor are they the environments that attract diving as a touristic exploit. They are, however, frequently the environments for modern modification. Coastal protection, flood control, dredging, construction of shipping channels, development of harbour structures are just a few instances (Adams, Van Holk and Maarleveld 1990). That makes them important environments in the public domain and a specific area of attention for the public responsibility of archaeological heritage management.

Moreover, these are the environments from which the most important archaeological discoveries may be expected. That at any rate is the conviction of the present author based on his experience in the Netherlands. What is deemed most important is at least partly a matter of subjectivity. The arguments in the present instance are twofold. One is the objective criterion of preservation of integrity and of otherwise not preserved artefacts and material, especially in organic categories. The other argument is the conviction that it is especially in such environments that discoveries can be expected that do not simply confirm what we already know from historical sources. No, it is exactly those discoveries that for instance can be inferred from the overviews for different periods that have been presented by Dr Bisht (2004: 19-34) and Professor Chakravarti (2004: 35–56). In channel deposits of estuarine areas evidence will be preserved that really may add to our views of world history, in which the Arabian Sea, the Indian Ocean and the Bay of Bengal have played such pivotal roles, both in Prehistoric and in Historical times. The public responsibility alluded to in the first part of this contribution is therefore especially important in those environments.

The first aspect to be dealt with needs stressing in view of developments world-wide. It relates directly to best and worst practices and to wise as well self-defeating choices. Such self-defeating choices are often instigated by western companies and have infested countries in the Indian Ocean Region as well as in South-eastern Asia. The second aspect is one where experience in the Netherlands may prove particularly relevant for India. It is in the Netherlands where most of the experience of the present contributor lies.

NOTES

1. The British Ministry of Defence negotiates the salvage of a deep wreck off Gibraltar on the assumption that the archaeological deposit indeed relates to this specific ship. Considerations of mutual responsibilities do not seem to be taken into account nor the political commitment expressed with regard to the operational rules of the 2001 UNESCO Convention on the Protection of the Underwater Cultural Heritage.
2. Freely rendered from 'Murphy's Law', attributed to Edward A. Murphy Jr., U.S. Air Force, 1949.

BIBLIOGRAPHY

ADAMS, J., VAN HOLK, A.F.L. and MAARLEVELD, Th.J. (1990) Dredgers and Archaeology, *Ship Finds from the Slufter*, Alphen aan den Rijn.

BASS, G.F. (1967) *Cape Gelidonya: A Bronze Age Shipwreck*, Transactions of the American Philosophical Society, New Series, Vol. 57, Pt. 8, Philadelphia.

BASS, G.F., PULAK, C., COLLON, D. and WEINSTEIN, J. (1989) The Bronze Age Shipwreck at Ulu Burun: 1986 Campaign, *American Journal of Archaeology* 93.

BISHT, R.S. (2004) Ships and Shipping in Ancient India, In: Alok Tripathi (ed.) *Proceedings of the International Seminar on Marine Archaeology*, Organising Committee of the ISMA, New Delhi.

Braudel, F. (1969) *Écrits sur l'histoire*, Paris.

Carswell, J. (2004) The Arab Seagoing Traditions in the Indian Ocean (8th to 15th Century AD), In: Alok Tripathi (ed.) *Proceedings of the International Seminar on Marine Archaeology*, Organising Committee of the ISMA, New Delhi.

Chakravarti, R. (2004) The Later Indian Trading and Power Systems in the Indian Ocean 4th to 12th Century AD, In: Alok Tripathi (ed.) *Proceedings of the International Seminar on Marine Archaeology*, Organising Committee of the ISMA, New Delhi.

Cousteau, J-Y., with F. Dumas (1953) *The Silent World*, London.

Deeben, J., Hallewas, D.P. and Maarleveld, Th.J. (2002) Predictive Modelling in Archaeological Heritage Management of the Netherlands: The Indicative Map of Archaeological Values (2nd Generation), *Berichten van de Rijksdienst voor het Oudheidkundig Bodemonderzoek* 45.

De Romanis, F. and Tchernia, A. (1997) *Crossings, Early Mediterranean Contacts with India*, New Delhi.

Ejstrud, B. (2007) Modelling Fetch Using GIS, *Maritime Archaeology Newsletter from Denmark* 22.

Franzén, A. (1960) *The Warship Wasa*, Stockholm.

Hocker, F. (ed.) (2006) *Vasa I: The Archaeology of a Swedish Warship of 1628*, Carl Olof Cederlund, Stockholm.

Kuppuram, G. and Kumudamani, K. (ed.) (1996) *Marine Archaeology - The Global Perspectives*, Sundeep Prakashan, Delhi.

Louwe Kooijmans, L.P. (ed.) (2001) Archeologie in de Betuweroute, Hardinxveld-Giessendam De Bruin, Een kampplaats uit het Laat-Mesolithicum en bet begin van de Swifterbant-cultuur (5500-4450 v. Chr.), Amersfoort.

Lyell, Ch. (1830-33) *Principles of Geology*, Being an Attempt to Explain the Former Changes of the Earth's Surface, London.

Maarleveld, Th.J. (1993) Between Frugality and Eclecticism - Some Thoughts on the Management of an Archaeological Cornucopia, *The Bulletin of the Australian Institute for Maritime Archaeology* 17/2.

Maarleveld, Th.J. (1995) Environmental Factors in Underwater Cultural Heritage Management - A Reflection on the Dutch Situation, In: O. Olsen, J. Skamby Madsen and F. Rieck (ed.) *Shipshape*, Essays for Ole Crumlin-Pedersen, Roskilde.

MAARLEVELD, Th.J. (1996) Underwater Heritage Management: Cultural and Legislative Perspective, *Analecta Praehistorica Leidensia* 26.

MAARLEVELD, Th.J. (1997) Nautical Archaeology in the Netherlands: Developments and Cultural Appraisal, In: W.J.H. Willems, H. Kars and D.P. Hallewas (ed.) *Archaeological Heritage Management in the Netherlands*, Fifty Years State Service for Archaeological Investigations, Assen/Amersfoort.

MAARLEVELD, Th.J. (1998) *Archaeological Heritage Management in Dutch Waters: Exploratory Studies*, Dissertation, Leiden.

MAARLEVELD, Th.J. and PEETERS, H. (2004) Can We Manage?, Chapter 14, In: N.C. Flemming (ed.) *Submarine Prehistoric Archaeology of the North Sea*, CBA Research Report 141, York.

MCKEE, A. (1982) *How We Found the Mary Rose*, London.

MONAGHAN, J. and BOUND, M. (2001) *A Ship Cast Away About Alderney, Investigations of an Elizabethan Shipwreck*, Alderney.

NIGAM, R., PATHAK, M.C. and HASHIMI, N.H. (1991) Consequences of Sea Level Rise due to Green House Effect for Coastal Archaeological Monuments, In: S.R. Rao (ed.) *Recent Advances in Marine Archaeology*, Society for Marine Archaeology, Dona-Paula.

PARKER, A.J. (1992) *Ancient Shipwrecks of the Mediterranean & the Roman Provinces*, BAR International Series 580, Oxford.

PATNAIK, A.K. (2004) Archaeological Investigation of *Princes Royal* off Bangaram Island – Introduction and Underwater Operations, In: Alok Tripathi (ed.) *Proceedings of the International Seminar on Marine Archaeology*, Organising Committee of the ISMA, New Delhi.

PATNAIK, A.P. (2003) *The Early Voyagers of the East*, The Rise in Maritime Trade of the Kalingas in Ancient India, Delhi.

RAO, S.R. (ed.) (1991) *Recent Advances in Marine Archaeology*, Society for Marine Archaeology, Dona-Paula.

RAY, H.P. and SALLES, J.-F. (ed.) (1996) *Tradition and Archaeology*, Early Maritime Contacts in the Indian Ocean, New Delhi.

RENFREW, C. (2000) *Loot, Legitimacy and Ownership: The Ethical Crisis in Archaeology*, London.

RIVAL, M. (1991) *La charpenterie navale romaine*, Paris.

SCHAMA, S. (1988) *The Embarrassment of Riches*, An Interpretation of Dutch Culture in the Golden Age, Berkeley.

SCHIFFER, M.B. (1987) *Formation Processes of the Archaeological Record*, Albuquerque.

SHA, L.P. (1990) Preservation Potential of Ebb-tidal Delta and Inlet Systems in Response to Sea-level Rise: Examples from the Dutch Wadden Sea, In: L.P. Sha (ed.) *Sedimentological Studies of the Ebb-tidal Deltas Along the West Frisian Islands, The Netherlands*, Utrecht.

SWAMY, L.N. (2000) *Maritime Contacts of Ancient India, with Special Reference to West Coast*, New Delhi.

TRIPATHI, Alok (1996) An European Shipwreck off Bangaram Island (India) in Arabian Sea, In: G. Kuppuram and K. Kumudamani (ed.) *Marine Archaeology - The Global Perspectives*, Sundeep Prakashan, Delhi.

TRIPATHI, Alok (2004) Archaeological Investigation of *Princes Royal* off Bangaram Island – Excavation and Identification, In: Alok Tripathi (ed.) *Proceedings of the International Seminar on Marine Archaeology*, Organising Committee of the ISMA, New Delhi.

TRIPATHI, Alok (ed.) (2004) *Proceedings of the International Seminar on Marine Archaeology*, Organising Committee of the ISMA, New Delhi.

UNESCO (1972) *Underwater Archaeology: A Nascent Discipline*, Paris.

VAN DER SPEK, A.J.F. (1994) *Large-scale Evolution of Holocene Tidal Basins in the Netherlands*, Dissertation, Utrecht.

VAN LEUSEN, M. and KAMERMANS, H. (ed.) (2005) *Predictive Modelling for Archaeological Heritage Management: A Research Agenda*, Nederlandse Archeologische Rapporten 29 [illegible] ersfoort.

Delegates

SPEAKERS

Dr R.S. Bisht

Dr R.S. Bisht started his career as an archaeologist in August 1968 in the Department of Archaeology and Museum, Punjab where he excavated the multi-cultural site of Sanghol. Later he joined the Department of Archaeology and Museum in Haryana and undertook excavation at Banawali, a Harappan site. He joined the Archaeological Survey of India in 1979 and has excavated many sites of archaeological importance, includin1g the majòr Harappan city at Dholavira in Kachchh.

He has published a number of research papers and popular articles in national and international journals. He has visited several countries to present papers in conferences or to participate in official meetings and has delivered lectures at academic institutions. [He retired from ASI as Joint Director-General.]

Professor John Carswell

Professor John Carswell taught at the American University of Beirut (1956-76); and was Visiting Professor (1997) and Fellow (1999) at the School of Oriental and African Studies, London. He was Curator of Oriental Institute Museum, and Director of the University of Chicago's Fine Art Museum (1988-97). He is a member of Council of the Royal Society for Asian Affairs, London.

His special interest is the export of Chinese porcelain to the Islamic world. He has conducted extensive surveys and excavation in India, Sri Lanka, Maldives, and more recently in the Gobi desert and Central Asia. His numerous publications include a new edition of G. Hourani's

Arab Seafaring and *Blue and White* with details of Red Sea wreck containing Yaun porcelain.

Professor Ranabir Chakravarti

Professor Ranabir Chakravarti has been an academician all through his professional career. He has been a Reader and Professor, Department of Ancient Indian History and Culture, University of Calcutta (1990-2002). Presently, he is a Professor of Ancient History at the Centre for Historical Studies, Jawaharlal Nehru University, New Delhi. He specialises in economic and social history of early India with a particular interest in the history of maritime trade of India in the Indian Ocean (700-1300 CE).

He has been especially pursuing the maritime trade history of early-medieval India on the basis of business letters of Jewish merchants of eleventh and twelfth century. He has published six books and more than hundred research articles in international journals both in India and abroad. He also regularly contributes book reviews to various journals and national dailies.

Rear-Admiral Joseph J. Krol

Rear-Admiral Joseph J. Krol graduated from the United States Naval Academy in June 1967. He completed Naval Nuclear Propulsion Training and Submarine Training in January 1969. His first shipboard assignment was in *USS Nathanael Greene* where he served in various division officer assignments. In December 1972, he reported to *USS Andrew Jackson* (SSBN 619) as Navigator and Operations Officer.

His other assignments include Navigation Branch Head in the Fleet Readiness and Training Branch at the Strategic Systems Project Office in Washington, DC, Executive Officer *USS Phoenix*, Command of *USS Norfolk* (1983-85), command of Pre-Commissioning Unit Oklahoma City (1985--87), command of Submarine Squadron EIGHT in Norfolk (1990-91), Chief of Staff of the Atlantic Fleet Submarine Force, Director, Deep Submergence Branch. He is currently serving as Assistant Deputy Chief of Naval Operations (Plans Policy and Operations). [He retired in June 2003 and joined National Nuclear Security Administration as Associate Administrator for the office of Emergency operations.]

Professor Andrew Lambert

Professor Andrew Lambert is a Laughton Professor of Naval

History in the Department of War Studies at Kings College, London. He has also taught at Royal Naval Staff College, Greenwich and the Royal Military Academy, Sandhurst. He is Honorary Secretary of the Navy Record Society, a fellow of the Royal Historical Society and Honorary Research Consultant, New Dictionary of National Biography Project. He is Director of the new Laughton Naval History unit housed in the Department.

His work focuses on the naval and the strategic history of British Empire since the Napoleon Wars. His work has addressed a range of issues, including technology, policy-making, regional security, deterrence, historiography, crisis management and conflict. He has lectured on aspects of his work around the world, Australia, the United States, Canada, Finland, Denmark and Russia.

Professor Thijs J. Maarleveld

Professor Thijs J. Maarleveld is Professor of Maritime Archaeology at the University of Southern Denmark. He studied History and Prehistory at the universities of Leiden and Amsterdam and learned to dive in 1973. He was attached to the cultural policy department of the Netherlands Ministry of Culture. He worked in heritage management in the Netherlands for 25 years, notably as head of the Maritime Heritage division of the National Service. He was appointed Head of the newly formed Department for Archaeology Underwater, Deputy Director of the Netherlands Institute for Ship and Underwater Archaeology and Head of the division of Maritime Heritage of the National Service for Archaeological Heritage.

In 1998 he took his doctoral degree at Leiden University. On a part time basis he taught nautical archaeology at Leiden University since 1994. He is one of the founding members of the International Committee on the Underwater Cultural Heritage of ICOMOS advising UNESCO on heritage issues.

Rear-Admiral K. Raja Menon

Rear-Admiral K. Raja Menon retired from the Indian Navy after a successful career in 1994. He was a pioneer submariner in the Indian Navy for which he was trained and worked with the Royal Navy. He subsequently commissioned two submarines in the USSR, and commanded a submarine squadron and four other ships. His later career included the important assignments of Head of Technology Transfer Group for building German designed submarines in India,

Deputy Director-General, Defence Planning Staff, New Delhi, Head of Submarine Acquisitions from USSR and Chief Instructor at the Defence Service Staff College, Wellington. He retired from the Navy as the Assistant Chief of Naval Staff (Operations).

He has extensively written for professional journals and published articles in the United States Naval Institute Proceedings and the British Naval Review. He has also written a few books such as *Maritime Strategy and Continental War, A Nuclear Strategy for India, 2000, The Indian Navy: A Photo Essay*, etc.

Commodore A.K. Patnaik

Commodore A.K. Patnaik, VSM is an alumni of National Defence Academy who joined the Indian Navy in 1971. He specialised in Clearance Diving in the year 1973 and has held various important diving appointments in the Navy since then including the Director of Special Operations and Diving at Naval Headquarters at New Delhi. During his diving career, he has had the singular distinction of being the first diver to undergo the pioneering experimental Saturation Dive conducted in India in 1977 by the Navy's Diving School at Kochi.

During his naval career, he has commanded *INS Abhay* (Seaward Defence Boat, 1975), *INS Nashak* (Missile Boat, 1980) *INS Sindhudurg* (Missile Corvette, 1989) and *INS Shakti* (Fleet Replenishment Tanker, 1995) and has also held important operational appointments ashore including that of Chief Staff Officer (Operations) of the Eastern Naval Command. He is a post-graduate from the Defence Services Staff College, Wellington (India) and has also served on the faculty of the college as the Senior Instructor (Navy). [He also served as NOIC, Calcutta and retired in 2006.]

Captain Jose Manuel Malhao Pereira

Captain Jose Manuel Malhao Pereira retired from the Portuguese Navy in 1998, where he served for more than 37 years mainly onboard ships, not only in continental Portugal's waters but also abroad. His knowledge of the sea and the prevailing conditions of life onboard, always made him deeply interested in investigating his predecessor's ways of living and conducting navigation. He has devoted major part of his life to the studies of the history of the Portuguese discoveries.

His main areas of interest are the ancient techniques and instruments of navigation, the sea routes and the correspondent pilots, and seamanship. His interest on these subjects resulted in the publication of many articles, participation in many seminars and conferences in Portugal, India, Brazil, USA, where he lectured, and the publication of more extensive works. He is member of the Academia de Marinha in Lisbon, where he is now the Vice Secretary of its Academic Council.

Dr Eric Rieth

Dr Eric Rieth was born in 1949 in Algeria (Alger) and has been living in France since June 1962. He studied medieval archaeology and maritime history at the Sorbonne University of Paris. The subject of his doctoral thesis submitted in 1978 to the Sorbonne University was the archaeological study of the 'flat bottom based' shipbuilding in north-west Europe, from ancient to modern times. He is a nautical and underwater archaeologist, who has been directing underwater excavation of wrecks at sea and inland waters since 1971. He also has a great interest in the ethnographical evidences as source of comparison and interpretation of archaeological data.

He is now Director of research at the National Center of Scientific Research (CNRS). He teaches nautical archaeology at the Institute of Art History and Archaeology, Sorbonne University Paris. He is chief of the ship archaeology department at the National Maritime Museum, Paris.

Dr Alok Tripathi

Dr Alok Tripathi joined the Archaeological Survey of India in 1987 and in the same year he was selected as the first Indian archaeologist to undergo systematic training in marine archaeology. After his training in India, he visited France and Spain where he obtained formal training and participated in several underwater archaeological excavations in Mediterranean Sea and Atlantic Ocean.

He has participated in a number of underwater archaeological excavations in India and abroad. He has also delivered a number of lectures on Marine Archaeology, Field Archaeology, Rock Art, Indian Art and Architecture, etc. in universities, museums and other academic institutions in India as well as abroad. He has completed his Ph.D. on *Archaeology of Indian Shipping and Shipbuilding Technology*. He has written and edited ten books on Marine Archaeology, Remote Sensing, art and architecture, legislations, etc. and has published more than 60 research papers on archaeology in reputed journals and research publications.

He is in-charge of the Underwater Archaeology Wing of the Archaeological Survey of India and has directed underwater archaeological excavations in the Arabian Sea and the Bay of Bengal.

DISCUSSANTS

Professor B. Arunachalam

Professor B. Arunachalam did his graduation in geography from the Presidency College, Chennai and has 40 years of postgraduate teaching and research experience in the Department of Geography, University of Mumbai from where he retired as the Head of the Geography Department.

He has published over 100 research papers in reputed academic journals and has completed research projects on Lakshadweep traditions of navigation, indigenous traditions of Indian navigation, etc.

Captain Peter Hore

Captain Peter Hore retired from the Royal Navy as Head of the Defence Studies during the British Government's Strategic Defence Review, the rewriting of the British Maritime Strategy, and the launch of a new concept of operations, the Maritime Contribution to Joint Operations. During the Falklands war he was the Logistics Controller on Ascension Island and he has headed both the Royal Navy's Applied Research Programme and its Non-Technical Research Programme.

He is now the associate editor of Warships International Fleet Review, op-ed writer for Newsday in New York and obituarist for the Daily Telegraph in London.

Shri M.C. Joshi

Shri M.C. Joshi joined the Archaeological Survey of India in 1956 and retired as the Director-General in 1993. After retirement he joined the Indira Gandhi National Centre for the Arts as the Member Secretary and served for seven years. Shri Joshi specializes in historical archaeology, ancient cultural tradition, Indian art and architecture including that of the Indo-Islamic period.

Shri Joshi is a widely traveled person and has been delivering lectures on different topics of archaeological and cultural interest in various institutions of learning in India and abroad. [He passed away in January 2007.]

Dr Jean Marie Lafont

Dr Jean Marie Lafont is a member of the French Education Service. He completed his Ph.D. in Greek Archaeology from the University of Lyons in 1977 and D.Litt. from the University of Sorbonne, Paris for his research on *La presence francaise dans Ie royaume sikh du penjab 1822-49.*

He has been teaching at the University of Benghavi (Libya), University of the Punjab, Lahore (Pakistan), University of Lyons (France) and University of Delhi. He was attached to the Institute National des Langues et Civilisations Orientales (INALCO), Paris from 1998-2002. He is on deputation to the University of Delhi.

Professor Roddam Narasimha

Professor Roddam Narasimha has been a distinguished academician, who has been a visiting professor to a number of universities in India as well as abroad. During his long career, he has held various important positions and is presently the Director of the National Institute of Advanced Studies, Bangalore and Chairman of the Engineering Mechanics Unit, Jawaharlal Nehru Centre for Advanced Scientific Research, Bangalore. His research interests mainly pertain to aerospace technology, fluid dynamics and related problems in the atmosphere. He is a recipient of the Karnataka Rajyotsava (1986) and Padma Bhushan (1987) awards.

He has been an Editor of 14 books, 200 papers, and numerous technical and policy reports. He is a member of the editorial board of numerous publications such as *Experiments in Fluids, Research, Sadhana,* Current Science, etc.

Professor Om Prakash

Dr Om Prakash is professor of Economic History at the Delhi School of Economics, University of Delhi. He is a fellow of the Royal Netherlands Academy of Arts and Sciences, Amsterdam. He is a specialist in the Indian Ocean trade of the early modem period and has published extensively on this subject.

He has held visiting positions in many Institutes and Universities abroad including Harvard University, University of Virginia, Curtin University of Technology, Netherlands Institute of Advance Study, Leiden University, Heidelberg University and Maison des Sciences de l' Homme, Paris.

Shri M. Rasgotra

Shri Maharajakrishna Rasgotra was member of the Punjab Education Service till 1949, when he entered the newly constituted Indian Foreign Service. He has worked at the United Nations and at Indian Embassies in Washington and Kathmandu. His Ambassadorial career began in 1967 in Morocco followed by assignments in USA, UK, Nepal, the Netherlands and France.

He was the Foreign Secretary to the Government of India from 1982 to 1985. After his retirement he served as the India's High Commissioner at London. In recognition to his life long service to the nation and achievements of high distinction in diplomacy, he was decorated with the Padma Bhushan in 2002.

Mrs Lotika Varadarajan

Mrs Lotika Varadarajan has had a varied background. She spent her early formative years in tribal Assam and pursued her higher studies at the Universities of Delhi and Mumbai, Chulalongkorn University, Bangkok and Newnham College, University of Cambridge, UK.

Over the years, she has developed an interest in ethno-history with special emphasis on indigenous knowledge systems. This has involved considerable periods of fieldwork in secluded parts of the country including Lakshadweep and the Andaman and Nicobar Islands. She has authored seventy articles and ten books among which may be mentioned *India of the Seventeenth Century, Memoirs of Francois Martin, Sewn Boats of Lakshadweep, The Rahmani of MP Kunhi Kunhi Malmi of Kavaratti, Journey in Science, Technology and Culture - Indo-Portuguese Experiences.*

MODERATORS

Vice-Admiral M.P. Awati

Vice-Admiral M.P. Awati, PVSM, VrC joined the Royal Indian Navy in May 1948. He specialised in signal communication from *HMS Portsmouth* (1953-54) and has held various important appointments in the Navy including Signal Communication Officer, *INS Delhi,* Executive and Cadet training officer, *INS Krishna,* Command of *INS Betwa, INS Tir, INS Kamorta* and *INS Mysore*. His other appointments include Chief Instructor DSSC, Commandant NDA, Chief of Naval Personnel, etc. He was the Commander-in-Chief, Western Naval Command before his retirement.

After retirement he has been associated with many non-government organisations and is currently the Chairman of Ecological Society, Pune. He has been the member of American Bureau of Shipping (1986-87), Llyod's Register of Shipping (since 1989), Director Toolani Group of Companies (since 1988), etc. He has written more than 100 articles and monographs on defence and strategic issues.

Shri Naresh Chandra

Shri Naresh Chandra is a postgraduate from the Allahabad University. After a brief stint as a lecturer in mathematics at the Allahabad University, he joined the Indian Administrative Service in 1956. He has held the posts of Secretary, Industries and Mines (1973), Chairman, State Electricity Board (1975), Finance Secretary (1985), Chief Secretary in the Government of Rajasthan.

In Government of India, he has held the posts of Joint Secretary, Ministry of Industry (1977-81), Secretary, Water Resources (1987-89), Defence Secretary (1989), Home Secretary (1990) and Cabinet Secretary (1990-92). After retirement he has served as the senior advisor to the Prime Minister of India (1992-95) and Ambassador of India to the USA (1996-2001).

Admiral J.G. Nadkarni

Admiral J.G. Nadkarni, PVSM, AVSM, NM, VSM joined the Royal Indian Navy in 1949 after completing 3 years of training on board Dufferin. He received training in UK (1949-53) during which he served onboard aircraft carriers and destroyers. He specialised in Navigation and Direction and was the Navigating Officer of the cruiser *INS Delhi* for over 4 years. He is a graduate of the Defence Services Staff College and the Naval War College, Newport, Rhode Island, USA.

He has held the command of *INS Talwar and INS Delhi.* His important shore appointments include Chief Instructor, Defence Services Staff College at Wellington, the Chief of Staff, Western Naval Command and the Senior Directing Staff, National Defence College, New Delhi. He is the founder member of the Centre of Advanced Strategic Studies and is on the advisory boards of many educational institutions. He is a frequent contributor to newspapers and magazines on Defence matters.

Vice-Admiral M.K. Roy

Vice-Admiral M.K. Roy, PVSM, AVSM was commissioned from the Royal Naval College, Dartsmouth in Britain. His early sea service was in Battleships, Cruisers and Minesweepers of the Royal Navy on the Murmansk route prior to independence and later he joined the Naval Air Arm. His operational sea commands include two Air Squadrons, *INS Godavari* (Destroyer), *INS Bramhaputra* (Frigate) and *INS Vikrant* (Aircraft Carrier). He retired after 40 years of naval service as the Commander-in-Chief of the Eastern Naval Command.

He has also authored a few books such as *War in the Indian Ocean, South Asia after the Cold War, Regional Cooperation in the Indian Ocean*, etc. He is an avid contributor to a spectrum of professional journals and is on several committees such as IDSA, USI, Centre of Policy Research, South African Institute of International Affairs, etc.

Admiral V.S. Shekhawat

Admiral V.S. Shekhawat, PVSM, AVSM, VrC grew up in a military environment having spent his childhood in various cantonments of undivided India. He joined the Indian Navy in 1956 and held various assignments in fleet ships. A pioneer of the submarine arm, he did his basic submarine training in UK in 1963-64 and subsequently in the former USSR in 1966-67. He commissioned the submarines *Kalvari* (1967) and *Karanj* (1969). He has also commanded ASW Frigate *Himgiri* during 1981-82.

During his years in service he has held various important appointments which include Assistant Chief of Naval Staff (Operations) at Naval Headquarters (1984), Flag Officer Maharashtra Area (1987-88), Commander-in-Chief, Eastern Naval Command (1990-92), etc. He took over as the Chief of the Naval Staff on 01 Oct 1993. Admiral Shekhawat was the member of the National Security Advisory Board (1999-2001).

Admiral R.H. Tahiliani

Admiral R.H. Tahiliani, PVSM, AVSM joined the Royal Indian Navy in March 1948. He took over as the Chief of the Naval Staff in Dec 1984 and retired three years later.

Admiral Tahiliani was the Governor of Sikkim from 1990 to 1994. During his years of retirement before and after being the Governor of Sikkim, he has been involved in social work with Servants of Peoples Society. He is currently the Chairman of two senior secondary schools catering mainly the weaker sections of the society. He is also the Chairman of Transparency International India and trustee of Memorial Fund for Disaster Relief in India.

Index

Abbasid Caliphate, 53, 58
Abdulla Wassaf, 68
abhayasasana, 62
Abraham bin Yiju, 61, 71, 80
Abu Bakar, 160
Abyssinia, 222
Achernar, 260
Adam's bridge, 42
Addu Atoll, 222, 233
Aden, 58, 59, 60, 61, 65, 66, 67, 68, 69, 71, 72, 80, 122, 124, 130, 221, 222, 225, 227, 238
Adhara, 268
Admiral
 Cheng Ho, 22, 61, 64, 77
 Zeng He, 61
 Paris, 19, 20, 115, 189, 208, 211, 218
aeroplane, 130, 233
Afghanistan, 226, 227
Africa, 7, 9, 50, 59, 78, 80, 91, 97, 103, 108, 146, 164, 225, 230, 231, 236, 237, 249, 280, 298
African, 7, 59, 66, 225, 227, 273, 280
Afro-Asian, 9
aghbab, 52
Agina-mala Samudda, 41
Agni-Mala Samudra, 41
Agnimali, 45
agricultural, 50, 58, 59, 60, 63, 79, 102
agriculture, 39, 50
aharana, 70
Ahmed bin Majid, 17
Ahmednagar, 127
airlift, 155, 167, 168
Aix-en-Province, 216
Ajanta, 45, 70, 87
Akbar, 98, 134
Akkad, 51
Al-askar, 70
Al-bahr-al-Hind, 52
Al-Beruni, 57, 59, 79
Al-Divi, 59
Al-Fustat, 53, 67, 71, 79, 80
Al-Idrisi, 57, 63, 79, 80, 97
Al-Mahdiyya, 61, 67, 80
Al-Manjrur, 59, 61, 66, 67, 72, 74, 80
Al-Marvazi, 57
Al-Masudi, 57, 60, 76
Al-Mulaybar, 61
Al-Muqaddam, 72
alabaster, 35
Alarm, 178
Alderney, 313, 328
Alexander, 3, 44, 281, 312
Alexandria, 14
Algerian, 115
Ali, 60, 212, 215, 217, 275
 Raja of Cannanore, 212
Alliya, 60
altitude, 122, 123, 263, 268, 274, 279, 282

Amalfi, 280
American
    frigate, 116
    terminology, 314
amphora, 315
Amsterdam, 124
Amta, 67
anchor, 40, 98, 151, 154, 155, 157, 171, 173, 175, 176, 202, 216
anchorage, 65, 151, 202, 222, 275
ancient
    coins, 17
    lakes, 8
    literature, 13, 17
    ports, 8
    settlements, 8
    sites, 8
    text, 4
Andalusia, 104
Andaman and Nicobar, 41
Andhra, 21, 62, 63, 65, 78, 134, 210
    coast, 62
Anglo-American, 234
Angria's fleet, 213
Angrias, 212
animal herding, 39
Antakhi, 42
Antarctica, 50
Antioch, 42
antiquarian remains, 16
antiquity, 7, 16, 43, 140, 159, 163, 164, 165, 171, 281
Anuradhapura, 100, 102, 104, 107
Apara, 53
aphaterion, 56
Arab (s), 5, 17, 18, 19, 20, 45, 52, 53, 55, 58, 59, 60, 61, 63, 66, 69, 76, 80, 81, 95, 96, 97, 98, 104, 105, 107, 108, 109, 115, 122, 130, 131, 248, 251, 255, 259, 260, 264, 274, 277, 279, 280, 281, 286, 291, 300, 311, 327
    accounts, 55, 59, 60, 64
    authors, 63
    horses, 108
    mariner, 17, 105
    merchants, 60, 76
    sailing traditions, 18
    vessels, 108
Arabia, 9, 67, 97, 107, 146, 164, 298
Arabian Sea, 9, 17, 41, 52, 53, 55, 73, 75, 77, 109, 146, 159, 163, 164, 244, 295, 326, 329
Arakan, 63
archaeological
    deposit, 313, 326
    evidence, 4, 7, 29, 39, 55, 181
    potential, 318
    research, 13, 310, 320
    significance, 147, 148, 164
    sites, 12, 16, 30, 35
    sources, 95
Archaeological Survey of India, 3, 4, 8, 11, 15, 145, 147, 148, 150, 163, 164, 309, 317
archaeologists, 4, 8, 12, 35, 179, 181, 185, 311, 314
archaeology, 4, 8, 11, 12, 15, 16, 40, 100, 153, 185, 243, 307, 308, 309, 310, 311, 312, 313, 314, 315, 316, 317, 319, 320, 322
    marine, 4, 5, 7, 9, 11, 12, 13, 14, 17, 145, 150, 243, 299
archipelago, 146, 163, 233, 238, 272, 302
archipels of Gilbert, 190
architectural plans, 199
architecture, 12, 14, 17, 19, 20, 34, 119, 202, 203, 205, 210
archives, 9, 146, 147, 211, 215, 216
Ardeshir Cursetji, 137
Ardrapura, 59, 73
Arikamedu, 100
Aristotle, 3
art, 12, 14, 17, 20, 30, 33, 97, 123, 127, 138, 165, 247, 248, 311
artefacts, 7, 12, 17, 146, 147, 148, 149, 150, 153, 154, 155, 163, 165, 167, 169, 317, 325
Artemise, 191, 194, 208

Arthasastra, 42
artillery, 212, 213
Aryabhata, 281
Aryans, 39, 40, 41, 43
ascorbutic, 129
Ashmolean Museum, 99
Ashwin, 41
Asia, 7, 17, 29, 50, 53, 55, 56, 57, 64, 67, 74, 75, 76, 77, 78, 90, 91, 96, 100, 102, 157, 238, 272, 282, 291, 292, 293, 294, 298, 300, 301, 302, 307, 326
Asiatic Society, 208, 213, 264, 277
Assaye, 113
astrolabe, 123, 280
astronomy, 19, 130, 140, 281
Atlantic, 4, 17, 50, 51, 96, 105, 129, 234, 237, 238, 247, 255, 274, 275, 279, 280, 293, 295, 296
Atlantic Ocean, 4
atolls, 146, 163
Australia, 29, 221, 222, 225, 227, 229, 231, 298
Australasian troops, 230
avartana, 41
Avdanas, 45
Aydhab, 65, 67, 69, 80
azimuthal compass, 281

Bab-el Mandap, 130
Babul, 120
Babylon, 44
Badar, 45
Badradweep, 45
Baghdad, 58, 97
Baghdadi, 60
Baghela, 77
Baghla, 21
bahars, 71
Bahr-al-Harkal, 52, 53
Bahr-al-Larvi, 52
Bahrain, 33, 39
bahri, 68
Balaputradeva, 64
Balipattana, 60, 73, 74
balisti, 282
ballast, 66, 175
stones, 175
Baltic, 229
Baluchi hills, 30
Baluchistan, 34
Bangaram Island, 4, 146, 147, 150, 163, 309, 329
Bangladesh, 52, 56, 63, 189, 295
bar shot, 127
Barus, 59, 74
Baruz, 72
Barygaza, 97
basket-boat, 192
Batavia, 222
Batel, 21
battles, 9, 16, 50, 113, 138, 300
battleship (s), 119, 129, 229, 234, 235
Baveru Jataka, 44
Bay of Bengal, 38, 41, 50, 52, 53, 56, 61, 62, 63, 65, 74, 75, 77, 78, 107, 108, 109, 236, 266, 294, 326
beaches, 18, 30, 147, 203
beads, 30, 33, 35, 103, 104
Bedouin, 18, 19
Beirut, 99
Belgium, 199
Bengal delta, 57, 63, 65
Bering Straits, 29
Betore, 65
Beypore, 21, 118, 120
Bhagavad Gita, 130
Bharator, 134
Bharuch, 57, 58, 59, 72, 97
Bharwamukh, 45
Bhatkal, 8
Bhoja, 20, 68
Bhrgukachchh, 55
Bhujyu, 41
biblical flood, 44
Bijapur, 55, 127
bilges, 118
bilij, 71
blocks, 121
Blue Plan, 293

Blue River, 64
blue water, 211, 212, 213, 215, 218, 303
boat, 192, 193
 basket, 192
 Flat-bottomed, 203
 -find, 319
 motorised, 151
 of river Ganges, 193
 sailing, 268
 traditional Indian, 193
Bodhisatva, 44
Bolan river, 30
Bombay, 21, 99, 125, 137, 179, 199, 202, 208, 212, 213, 217, 221, 222, 225, 226, 227, 228, 236, 244, 293
 coast, 21
Borneo, 190
Bougainville, 190
Bowl (s), 169, 170
Brahma Siddhantha, 130
Brahmagupta, 130
Brahmanical, 49, 75, 133
Brahmins, 140
Brazil, 248, 273
bricks, 17, 169, 175
Brihadeeswara temple, 132
Britain, 22, 101, 119, 127, 138, 141, 178, 190, 222, 223, 224, 231, 232, 233, 234, 235, 238, 277, 293, 298, 300, 301, 303, 318
British, 9, 24, 101, 113, 114, 125, 126, 134, 137, 138, 139, 140, 173, 178, 179, 211, 212, 213, 215, 217, 221, 222, 223, 224, 225, 226, 227, 229, 230, 231, 232, 233, 234, 235, 236, 237, 238, 244, 293, 294, 295, 298, 299, 324, 326
 diplomacy, 229
 Empire, 224, 299
 maritime hegemony, 224, 232
 policy-makers, 224
 power, 224
bronze, 33, 131, 132, 148, 164, 177, 179
 bell, 179
Bronze Age, 33, 311, 326
brown glazed ware, 148, 157, 169, 170, 179
Buddha, 70, 159
Buddhagupta, 57
Buddhist, 41, 49, 56, 101, 302
 literature, 41
Burma, 91, 222, 225, 233, 237, 238, 294
burrama, 69
Byzantine rulers, 54

Cairo, 53, 67, 79, 130
calamities, 16
Calcutta, 125, 138, 181, 208, 213, 227, 228, 236, 265
Calicut, 61, 77, 98
Caliph Mansur, 281
Cambridge, 79, 127, 283
camphor, 67
Cana, 97
Candrapura, 60, 73
Cannanore, 212
cannon, 22, 126, 127, 131, 132, 171, 172
 making, 131
Canopus, 260
canvas, 121
Cape
 Comorin, 97, 205
 Gelidonya, 311, 326
 of Good Hope, 222
 Station, 230
 Town, 222, 227
 Verde Islands, 227
Capella, 259
cargo, 7, 21, 33, 61, 68, 72, 100, 157, 292, 319
carnelian, 35, 103
Carolines, 190
Carronade, 126
cartographic tradition, 281
catamaran, 21
catimaron, 192, 199, 205
Cattagrama, 63
celestial
 bodies, 250, 251, 273

hemisphere, 250
sphere, 250, 274
Cemulya, 60, 73
Central Asia, 33, 185
Central Treaty Organization, 294
Cesare de Fedrici, 14
Ceylon, 53, 203, 208, 221, 224, 236, 294
Chachnama, 55
Chalukya (s), 55, 58, 62
of Badami, 55
Chandore, 60
Chanhu-daro, 35
Charaka, 130
charts, 44, 123, 124, 129, 147, 190, 191, 196, 208, 280, 282
Chau ju Kua, 57
Chaul, 55, 60, 214, 216
chelinque, 203
Chennai, 61
Chetlat, 286
Chicago, 101, 283
Chief of the Naval Staff, 3, 9, 11, 15, 23, 42, 297, 298
China, 4, 96, 100,
Sea, 50
War, 223
ware, 171
Chinese, 5, 19, 45, 53, 54, 56, 61, 64, 69, 71, 75, 77, 79, 98, 99, 100, 102, 104, 108, 109, 122, 223, 248, 249, 255, 280, 281, 282, 303
ceramics, 75
crown, 77
Junks, 98
pilgrim, 56
porcelain, 99
pot, 100
Chittagong, 52, 63
chlorite, 35
Chola, 61, 62, 77, 96
dynasty, 62
power, 61
rulers, 61
territories, 62
throne, 62
Christianity, 9
Christopher Columbus, 114
chronometer, 123
Chryse Chersonesys, 56
Chryse Chora, 56
Chuel, 8
Churchill, 234, 235, 236, 237, 294
circumnavigation, 190, 191
civilization, 12, 17, 19, 29, 33, 40, 50, 279
non-European, 139
clinch ring, 173
Clive, 138, 212
cloth, 38, 65, 160
Coast
Defence ships, 228
Bombay, 21
Coromandel, 56, 62, 194, 199, 203, 208, 209, 286
east, 14, 50, 55, 61, 98, 107, 108, 109, 286
Indian, 4, 8, 13, 114, 207, 224, 227, 236
Konkan, 55, 68, 76
Makran, 55, 58, 134
Malabar, 9, 68, 74, 107, 164, 202, 205, 206, 212, 213, 214, 217
of China, 56
of Golconde, 192
of Philippines, 190
Sumatran, 109
coastal, 13, 39, 45, 55, 56, 59, 60, 62, 63, 68, 71, 72, 73, 74, 75, 76, 78, 97, 108, 125, 134, 203, 205, 211, 215, 226, 249, 320, 324
navigation, 249
offensive operations, 226
protection, 325
voyage, 59, 72
coasting, 97, 122
coastline, 4, 280, 293, 294, 297
Cochin China, 211
Cocos Islands, 230
Cold War, 294
collections, 79, 317

Colombo, 99, 104, 222, 228, 236, 237, 293
Colonial
  Defence Committee, 229
  power, 126
Comilla, 52, 56, 64
commodities, 57, 62, 65, 67, 71, 75
communication, 17, 64, 65, 149, 151, 155, 165, 169, 229, 233, 244, 282
communities, 16, 19, 50, 75, 77, 134
Company rule, 225
Company's ships, 212
compass, 102, 120, 122, 123, 247, 251, 255, 256, 257, 259, 260, 261, 272, 273, 274, 280, 281, 282, 286
  azimuthal compass, 281
  floating, 19
  magnetic, 122, 272, 286
  Star, 286
  stellar, 255
composite sloops, 228
compressors, 151
Congreve, 138
construction, 5, 19, 20, 29, 70, 76, 77, 115, 116, 121, 157, 177, 196, 199, 206, 212, 214, 325
continents, 5, 29, 132, 208, 253, 318
copper, 33, 34, 35, 43, 67, 72, 97, 131, 148, 157, 165, 169, 173, 174, 175, 176, 177, 178, 185
  Age, 32
  bolts, 178
  sheet, 165, 174, 176, 178
coral
  island, 147
  reef, 147, 149, 154, 164
  Sea, 237
cordage, 121
Coringui, 192
Coromandel, 56, 62, 78, 199, 202, 203, 205, 212, 286
  coast, 56, 62, 199, 203, 286
Cosmas Indicopleustes, 55, 60
country crafts, 20
cowry shells, 66
crafts, 20, 30, 33, 38, 45, 63, 70, 115, 149
craftsmanship, 13, 141
Cranganore, 74
crocodile, 33, 102
Cronstadt, 226
cross staff, 123, 266, 276
crossbeams, 203
crossed lashings, 207
cruiser, 228, 229, 230
Cuddalore, 217, 286
cultural
  activities, 49
  deposit, 32
  empire, 34
  Landscapes, 320, 321
  life, 49
  practices, 49
  Resource Management, 314
culture, 12, 14, 16, 17, 19, 20, 29, 30, 40, 139, 141, 146, 163, 164, 319, 323
current meter, 148

da'yra, 19
Dahabattan, 71
Dahlak, 130
Dahomey, 66
Daibul, 55, 85
Damascus, 99, 126, 131
Danish, 89, 120, 321
Dar-es-Salaam, 230
Darwin, 235
Debal, 55
Deccani kingdoms, 127
deck, 126
Declaration of Paris, 225
decompression illness, 151
defensive policy of the Empire, 229
Delhi, 81, 134, 149, 216, 227, 229, 244, 295, 297, 317, 329
Deltas, 324, 329
Denmark, 321, 322, 324, 327
Desert
  Shield, 295

Storm, 295
Deserters, 129
design, 5, 78, 115, 116, 117, 119, 127, 130, 169
Deva-pala, 64
Devnagari, 185
Devparvata, 64
Dewar, 134
DGPS, 148
Dhaka, 53, 65, 90
Dharma Shastras, 133
dharmasthanam, 76
Dhofar, 18
Dholavira, 38
dhoti, 21
Dhow, 18, 19, 20, 21, 98, 115, 117
diamond mines, 65
Diarc, 140
Dijbuti, 297
Dilmun, 34, 35
Diphda, 268
Director-General, 3, 7, 11, 15, 164, 185
dirgha, 68
dishakak, 33
dishes, 170
displacement, 116, 119, 121
Diu, 59
Dive Support Vessel, 23, 150
divers, 4, 72, 147, 148, 149, 150, 152, 154, 155, 157, 164, 165, 168, 169, 177, 181
saturation, 149
diving
operations, 149, 151, 152, 155
School, 148
-bells, 3
-suit, 4
Diyarbakir, 97
dockyard, 35, 228, 232
documentation, 44, 54, 89, 90, 150, 167, 169, 175, 189, 190, 194, 196, 197, 199, 201, 207, 316
domestication, 29
Dravidians, 40
drawings, 20, 115, 116, 117, 129, 130, 131, 132, 172, 194, 196, 198, 207, 211
dredging, 323, 325
drun, 40
dry-docks, 228
Duke of Wellington, 138
Dumont d' Urville, 190
dunes, 324
Durban, 227
Dutch, 9, 51, 78, 89, 90, 120, 178, 221, 224, 234, 248, 291, 299, 325, 329
East Indian Companies, 51
dvara, 65
Dvarahataka, 65
Dvaravati, 57
Dvipalaksham, 163
dvipantara, 62, 68
Dwarka, 8, 42

earthen ware, 100, 104
East
Africa, 230
Asia, 78, 307
coast, 14, 50, 55, 61, 98, 107, 108, 109, 286
India Company, 125
Indies, 122, 225, 229, 232, 234, 235, 238
eastern
Ganga rulers, 62
ghats, 41
seas, 18
ebony, 97
echo-sounder, 148, 153
ecological system, 12
economical empire, 34
economy, 29, 30, 49, 51, 66, 79, 102, 139, 223, 229, 234, 311
ecosystem, 13, 149
education system, 14
Egypt, 32, 43, 44, 53, 57, 58, 65, 79, 103, 104, 222, 226, 230, 236, 275
Einstein, 140
Elephant, 35

Elephanta, 55
Emden, 222, 230, 233
Ems, 324
engineering drawing, 130
engineers, 4, 41, 137, 201
English, 14, 51, 52, 78, 89, 115, 120, 132, 186, 190, 212, 213, 215, 217, 248, 282, 283, 291
  garrison, 212
  merchants, 14
environment, 13, 16, 157, 168, 172, 197, 203, 205, 248, 319, 324, 325
epigraphs, 17
epigraphy, 45
equator, 251, 261
estuaries, 30, 325
estuarine, 18, 308, 319, 320, 326
etched carnelian beads, 35
Ethnographic Museum, 265
ethnology, 19, 309
Europe, 50, 78, 122, 123, 126, 127, 129, 130, 131, 132, 133, 139, 141, 178, 213, 214, 221, 234, 237, 280, 282, 291, 293, 294, 301, 311, 314
European (s), 19, 21, 24, 51, 78, 105, 115, 123, 130, 131, 177, 185, 282
  encounter, 137
  Navies, 174
  waters, 224
excavation (s), 4, 8, 9, 12, 13, 16, 17, 19, 22, 24, 58, 95, 96, 98, 100, 101, 145, 146, 148, 149, 150, 151, 153, 155, 157, 159, 163, 164, 165, 167, 168, 169, 170, 172, 173, 174, 175, 176, 177, 181, 185, 309, 311, 314, 324
  site, 4, 149
exchange centers, 63
Exclusive Economic Zone, 146
exotic spices, 67
expansion, 51, 60, 63, 64, 96, 213, 223
exploration (s), 3, 8, 16, 159, 190, 243, 247, 302, 309, 315, 317, 325
Extra-European, 189, 192, 194, 196, 197, 199, 202, 206, 208
Fa Hsien, 56, 85
Fadiyar, 72
Failaka, 33
Falkland Islands, 230
False Divi Point, 192
Fandarina, 61
fastenings, 20, 122, 173, 177, 178
Fatimid Caliphate, 53, 58
Favorite, 190, 191, 192, 193, 194, 209
feudalism, 51, 52, 79, 90
fisherman, 18, 51
Fishing, 18
flagship, 215, 216, 228, 229
flat-bottomed boat, 203
flax, 121
floating
  compass, 19
  dock, 231
fort of Revdanda, 214
fortifications, 3, 212, 213, 216
fortresses, 125, 227
fragrance material, 38
frame, 116, 121, 203, 206, 260
France, 22, 24, 119, 120, 127, 190, 196, 205, 208, 211, 213, 214, 215, 216, 233, 298, 318
Francois Edmond Paris, 189
Fremantle, 222, 225, 239
French, 9, 29, 51, 78, 89, 113, 114, 119, 120, 127, 131, 178, 186, 190, 191, 196, 208, 211, 212, 213, 214, 215, 216, 217, 218, 223, 224, 227, 228, 233, 244, 248, 273, 291, 299, 315
  contingent, 212
  diplomatic mission, 214
  experts, 215
  influence, 190
  navy, 191
frigate, 120, 127, 173, 190, 228, 297
Fua Mulaku, 19

Gallivats, 127
galvanic action, 173, 178
Ganapatideva, 62, 85
Ganga valley, 64

Gangetic plain, 41
Ganjam, 63
Genizah
  documents, 54, 67, 76
  letters, 66
  papers, 65
geographical, 18, 21, 50, 52, 53, 60, 123, 202, 206, 208, 280, 314, 322
geography, 18, 139, 322, 323
  medieval, 17
German Cameroons, 230
Germany, 101, 199, 232, 233
Gharapuri, 8
Gheria, 113, 212, 214
Ghogha velakula, 59
Gibraltar, 313, 326
Gilan, 97
global strategy, 221, 226, 232, 234
Gobi desert, 103
Godavari, 56, 297
gold, 18, 33, 34, 43, 61, 67, 68, 225
  beads, 33
Golkonda, 127
google, 38
Gopakapattana, 60, 82
gravimeter, 148
Great White Fleet, 293
Greece, 279
Greeks, 97, 130, 247, 248
green
  glass, 169, 175
  glazed sherd, 169
grinding stone, 165
Gujarat, 21, 44, 52, 55, 58, 59, 60, 65, 73, 76, 77, 90, 91, 185, 186, 275, 281
Gujarati, 44, 125, 131, 281, 285
Gulf
  of Gibraltar, 29
  of Khambat, 35
  of Mannar, 199
gun (s), 126, 212, 213, 214, 215, 239
  carriages, 126
gunsmith, 127
gun-vessels, 228
Hadramawt, 18
hajj, 44, 73
Hamburg, 265
handicrafts, 4
Harappa, 12, 17, 33, 38, 40
Harappan
  culture, 7, 40, 43, 319
  Empire, 34, 39
harbour (s), 191, 213, 222, 227, 253, 261, 266, 275, 319, 324
  structures, 325
Harikela, 52, 86
Harrison Chronometers, 123
Hastings, 138
hatta, 58, 65
hattika, 58
Hawaii, 232, 249, 277
Hawkins, 20
heavenly bodies, 19, 250, 251, 260
heritage, 4, 7, 8, 13, 148, 307, 309, 315, 316, 317, 318, 319, 320, 321, 323
  management, 320
  maritime, 4, 5, 11, 13, 14
high sea, 56
Himachal Pradesh, 38
Himalaya, 121
Himalayan, 121, 226
  timber, 121
Hindu, 21, 76, 97, 133, 205, 265, 281
Hippalus, 97
historian (s), 4, 8, 43, 49, 50, 51, 57, 63, 68, 113, 114, 130, 247
  Greek, 3
historic sites, 127
historical
  data, 308, 313
  importance, 7
historiography, 49, 51, 52, 57
history, 3, 4, 5, 7, 8, 11, 12, 13, 14, 15, 16, 17, 18, 19, 20, 22, 33, 49, 50, 51, 59, 62, 63, 103, 105, 113, 137, 139, 145, 146, 163, 164, 178, 179, 181, 189, 197, 201, 208, 213, 218, 222, 243, 244, 263, 266, 291, 294, 300, 309, 310, 326

human, 3, 145
HMS
Hercules, 295
Minden, 137
Prince of Wales, 235
Repulse, 235
Sussex, 313
Trincomalee, 137
hogging, 117
Holocene, 323, 324, 329
Hong Kong, 225, 228, 232, 234, 238
Honnavar, 8
Hormuz, 58, 59, 68, 73, 76, 108
Hornell, 20
Howrah, 65
Hsi-hai, 53
Hsuan Tsang, 55, 56, 79, 81
Hudud-al-Alam, 53, 57, 79
hull, 117, 127, 154, 157, 173, 176, 178, 202, 204
humanities, 7
Hurmuji-vahana, 59
Hyder Ali, 212, 213, 215, 216, 217
Hyderabad, 215
hydrographic, 123, 124, 125, 190, 191
hydrography, 114, 123, 124, 190, 191

Ibn Abi'l Kataib, 71
Ibn-al-Muqaddam, 72
Ibn-e-Awkal, 67, 87
Ibn-e-Batuta, 17, 57, 61, 63, 66, 71, 79, 96, 97, 98, 99
Ibn-e-Jubayr, 65, 69, 79, 80, 81
Ibn-e-Khuradbeh, 57
Ibn-e-Majid, 53, 69, 80, 105, 259, 260, 262, 264, 268, 273, 274, 275, 276
iconographic, 68, 199
Imperial
Government, 228
naval squadron, 226
incense, 38, 97
Independence, 295
India, 3, 4, 5, 7, 8, 9, 11, 12, 13, 14, 15, 16, 17, 19, 20, 22, 29, 32, 33, 34, 35, 38, 41, 42, 43, 44, 49, 50, 51, 52, 53, 54, 55, 56, 57, 58, 61, 63, 65, 66, 67, 71, 72, 73, 74, 75, 76, 77, 79, 81, 89, 90, 91, 95, 96, 97, 98, 99, 100, 101, 102, 103, 104, 105, 107, 108, 113, 114, 117, 120, 121, 122, 124, 125, 126, 127, 130, 131, 132, 133, 134, 137, 138, 139, 140, 141, 145, 146, 147, 148, 150, 157, 159, 163, 164, 165, 177, 179, 181, 185, 186, 189, 190, 192, 193, 194, 199, 205, 206, 208, 211, 212, 213, 214, 215, 216, 217, 218, 221, 223, 224, 225, 226, 227, 229, 230, 231, 233, 234, 235, 236, 237, 260, 276, 280, 283, 291, 292, 293, 294, 295, 296, 297, 298, 299, 300, 301, 302, 303, 307, 309, 311, 312, 314, 315, 316, 317, 318, 319, 322, 324, 325, 326, 329
Traders, 54, 84
Indian (s), 5, 7, 18, 42, 75, 81, 104, 113, 122, 132, 138, 141, 213, 255, 280, 298, 300, 301
cannon, 126, 127, 131
coast, 4, 8, 13, 114, 207, 224, 227, 236
coastline, 4, 89
designer, 121
Government, 223, 228
heritage, 4
indigo, 66
know-how, 214
mariners, 44, 114
museums, 127
navy, 4
pepper, 65, 80
ports, 57, 216
ships, 68, 70, 113, 121, 127, 186
troops, 229
warships, 126
workers, 214
Indian
Naval Ship Jamuna, 23, 148
Sharda, 296
Sujata, 148

Sukanya, 296
Vikrant, 295
Nireekshak, 23
Indian Ocean, 4, 5, 8, 13, 15, 17, 18, 19, 21, 22, 24, 50, 51, 52, 53, 54, 55, 57, 58, 64, 66, 69, 70, 71, 73, 74, 75, 76, 77, 78, 79, 89, 90, 95, 96, 97, 98, 105, 114, 115, 124, 129, 130, 146, 159, 163, 181, 189, 212, 213, 215, 216, 221, 222, 224, 225, 226, 227, 228, 229, 230, 231, 232, 233, 234, 235, 236, 237, 238, 243, 244, 247, 255, 260, 261, 264, 272, 274, 275, 277, 280, 282, 285, 286, 291, 292, 293, 294, 295, 296, 297, 298, 299, 300, 301, 302, 303, 307, 311, 326, 328
network, 51, 53, 64
region, 237, 301
indigenous tradition, 104
Indo-China, 229
Indonesia, 74, 91, 108, 224
Indra, 39, 40
Indus
delta, 55, 58
signs, 35
industrial revolution, 127, 132
industrialisation, 13
inscription (s), 53, 54, 57, 59, 60, 62, 64, 65, 68, 71, 74, 75, 76, 77, 79, 80, 81, 97, 102, 148, 157, 163, 185, 217
Devanagari, 185
interdisciplinary, 7
inter-island routes, 249
international trade, 33
inter-tropical areas, 248, 274
invasion, 55, 96, 223, 229, 235, 237, 281, 301
Iran, 32, 33, 35, 54, 55, 67, 231, 298
Iraq, 104, 130, 231
Iraqi, 73, 241
Irawaddy, 56
iron objects, 148, 169, 172
Isaq, 71
isba, 264, 266, 268, 275
Isfahan, 97
Isha, 71
I-shung-na-pu-lo, 57
Islam, 52, 58, 95, 99, 160
island (s), 5, 24, 77, 99, 146, 147, 148, 159, 163, 164, 249, 253
Andaman and Nicobar, 41
Bangaram, 4, 146, 147, 150, 163, 309, 329
Cape Verde, 227
Chetlat, 286
Cocos, 230
Elephanta, 55
Falkland, 230
Fua Mulaku, 19
King George, 228
Lakshadweep, 159
Maldives, 64, 66, 96, 98, 99, 100, 104, 108, 159
Marshal group, 302
Minicoy, 159
of Crete, 124
of Nilandu, 19
Pulao group, 302
Seychelles, 191
Solomon, 190
Sound, 228
Thursday, 228
Willingdon, 15
Isnapura, 57
Italian ports, 65
Italy, 199, 233, 234
Itsing, 56, 79

Jaffna peninsula, 100
Jagadu, 73
Jagaducaritam, 59, 73
jahaz, 69
Jahangir, 98
Jaina, 49, 80
Jalia, 134
jars, 170
jashujiyat, 69
Jatakas, 41, 302

Baveru, 44
Supparak, 44
Java, 50, 56, 64, 109, 221, 281, 302
Javanese, 75, 281
Jewish, 54, 57, 59, 60, 61, 65, 67, 69, 70, 71, 72, 73, 74, 76, 79, 80, 130
business letters, 57, 60, 61, 76
merchants, 60, 72, 79
Jews, 73, 131, 280
Jiajing period, 98
Jordan, 130
Joseph bin Abraham, 72
Judah bin Joseph Ha Kohen, 72
Judah Ha Kohen, 76
Junks, 61, 69, 71
Jutland, 104

Kachchh, 38, 39
Kachi plain, 30
Kadambas of Goa, 60, 80
Kadaram, 61, 77, 78
Kakatiya (s), 62, 63, 65, 85
Kalapani, 133
Kali
Age, 75
era, 33
Kalinga, 62, 63
Kalliene, 55
Kalyan, 55
Kamal, 122, 123, 247, 262, 265, 266, 269, 276, 279, 282, 285, 286
Kamalanka, 56
Kamar, 45
Kambaya, 58
Kambayat, 72
Kamina, 230
Kamkam, 59
Kanara, 59, 61, 76
Kanbaya, 58
Kanchipuram, 63, 185
Kanhoji Angre, 134
Kanja, 63
Kanniyakumari, 109, 122
Karachi, 225
kardar, 80
karimaradu, 18
karnadhar, 45
Karnataka, 8, 43, 159, 160
Kashmir, 38
Kashmiri era, 33
Kathiawad, 59, 60
Kautilya, 42, 303
Kavali palagahi, 285
Kaveri delta, 55, 61
Kaveripattinam, 55, 56, 57
Kedah, 61, 78
keel, 116, 120, 121, 173, 176, 206
Kerala, 19, 44, 61, 65, 74, 96, 126, 132, 134, 139, 160, 286
Khair, 120
Khalaf, 71, 72
Khalaf bin Isaac bin Bundar, 71
Khalila wa Dimna, 102
Kharepatan, 60, 73
khawar, 63
Kush mali, 45
Khurmali, 45
Khusru, 55
Kia-mo-lang-kia, 56
Kilwa, 66
Kish, 35, 58, 68, 69
Kitab-ul-Ajaib-ul-Hind, 57, 70, 79
Klings, 131
Kochi, 148, 150
kolepalagai, 282
Koli, 134
Kollam, 61
Kon Tiki, 19
Konigsberg, 230
Konkan, 55, 59, 60, 67, 68, 73, 76, 80, 134
coast, 55, 68, 76
ports, 60
Korea, 103, 293
Kotia, 21
Kshatriyas, 133
Ksiroda, 64
Kufic, 160
Kulam Mali, 61, 98
Kulottunga, 61, 62

Kulottungacholapattanam, 62
kundika, 102
Kunkan, 59
Kura-mala Samudda, 41
Kuwait, 99, 225, 295
Kyot, 134
Kyoto, 185
Kyushu, 124

Labuan, 225, 228
lagoons, 30
Lakshadweep, 4, 9, 24, 77, 146, 147, 159, 163, 164, 181, 186, 261, 279, 281, 282, 317, 323
  Islands, 4, 159
Lakshyalaksha, 77
Lanka, 13, 19, 41, 42, 50, 52, 53, 55, 56, 61, 63, 67, 71, 74, 77, 78, 95, 96, 98, 99, 100, 101, 104, 107, 108, 109, 186, 208, 256, 294, 311, 319
Laperouse, 190
Lata, 52
lateen
  sail, 115
  -rigged, 114
Latins, 247
lead, 4, 169, 174, 235, 260, 263, 321
Lebanon, 3, 38, 130
leeward, 117
legends, 15, 33
legislation, 311, 314
Leonardo da Vinci, 3, 130
Levant, 57, 67
lifting balloons, 172
lighthouse, 56
Lisbon, 124
lithographs, 199
Little Bear, 255, 262
Lloyds register, 178
lofting, 116, 118, 129
London, 81, 101, 134, 226, 227, 229, 232, 277, 283, 293, 299, 328
longitude, 123, 146, 163, 256, 261, 281
Lothal, 35
Lo-ts-lo-tsa, 61
Louvre, 199

Macedon, 281
Machwa, 21
Madagascar, 122, 236, 272, 280
Madapollum, 14
Madhumati, 60
Madmun, 71, 72
Madras, 81, 113, 125, 212, 217, 230
Madurai, 68
Magan, 34, 35, 38
Magellan, 105, 190
Maghrabin traveler, 17
magnetic
  compass, 122, 272, 286
  needle, 19, 279
magnetometer, 148
Mahabalipuram, 61
Mahabharata, 33, 42
Mahacampa, 57
mahanavika, 57, 69, 79
Mahar, 18
Maharashtra, 8, 81, 134, 160
Mahodadhi, 159
Mahodayapuram, 74
Mahruz, 72, 76
Mahumata, 60
Makamkam, 59
Makran, 55, 58, 134
  coast, 55, 58, 134
Malabar, 9, 59, 60, 61, 65, 66, 68, 72, 74, 80, 107, 120, 146, 164, 194, 199, 202, 205, 206, 208, 212, 213, 214, 217, 297
  coast, 9, 68, 74, 107, 164, 202, 205, 206, 212, 213, 214, 217
  yards, 120
Malacca, 64, 108, 109, 131, 222, 281, 292, 294, 296, 298
Maladeva, 77
Malaimandalam, 74
Malainadu, 68
Malay, 57, 91, 108, 229, 235
  peninsula, 56, 57, 61, 78

Maldives, 64, 66, 96, 98, 99, 100, 104, 108, 159
  Islands, 95, 264
  Middle, 19
Maldivian Muslim, 99
Male, 96, 99, 311
Malibar, 61
Mallah, 134
Mallam ni pothi, 125
Malpe, 8
Malwa, 58, 65, 299
Malyalam, 160
Malyalee, 21
management policy, 314
Manakkavaram, 78
mandapikas, 58
mandis, 58
Mangalapura, 59
Mangalore, 21, 59, 71, 72, 76, 113, 212, 213, 215, 216, 217
Mangrol, 59
Manibar, 61
Manigramam, 74
Manila, 234, 235, 298, 302
Manimekhalai, 56
Manimekhla, 41
man-of-war, 127, 211, 212, 213, 214, 215, 216, 218
mansions, 264
Mansura, 58
Mantai, 19, 96, 100, 101, 102, 103, 104, 107, 108, 109, 186, 319
Manu, 44, 133
map making, 280
Maraikkar, 74
marakkalam, 74
Maratha (s), 213, 214, 215, 218
  ports, 214
  warships, 126
Marco Polo, 14, 54, 57, 60, 62, 65, 68, 81, 122
mare nostrum, 21
Mari Indicum, 52
marine
  archaeological club, 9
  archaeology, 4, 5, 7, 9, 11, 12, 13, 14, 17, 145, 150, 243, 299
  encrustations, 170, 175
  engineering, 7
  growth, 154
  influence, 324
  products, 30
  resources, 29
  shells, 30
mariner (s), 16, 19, 38, 41, 42, 122, 123, 124, 146, 147, 185, 279
  master, 57, 70
maritime
  activities, 7, 8
  archaeological discoveries, 315
  expeditions, 77
  heritage, 4, 5, 11, 13, 14
  history, 4, 50, 51
  network, 5
  studies, 50, 309
  trade, 52, 58, 67, 76
  traditions, 4
  zone, 50, 77
Maritime Heritage Foundation, 11
Maritime History Society, 11, 15, 118, 134
markab, 69
markets, 58, 63, 66, 91, 96, 223, 292, 311
Marshal group of Islands, 302
Marshall, 190
Mary Rose, 312, 328
master mariner, 57, 70
masts, 40, 121, 131
Masula, 21, 203, 207, 209
mathematics, 19, 139, 281
Mauritius, 213, 216, 221, 223, 224, 228
mawasim, 53
medicinal plants, 38
medicines, 300
medieval geography, 17
Mediterranean, 4, 42, 54, 57, 58, 65, 66, 81, 96, 122, 124, 225, 227, 233, 234, 243, 279, 286, 311, 312, 315, 328

basin, 4
Meghna, 53, 64
Mehrgarh, 29, 30, 32
Meluhha, 34, 35, 38
mercantile activity, 96
merchandise, 68, 292
merchant (s), 41, 52, 54, 56, 57, 60, 61, 62, 63, 65, 66, 67, 71, 72, 73, 74, 75, 76, 77, 78, 79, 80, 81, 90, 97, 99, 139, 191
guilds, 80, 115
ships, 45, 174, 178, 230, 291
vessel, 56
meridian, 251, 257, 262, 281
Mesolithic, 325
Mesopotamia, 32, 33, 35, 38, 39, 44, 104, 230
metallurgy, 126, 131
method of construction, 19
Metropolitan, 101
Meuse, 324
middle ages, 59, 63, 64, 67, 79
Middle East, 114, 130, 234, 236, 237, 277, 291, 294
Maldives, 19
Midway, 237, 295
Minicoy, 159
Missouri, 295
Mocha, 97
models, 116, 119, 140, 199, 201, 202, 207, 208, 279, 320
modern cruiser, 228
Mohenjo-daro, 12, 30, 35
Mo-ho-chen-po, 57
Moka, 212
Mombasa, 122
Mongol, 130
Mongolia, 103
monsoon wind system, 53
monuments, 317, 318, 322
Moplahs, 131
Morocco, 99
Mortimer Wheeler, 39, 100
motorised boats, 151
Motupalli, 62
mound, 16, 98, 101
Mu allim, 17, 69
Mughal army, 126
Muhammad, 60, 67
Multan, 58, 66
Mumbai, 9, 11, 15, 23, 55, 60, 68, 72, 134, 148, 186
mundu, 21
Murshidabad, 57
musafirun-ul-Hind, 54
Muscat, 212
Musee National de la Marine, 192
Museum, 69, 75, 101, 182
Ashmolean, 99
Ethnographic, 265
Indian, 127
Kyoto, 185
Metropolitan, 101
National, 44, 104, 216
National Maritime, 194, 208
naval, 189, 199, 201, 208
muskets, 127, 131
Muslim, 130
Mutfili, 62
mutineers, 129
Muza, 97
Myanmar, 56, 61, 74, 78, 294
Mysore, 137, 138, 141, 212, 213, 215, 217
mysteries, 3, 4, 12
myths, 15

Nagapattinam, 61, 96
Nagapura, 73
Nahray bin Alan, 70
Nails, 19
nakhuda, 71, 72, 74, 76
Nal, 41, 42
Nal Malai, 41, 42
Nalmali, 45
Nala-mala Samudda, 41
Nalanda, 64
Nambiyar, 72
Nanadesi, 74
Napolean, 138

Napoleonic
  conflicts, 224
  wars, 120, 127, 138, 224
Narasimhavarman, 163
Narmada, 55, 72, 97
National Maritime Museum, 194, 208
National Museum, 44, 104, 216
natural resources, 14, 34
nau, 40, 42, 45, 68, 71
Nautical, 123, 202, 323, 328
  Almanac, 123
  ethnography, 189, 196, 205
  ethnologist, 20, 21
  tables, 129
  techniques, 201, 206
nauvittaka, 71, 73
Naval
  Base, 15
  blockade, 69
  expeditions, 42
  infrastructure, 213
  mastery, 222, 224
  museum, 189, 199, 201, 208
  stocks, 214
navigable, 63
navigation, 18, 19, 40, 95, 105, 107, 108, 114, 122, 123, 124, 125, 133, 147, 148, 203, 205, 225, 247, 248, 249, 260, 261, 262, 263, 274, 275, 280
navigational
  manual, 53
  skills, 4
navigators, 7, 18, 40, 109, 122, 125, 133, 250, 253, 255, 261, 266, 272, 276
navika, 64
navikakarmakara, 70
Navy, 4, 5, 9, 11, 14, 15, 16, 22, 42, 134, 145, 146, 147, 148, 150, 159, 163, 164, 172, 174, 178, 212, 213, 216, 222, 225, 226, 228, 229, 231, 232, 233, 244, 274, 291, 293, 294, 295, 296, 297, 298, 303, 309, 312, 315, 317, 318, 319
necklaces, 30
Neel, 41, 42
Neel Malai, 41, 42
Nejd, 17
Neolithic, 29, 30
  pre-pottery, 29
Neptune, 134, 248
Netherlands, 199, 235, 308, 313, 318, 319, 320, 321, 322, 323, 324, 325, 326, 329
New Guinea, 124, 190
New York, 101, 134, 283
New Zealand, 190, 208, 229, 237, 249, 298
Newton, 138, 140
Niarchus, 44
Nicolo Conti, 122
Nigerian, 285
Nila-mala Samudda, 41
nilloham, 70
nivartana, 41
niyana, 41
Noakhali, 52, 56
non-European civilization, 139
North Pole, 250, 256, 259
Nuruddin Firuz, 76
nyayana, 41

Oak trees, 119, 120
observatory, 123
occupational deposit, 29
oceanographers, 8
oceanographic, 321
octant, 123
Old Goa, 60, 69
Oman, 18, 29, 33, 34, 35, 38, 44, 97, 98, 225
Omani, 60, 285
opium, 299
Organic materials, 325
Organising Committee, 3, 23, 24, 329
Oriental Institute, 101
Orissa, 63, 78, 134
ornaments, 30, 34, 35
outrigger, 202, 207
Oxford, 99, 328

Pacific, 17, 50, 96, 104, 105, 231, 232, 235, 236, 237, 238, 247, 248, 249, 272, 274, 277, 294, 295, 296, 297, 298, 302
Pagan, 74
paintings, 17, 45
Pakistan, 30, 189, 294, 295, 296, 298, 301, 302, 303
Pala king, 64
Palembang, 61, 78
Palestine, 67, 230
Pali, 49, 281
Palk Strait, 29, 52
Pallava (s), 61, 163
Panatalayani, 61
Panchatantra, 130
Pandyan, 65, 68
Pandyas, 68
Panini, 139
Paravan, 134
parayana, 41
Parikshit, 33
Parumdurai, 108
Parys Mine Company, 178
paschima, 53
passengers, 68, 69, 70, 71
Pasupatacharya, 76
Patamar, 21, 206, 207, 208
pattana, 58
pattanams, 58
pattanaswami, 72
Pearl Harbour, 235
Peat, 324
Pegu, 57
pemta, 63
pendants, 30
pentha, 63
Persia, 222
Persian (s), 19, 33, 34, 35, 41, 44, 50, 51, 52, 53, 54, 55, 57, 58, 59, 60, 61, 66, 68, 71, 79, 91, 96, 97, 98, 102, 108, 109, 122, 127, 181, 186, 221, 225, 226, 228, 230, 233, 237, 248, 249, 255, 275, 280, 295, 298
  court, 122
  Gulf, 33, 34, 35, 41, 44, 50, 51, 53, 55, 58, 59, 60, 61, 66, 91, 96, 97, 98, 102, 181, 186, 221, 225, 226, 228, 230, 233, 237, 275, 280, 295, 298
Perumal Cheraman, 160
peth, 63
Philippines, 235, 298
philosophy, 130, 139, 140, 309, 310, 316, 317, 318
photography, 149, 151, 153, 310
Pink Plan, 293
pintha, 63
Pitris, 133
planks, 66, 79, 121, 122, 165, 173, 176, 177
plava, 40
Pliny, 52, 86
Poksha, 45
pole-star, 122, 255, 257, 261, 262, 266, 272, 275, 281
Pollonaruwa, 100
Pondicherry, 190, 191, 217, 286
Poompuhar, 8, 55
porcelain, blue on white, 170
port (s), 8, 16, 40, 42, 53, 54, 55, 57, 58, 59, 60, 61, 62, 63, 64, 65, 66, 67, 68, 69, 71, 73, 75, 77, 79, 96, 97, 107, 108, 109, 114, 122, 125, 185, 186, 191, 214, 215, 216, 224, 229, 286, 297
  ancient, 8
  cities, 13, 125
  city of Tyre, 3
  Konkan, 60
  Maratha, 214
  of Chaul, 214
  of Ghogha, 59
  of Khambat, 58
  of Quanzhou, 75
  of Samyana, 59, 80
  of the call, 207
  officers, 18
  Sumerian, 7
Port-Louis, 216

Porto Novo, 191
Portolan chart, 280
Portugal, 24, 279
Portuguese, 8, 21, 22, 51, 53, 64, 65, 77, 78, 89, 95, 99, 105, 113, 131, 159, 186, 212, 215, 247, 248, 255, 273, 274, 275, 277, 279, 280, 282, 283, 291, 299, 300
Pota, 45
Potavanij, 44
Potavanik, 45
pottery, 5, 17, 29, 30, 34, 100, 102, 103, 104, 155, 169, 170, 175, 177, 185
 making, 30
Prakrit, 49, 70
pratarani, 40
pre-European, 17, 18
Prehistoric, 313, 326, 328
pre-Islamic, 95, 97, 104
preservation, 13, 102, 103, 150, 155, 157, 173, 320, 321, 322, 323, 324, 325
Princes Royal, 145, 146, 157, 177, 179, 329
Prome, 56
propulsion, 114, 203, 323
Ptolemy, 140, 281
public responsibility, 307, 317, 318, 325, 326
Puhar, 55
Pulakesin, 55
Pulao, 302
Pune, 213, 214
 Darbar, 214
punnai, 18
Puranic literature, 44
puras, 39
Puri, 55, 79
purva, 53
Puspak, 42

Qays, 58
Queen Elizabeth, 115
Quilon, 61, 96, 98, 107
Quran, 76, 160

Raja Rammohan Roy, 138
Rajapur, 214
Rajaraja, 61, 77
Rajarajendra, 163
Rajarajeswara, 163
Rajasthan, 34, 58
Rajendra, 61, 77, 78
Rakhigarhi, 38
Raktamrittika, 57
Ram, 13, 42, 77
Ramayana, 42, 302
Rameshwaram, 41
Ramisht, 72, 87
Ranger, 295
Rangoon, 225, 236
Rastrakutas, 76
Ratnagiri, 134
Ratnakar, 159
Ravana, 42
recession, 29, 39
re-compression, 149
Red Sea, 38, 41, 50, 51, 54, 55, 58, 61, 65, 66, 67, 69, 80, 91, 96, 97, 98, 108, 130, 222, 230, 280, 298
 network, 58
red ware, 55
reeds, 44
reefs, 9, 13, 99, 100, 146, 147, 163, 164, 190, 323
relics, 16
religion, 18, 133
Renaissance, 130, 311
reservoir, 35, 102
Rhine, 308, 324
Rhineland, 103
Rhino, 35
rhumbs, 259, 264, 274, 280
Ricciardi, 20
rig, 18, 19, 20, 122
Rigveda, 39, 40, 41, 43
rival fleet, 225
river (s), 8, 16, 40, 43, 308, 324
 Bangshi, 65
 beds, 13
 Blue, 64

Bolan, 30
-craft, 68
Ems, 324
Godavari, 56, 297
Meghna, 53, 64
Meuse, 324
Narmada, 55, 72, 97
Rhine, 308, 324
Rupnarayan, 56
Sarasvati, 59
Scheldt, 324
Riverine, 324
channel, 63
rockets, 137, 138
Roma, 42
Roman (s), 130
Empire, 51, 54, 57
Rome, 42, 279
Roosevelt, 234, 293
rosewood, 44
Rotary querns, 174
Royal
Cemetery, 35
Navy, 225
ruler of Pagan, 62
Russian, 223, 225, 226, 229, 230, 231

sagging, 117
Sahyadri range, 60
Sailendra ruler, 64
sailing
boat, 268
craft, 5, 115
ship, 114, 121, 126, 177, 199
-vessels, 18, 21
sailor, 51, 70, 98, 125, 127, 247, 249, 264, 274, 282, 286, 323
sails, 40, 44, 45, 70, 114, 121
Saimur, 60
Saint-Malo, 215
Saiva, 59, 76
Saivism, 76
Salva, 42
Samandar, 63, 65
samanya, 68
Samatata, 56, 57
samudra, 41, 43, 63
Samudri
of Calicut, 22
of Kozhikode, 113
samundar, 63
samyatrikas, 62, 68
San Diego, 296
sand barriers, 324
sandal, 67
Sandwip, 63
Sangam, 13, 41, 55
Age, 13
literature, 41
Sanjan, 59, 60, 67, 76, 80
sanku, 282
San-mo-ta-ta, 56
Sanskrit, 20, 40, 44, 45, 49, 53, 57, 58, 59, 60, 68, 69, 70, 71, 72, 75, 76, 77, 81, 159, 160, 281, 282
sanyayana, 41
Sarasvati, 59
Sargon of Akkad, 34, 35
Sasanid Empire, 54, 55
Saurashtra, 35, 38, 39, 120
Scheldt, 324
sciences, 7
scientists, 122, 137, 140
SCUBA, 165, 311, 315
sculptures, 68
Sea, 3
Arabian, 9, 17, 41, 52, 53, 55, 73, 75, 77, 109, 146, 159, 163, 164, 244, 295, 326, 329
battle, 68, 113, 129, 222, 229, 234, 295, 296
China, 50
Coral, 237
eastern, 18
floors, 9
Red, 38, 41, 50, 51, 54, 55, 58, 61, 65, 66, 67, 69, 80, 91, 96, 97, 98, 108, 130, 222, 230, 280, 298
route, 8, 122, 127
South China, 4

seabed, 12, 154, 155, 164
seafaring, 5, 13, 14, 16, 18, 24, 33, 41, 73, 79, 81, 95, 104, 279, 280, 319
  activities, 13, 24, 81
  nation, 14, 319
  traditions, 5
seal, 33, 35
sea-level
  changes, 13, 310
  fluctuations, 16
sea-lines, 17
seals, 17, 33, 34
seamanship, 17, 95, 105, 133
seamen, 18, 19, 134, 189, 202, 260, 274
seapower, 21, 114
search, 7, 8, 16, 20, 121, 178, 179, 312, 317
sea-route to India, 226
seasonal wind, 18
secrets of the sea, 3
sedimentation, 324
sediments, 9, 319, 324
Seringapatam, 113
Serndib, 63
setting of stars, 253
settlements, 8, 29, 40, 60, 63, 191
sewn planks, 19, 203
sextant, 123
Seychelles, 191
shaffara, 69
Shah Bandars, 18
Shah Jahan, 98
Shahriyar, 60, 70, 79, 85
Shanghai, 124
Shariyarahara, 60
sheathing, 157, 174, 177, 178
shells, 29, 30, 32, 66
sherd (s), 17, 100, 102, 169, 171
Shi-li-chi-ta-lo, 56
ship, 13, 19, 20, 21, 23, 40, 41, 42, 44, 45, 51, 53, 59, 68, 69, 70, 71, 72, 76, 79, 115, 116, 117, 118, 119, 120, 121, 122, 126, 129, 130, 131, 133, 148, 149, 150, 151, 154, 155, 157, 164, 165, 169, 170, 171, 172, 173, 174, 175, 176, 177, 178, 179, 181, 191, 206, 212, 214, 215, 218, 232, 249, 261, 273, 294, 295, 319, 320, 326
  design, 115
  forms, 19
  merchant (s), 41, 52, 54, 56, 57, 60, 61, 62, 63, 65, 66, 67, 71, 72, 73, 74, 75, 76, 77, 78, 79, 80, 81, 90, 97, 99, 139, 191
ship's hull, 157
shipbuilder (s), 7, 19, 20, 51, 117, 193, 202, 214
shipbuilding, 4, 13, 14, 18, 19, 20, 41, 66, 69, 104, 114, 115, 119, 120, 121, 132, 206, 214, 216, 218, 247, 315
shipowners, 18, 71, 73, 74, 191, 214
shipping, 14, 43, 44, 45, 53, 65, 68, 69, 70, 71, 73, 74, 75, 80, 96, 98, 100, 107, 178, 221, 224, 225, 230, 231, 234, 235, 236, 237, 244, 292, 301, 302, 323, 325
  centre, 14
  industry, 14, 45, 292
shipwrecks, 4, 7, 8, 9, 12, 13, 41, 70, 72, 146, 147, 150, 163, 186, 190, 322
shipwright (s), 18, 20, 115, 119, 130, 193
shipyards, 116, 132
Shivaji, 134
Shraddha, 133
Siam, 223
Siberian forests, 120
Sibor, 55, 60
Sicily, 237
side scan sonar, 148, 153, 154
Sidis, 212
Sikh wars, 113
Silahara (s), 60, 80
  ruler, 76
Silandib, 63
Silan-hai, 53
Silappadikaram, 56
silk, 5, 67, 90, 96

Silted tidal basins, 325
silver, 34, 43, 67, 129, 178
Sind, 55
Sindabur, 60, 82
Sindan, 59
sindhu, 43
Singapore, 221, 222, 225, 228, 229, 230, 231, 232, 233, 234, 235, 236, 237, 238
Sinhalese, 101, 159
Siraf, 53, 58, 72, 98
Sirafi, 60
Siraj-ud-dowlah, 127
site-formation, 321
siyapata, 70
sloops, 228, 229, 233
sluice gates, 35
Sobraon, 113
societies, 13, 21, 91, 163, 322
sociology, 139
Socotra, 124, 130, 261, 273
soldier, 127
Solomon, 190
Somalia, 295
Somnath, 59, 60, 76
Song dynasty, 100
Sopara, 8, 186
South China Sea, 4, 235
Southeast Asia Treaty Organization, 294
Southern Naval Command, 148, 296
Soviet Union, 294
Spain, 104
Spanish, 178, 207, 248
spars, 121
spices, 5, 61, 65, 244
square lashings, 207
square-rigged
  ships, 22
  vessel, 109
squatter, 120
Sri Lanka, 55, 56, 67, 71, 96, 98, 100, 104, 107, 109
Srikshetra, 56
Srirangapattna, 138
Sristhanaka, 59, 73
Srivijaya, 61, 78
St. Petersburg, 226
Stambatirtha, 58
Stambhaka, 58
standing rigging, 202
star (s), 41, 249, 250, 251, 253, 255, 256, 257, 259, 260, 261, 262, 268, 272, 273, 274, 280, 281, 286, 287
  compass, 286
steamships, 53, 221, 226
stellar, 255
stem, 19, 120, 206, 286
stern, 120, 206, 272
stoneware, 102, 104
storage jars, 8, 169, 170
storm, 129, 275
Strait
  of Malacca, 296
  of Hormuz, 297
subcontinent, 8, 50, 67, 90, 99, 104, 221, 223, 225, 226, 227, 261, 291, 318
submarine, 95, 164, 168, 221, 226, 231, 238, 297
Submerged
  cities, 7
  ports, 15
  terraces, 154
submergence, 8
submersible dive-bell, 149
Sudkawan, 63
Sulaiman, 57, 79, 80, 81, 276
Sultanpettah, 138
Sultans of Constantinople, 14
Sumatra, 61, 74, 78, 109, 122, 281
Sumatran coast, 109
Sumer, 51
Sumerian ports, 7
Sunda Straits, 222
Sung dynasty, 57
Sunganadavirttachola, 62
Sunken
  archaeological sites, 8
  cultural heritage, 7
superlite helmets, 151

Supparak, 44
Sur, 18
Surat, 75, 215
surgeon, 129
survey
 of wrecks, 4
 preliminary, 23, 148
 team, 148
Sus siti, 71
Susa, 35
Suvarnabhumi, 56
Suvarna-dvipa, 56
swords, 66
Sydney, 221, 227, 230, 232, 238
Syria, 42, 130
Syrian Christian, 55

Tajjikas, 60
Takua Pa, 74, 77
Talai Mannar, 107
Talaittakkolam, 77
Tamil, 101
 epics, 56
 inscriptions, 61, 74, 75
 literature, 13
 merchants, 61, 74
Tamil Nadu, 8, 21, 108, 134
Tamluk, 56
Tamralipta, 56, 57, 63
Tana, 59, 68, 72, 81
tandel, 69
Tang
 dynasty, 102
 splashed ware, 102
 white ware, 102
Tangiers, 96
tarani, 40
Tasmania, 124
Tatari horses, 67
Tavaccakkaravartikal, 75
teak, 18, 38, 66, 80, 118, 120
techniques of navigation, 247
technology, 5, 13, 19, 20, 103, 105, 114, 126, 127, 129, 131, 137, 138, 139, 141, 319
Tectonic subsidence, 324
Telengana, 63
temple, 35, 76, 132, 163
terraces, 148, 164, 165, 167
terracotta
 amulet, 33
 ring-wells, 8
territorial waters, 146
Thailand, 61, 74, 91, 298
Thane, 14
Thanjavur, 108, 160, 163
theatre of war, 230
Thoni, 21
Thor Hayerdahl, 19
through bolts, 173, 176
Thursday Island, 228
tidal flat, 308
tides, 16, 21, 35, 323
Ti-hua-kie-lo, 61
timber (s), 18, 20, 38, 97, 116, 120, 121, 315
Tinbu, 74, 76
Tipu Sultan, 114, 137, 212, 215, 218
Tipu's navy, 212
Tokyo, 294
To-lo-po-ti, 57
tomb, 97, 104
tombstone, 104
Tondaimandalam, 61
Torres Straits, 222
Toulon, 191, 218
trade, 4, 5, 7, 9, 13, 16, 17, 18, 19, 21, 22, 33, 34, 35, 38, 39, 41, 42, 44, 50, 51, 52, 54, 55, 57, 58, 62, 66, 67, 71, 72, 73, 75, 76, 78, 79, 80, 89, 90, 95, 96, 97, 99, 100, 101, 102, 107, 108, 109, 130, 146, 164, 171, 178, 181, 191, 214, 215, 222, 223, 224, 225, 226, 227, 228, 229, 231, 232, 237, 238, 243, 244, 280, 291, 292, 298, 300, 302, 314, 319, 323
 Muslim, 130
 overseas, 51
 route, 5

traditional
    data, 4
    Indian boats, 193
traditions, 4, 5, 12, 14, 18, 19, 75, 115, 123, 186, 189, 190, 196, 205, 208, 280, 286, 310, 311, 314, 315, 316, 317
    maritime, 4
Tranquebar, 8
transgression, 29
transshipment, 54, 61
Travancore, 126, 132
travellers, 13, 14, 198
treasure (s), 3, 22
    -hunters, 325
Treaty of Paris, 213
Treaty of Vienna, 224
Trincomalee, 193, 213, 217, 222, 228, 237
Tudor warship, 313
Tugra, 41
Tunisia, 57, 61, 67, 80, 104
Tunisian, 71
Turkey, 230
Turkish gun makers, 127
turquoise-glazed, 104
Tuticorin, 21, 65, 120
Tyre, 3

Ulu Burun, 311, 326
underwater, 3, 7, 8, 12, 13, 16, 17, 22, 100, 117, 145, 149, 150, 151, 152, 153, 155, 157, 165, 181, 307, 309, 310, 311, 315, 316, 317, 318, 319, 321, 322, 325
    archaeological heritage, 317
    archaeology, 8
    Archaeology Wing, 163, 164, 182
    cultural heritage, 309
    exploration, 3, 13, 16
Union Territory Administration, 147
United States, 24, 101, 232, 233, 234, 292, 294, 295, 296, 297, 298, 301, 302, 303
    of America, 24, 302
    Navy, 293, 294, 295, 296
unnata, 68
upwind, 117
Uranshin, 63
urban
    areas, 63
    life, 52
urbanisation, 13, 32, 33, 38, 43, 44
Ushas, 41
USS
    Carl Vinson, 296
    Tarawa, 295
Uttar Pradesh, 58

Vaalon, 134
vahitra, 68
Vaisnavism, 76
Vallarapatnam, 72
Valparai forest, 120
Vanga, 53, 65
Vangasagara, 53, 65, 79
vanijya, 50
Vanikoro, 190
Vankalai, 100
Varaharamihira, 281
Varuna, 41, 157
Vasa, 312, 327
Vasco da Gama, 17, 77, 122, 124, 265, 274, 285
Vasishta, 41
Vastupalamahatmayam, 59
Vayalur, 163
Vedas, 40, 44, 133
vedic
    ceremony, 133
    literature, 38, 43
velakula, 58
velapuras, 58
Venetian, 14, 115
Vengi, 56, 62
    kingdom, 62
verbalisation, 19
Versailles, 214, 217, 218
Viceroy of India, 228
videography, 149, 151, 153

Vietnam, 57, 295, 301
  war, 295, 301
Vijaydurg, 8, 212, 216
Villiers, 20
Viman, 42
Vincennes, 216
vishesa, 68
Visvanatha, 77
Visvarupa, 77
vitta, 71
VOC ships, 179
voyages, 7, 18, 41, 55, 60, 61, 63, 65, 68, 73, 74, 75, 76, 77, 79, 107, 179, 249, 253, 255, 260, 272
Vrddhavasistha Siddhanta, 282

Wadias, 137
wakil-ul-tujjar, 71
Wandiwash, 113
Warangal, 62, 63
warships, 121, 215, 223, 224, 225, 228, 229, 230, 231, 232, 237
Washington, 138, 218, 232, 293, 294, 296
wastelands, 13
Wat Boroma That, 74
water bodies, 8
water-dredge, 150
weapon (s), 45, 66, 125, 127, 132, 133
  technology, 114
weights, 33, 126, 177
Wellington, 297
West Asia, 7, 32, 35, 43, 66, 67, 90, 91, 141
West Bengal, 56, 57, 63
western traders, 14
wharf, 56
wheels, 40
William Congreve, 138
Willingdon Island, 15
Winger, 140
Wolf, 231
wood, 38, 66, 80, 103, 118, 119, 120, 121, 157, 165, 169, 173, 213, 214, 253, 255, 264, 265, 266, 272, 276
woodborers, 176, 178
wooden
  pegs, 207
  planks, 70, 75, 165, 176
World
  Heritage, 309, 316, 317
  War I, 230, 293
  War II, 100, 238, 291, 293, 294
wreck site, 148, 153, 164, 170, 171, 179, 181

Xebec, 115

Yadavas of Devagiri, 60
Yajnavalkya, 133
Yamanadvipa, 57
yanapatras, 68
Yavana (s), 42, 56
Yemen, 97, 297
Yen-niona-chen, 57
yogataras, 281
Yuktikalpataru, 20, 68, 185

Zalari, 134
Zanz, 59
Zanzibar, 59
Zulu War, 227